AF574023

Catalysis of Organic Reactions

CHEMICAL INDUSTRIES

A Series of Reference Books and Textbooks

Consulting Editor
HEINZ HEINEMANN
Heinz Heinemann, Inc.,
Berkeley, California

Volume 1: Fluid Catalytic Cracking with Zeolite Catalysts, *Paul B. Venuto and E. Thomas Habib, Jr.*

Volume 2: Ethylene: Keystone to the Petrochemical Industry, *Ludwig Kniel, Olaf Winter, and Karl Stork*

Volume 3: The Chemistry and Technology of Petroleum, *James G. Speight*

Volume 4: The Desulfurization of Heavy Oils and Residua, *James G. Speight*

Volume 5: Catalysis of Organic Reactions, *edited by William R. Moser*

Volume 6: Acetylene-Based Chemicals from Coal and Other Natural Resources, *Robert J. Tedeschi*

Volume 7: Chemically Resistant Masonry, *Walter Lee Sheppard, Jr.*

Volume 8: Compressors and Expanders: Selection and Application for the Process Industry, *Heinz P. Bloch, Joseph A. Cameron, Frank M. Danowski, Jr., Ralph James, Jr., Judson S. Swearingen, and Marilyn E. Weightman*

Volume 9: Metering Pumps: Selection and Application, *James P. Poynton*

Volume 10: Hydrocarbons from Methanol, *Clarence D. Chang*

Volume 11: Foam Flotation: Theory and Applications, *Ann N. Clarke and David J. Wilson*

Volume 12: The Chemistry and Technology of Coal, *James G. Speight*

Volume 13: Pneumatic and Hydraulic Conveying of Solids, *O. A. Williams*

Volume 14: Catalyst Manufacture: Laboratory and Commercial Preparations, *Alvin B. Stiles*

Volume 15: Characterization of Heterogeneous Catalysts, *edited by Francis Delannay*

Volume 16: BASIC Programs for Chemical Engineering Design, *James H. Weber*

Volume 17: Catalyst Poisoning, *L. Louis Hegedus and Robert W. McCabe*

Volume 18: Catalysis of Organic Reactions, *edited by John R. Kosak*

Volume 19: Adsorption Technology: A Step-by-Step Approach to Process Evaluation and Application, *edited by Frank L. Slejko*

Volume 20: Deactivation and Poisoning of Catalysts, *edited by Jacques Oudar and Henry Wise*

Volume 21: Catalysis and Surface Science: Developments in Chemicals from Methanol, Hydrotreating of Hydrocarbons, Catalyst Preparation, Monomers and Polymers, Photocatalysts and Photovoltaics, *edited by Heinz Heinemann and Gabor A. Somorjai*

Volume 22: Catalysis of Organic Reactions, *edited by Robert L. Augustine*

Additional Volumes in Preparation

Catalysis of Organic Reactions

edited by

Robert L. Augustine

Department of Chemistry
Seton Hall University
South Orange, New Jersey

MARCEL DEKKER, INC. New York • Basel

Library of Congress Cataloging in Publication Data
Main entry under title:

Catalysis of organic reactions.

(Chemical industries ; 22)
Papers from the Tenth Conference on the Catalysis of Organic Reactions, held May 7–9, 1984, at Williamsburg, VA.
Includes index.
1. Chemistry, Organic--Synthesis–Congresses.
2. Catalysis–Congresses. I. Augustine, Robert L., [date]. II. Conference on the Catalysis of Organic Reactions (10th : 1984 : Williamsburg, VA.) III. Series: Chemical industries ; v. 22.
QD262.C35 1985 547.1'395 85-10242
ISBN 0-8247-7263-6

MARCEL DEKKER, INC.
270 Madison Avenue, New York, New York 10016

Current printing (last digit):
10 9 8 7 6 5 4 3 2 1

PRINTED IN THE UNITED STATES OF AMERICA

Preface

About twenty years ago a group of individuals who were involved on a day to day basis with the application of catalytic processes to organic synthesis came together to establish a forum for the discussion of the practical aspects of catalysis in organic chemistry. The result was a conference held in June, 1966 on Catalytic Hydrogenation and Analogous Pressure Reactions,[1] sponsored by the New York Academy of Sciences and co-chaired by Joseph M. O'Connor of CIBA(presently CIBA-Geigy), Morris Friefelder of Abbot, and Melvin A. Rebensdorf of Upjohn. This conference was very successful and two years later in June, 1968, the Second Conference on Catalytic Hydrogenation and Analogous Pressure Reactions[2] was held, chaired by Joseph O'Connor and again sponsored by the New York Academy of Sciences. The Third Conference[3] (June, 1970), chaired by Melvin A. Rebensdorf and the Fourth Conference[4] (September, 1972), chaired by Paul N. Rylander of Engelhard, were also sponsored by the New York Academy of Sciences despite the Academy's general policy of not sponsoring series of conferences in the same area, a fact which attests to the value of these conferences.

At this stage, however, continuing Academy support for further conferences was not forthcoming so a small group of active participants in these previous meetings worked to independently sponsor the next conference. This group called itself The Organic Reactions Catalysis Society and in the discussions leading to the formulation of a general policy for future meetings it became clear that the titles used for the first four conferences did not accurately describe the groups' changing interests. Thus, the Fifth Conference on Catalysis in Organic Synthesis[5], co-chaired by Paul Rylander and Harold Greenfield of Uniroyal Chemical was held in Boston in April, 1975. A result of these conferences was the merging of the interests of the organic chemists with those of the catalytic chemists. Further evidence of this development came in 1975 with the affiliation of the Organic Reactions Catalysis Society (ORCS) with the North American Catalysis Society. The Sixth Conference on Catalysis in Organic Synthesis[6], chaired by Gerard V. Smith of Southern Illinois University was held in Boston in May, 1976 in order to have the ORCS biennial conferences held in those years in which the biennial North American Catalysis Society National Meetings were not held.

The Seventh Conference on Catalysis in Organic Synthesis[7], chaired by William H. Jones of Merck Sharp and Dohme, was held in Chicago in June, 1978. Throughout this time the interests of the Society were expanding to include chemical processes and catalyst development. The Eighth Conference, chaired by William R. Moser of Worcester Polytechnic Institute and held in New Orleans in June, 1980, reflected this with a change in the conference title to Catalysis of Organic Reactions.[8] The Ninth Conference on Catalysis of Organic Reactions,[9] chaired by John R. Kosak of du Pont, was held in Charleston, SC in April, 1982. And that brings us to the present, the Tenth Conference on Catalysis in Organic Reactions held on May, 7-9, 1984, in Williamsburg, VA, which is the topic of this volume.

It is obvious that the Organic Reactions Catalysis Society could not have reached this milestone without the enthusiastic support of a lot of people. Particular thanks must go to our "Founding Fathers" who worked so hard in the 1960's to get the first conferences established: Dale Blackburn of Smith Kline and French, Jack Campbell of Lilly, Morris Friefelder of Abbott, William Jones of Merck Sharp and Dohme, Joseph O'Connor of CIBA (now CIBA Geigy), William Pearlman of Parke-Davis (now Warner Lambert), Melvin Rebensdorf of Upjohn, Paul Rylander of Engelhard, William Selby of G.D. Searle, and David Wagner of Hoffman-LaRoche. Thanks must also go to the New York Academy of Sciences for its support during the formative years of the Society and to all of the Officers and Directors who have served over the years to develop the Society into what it is today.

We cannot forget that without participants and speakers there would be no conferences and I would particularly like to thank our present speakers for taking the time out of their busy schedules to present aspects of their work and for being so cooperative in providing manuscripts of their talks so readily as to facilitate the publication of these proceedings.

The financial support of this conference from the following corporations is gratefully acknowledged: Air Products and Chemicals, Allied Corporation, Chemical Data Systems, Degussa, E.I. du Pont de Nemours and Company, General Electric Company, W.R. Grace and Company, Hoffman-LaRoche, Merck Sharp and Dohme, The Procter and Gamble Company, Sandoz, G.D. Searle, Shell Development Company, Smith Kline and French, Standard Oil Company (SOHIO), Uniroyal Chemical, and Warner Lambert Company.

The excellent job done by Jean DeRosa in preparing all of the manuscripts in this volume for publication is greatly appreciated.

References

1. Catalytic Hydrogenation and Analogous Pressure Reactions (J.M. O'Connor, ed.) Ann. N.Y. Acad. Sci., 145, Art. 1, 1-206 (1967).
2. Second Conference on Catalytic Hydrogenation and Analogous Pressure Reactions (J.M. O'Connor, ed.), Ann. N.Y. Acad. Sci., 158, Art. 2, 439-588 (1969).
3. Third Conference on Catalytic Hydrogenation and Analogous Pressure Reactions (M.A. Rebensdorf, ed.), Ann. N.Y. Acad. Sci., 172, Art 9, 151-276 (1970).
4. Fourth Conference on Catalytic Hydrogenation and Analogous Pressure Reactions (P.N. Rylander, ed.), Ann. N.Y. Acad. Sci., 214 (1973).
5. Catalysis in Organic Synthesis, 1976 (P.N. Rylander and H. Greenfield, eds.), Academic Press, New York, 1976.
6. Catalysis in Organic Synthesis, 1977 (G.V. Smith, ed.), Academic Press, New York, 1977.

7. *Catalysis in Organic Synthesis* (W.H. Jones, ed.), Academic Press, New York, 1980.
8. *Catalysis of Organic Reactions* (W.R. Moser, ed.) Marcel Dekker, Inc., New York, 1981.
9. *Catalysis of Organic Reactions* (J.R. Kosak, ed.), Marcel Dekker, Inc., New York, 1984.

Robert L. Augustine

Contents

Contributors

R.G. Austin*, Exxon Chemical Company, 4 Pearl Court, Allendale, New Jersey

Keith M. Bailey, Department of Chemical Engineering, University of Colorado, Box 424, Boulder, Colorado

Werner Brijoux,Max-Planck-Institut für Kohlenforschung, P.O. Box 01 13 25, D - 4330 Mülheim a.d. Ruhr

Helmut Bönnemann, Max-Planck-Institut für Kohlenforschung, P.O. Box 01 13 25, D - 4330 Mülheim a.d. Ruhr

D.R. Bryant, Union Carbide Corporation, Technical Center, South Charleston, West Virginia

M.J. Burk, Department of Chemistry, Yale University, New Haven, Connecticut

Robert L. Burwell, Jr., Department of Chemistry, Northwestern University, Evanston, Illinois

J.N. Cawse, General Electric Company, Plastics Business Group, Pittsfield, Massachusetts

---

Current Affiliation:
*Exxon Chemical Company, Chemical Technology Center, P.O. Box 5200, Baytown, Texas

G.R. Chambers, General Electric Company, Corporate Research and Development, P.O. Box 8, Schenectady, New York

C.H. Chan, Allied Corporation, Petersburg, Virginia

R.H. Crabtree, Department of Chemistry, Yale University, New Haven, Connecticut

A.J. Dennis, Davy McKee Research and Development, Stockton-on-Tees, Cleveland, U.K.

R.P. Dion, Department of Chemistry, Yale University, New Haven, Connecticut

Michael P. Doyle*, Department of Chemistry, Hope College, Holland, Michigan

A. Eisenstadt, SRI International, Menlo Park, California

John L. Falconer, Department of Chemical Engineering, University of Colorado, Box 424, Boulder, Colorado

William B. Fisher, Allied Corporation, Petersburg, Virginia

M.F. Fredericks, Nicolet XRD Co., Fremont, California

C.M. Giandomenico, SRI International, Menlo Park, California

Paul D. Gochis, Department of Chemical Engineering, University of Colorado, Box 424, Boulder, Colorado

D.W. Goodman, Surface Science Division, Sandia National Laboratories, Albuquerque, New Mexico

Jack Halpern, Department of Chemistry, The University of Chicago, Chicago, Illinois

Anders Hansson, W.R. Grace & Co., Washington Research Center, 7379 Route 32, Columbia, Maryland

G.E. Harrison, Davy McKee (London) Ltd., London, U.K. NW1 2 PG

D.R. Herrington, The Standard Oil Company (Ohio), 4440 Warrensville Center Road, Cleveland, Ohio

A.S. Hirschon, SRI International, Menlo Park, California

Current Affiliation:
*Department of Chemistry, Trinity University, San Antonio, Texas

E.M. Holt, Department of Chemistry, Yale University, New Haven, Connecticut

Yuzo Imizu*, Department of Chemistry, Kansas State University, Manhattan, Kansas

Felek Jachimowicz, W.R. Grace & Co., Washington Research Center, 7379 Route 32, Columbia, Maryland

Kenneth J. Klabunde, Department of Chemistry, Kansas State University, Manhattan, Kansas

R.M. Laine, SRI International, Menlo Park, California

M.E. Lavin, Department of Chemistry, Yale University, New Haven, Connecticut

R.C. Lindberg, General Electric Company, Plastics Business Group, Pittsfield, Massachusetts

Jack H. Lunsford, Department of Chemistry, Texas A&M University, College Station, Texas

Tobin J. Marks, Department of Chemistry, Northwestern University, Evanston, Illinois

R.C. Michaelson, Exxon Chemical Company, 4 Pearl Court, Allendale, New Jersey

S.M. Morehouse, Department of Chemistry, Yale University, New Haven, Connecticut

R.S. Myers, Exxon Chemical Company, 4 Pearl Court, Allendale, New Jersey

C.P. Parnell, Department of Chemistry, Yale University, New Haven, Connecticut

Ramesh N. Patel, Exxon Research and Engineering Company, Route 22 East, Annandale, New Jersey

Ronald Pierantozzi, Corporate Science Center, Air Products and Chemicals, Inc., P.O. Box 538, Allentown, Pennsylvania

Norbert Pingel, Max-Planck-Institut für Kohlenforschung, P.O. Box 01 13 25, D - 4330 Mülheim a.d. Ruhr

---

Current Affiliation:
*Kitami Institute of Technology, Kitami, Hokkaido 090, Japan

Dennis P. Riley*, The Procter & Gamble Company, Miami Valley Laboratories, Cincinnati, Ohio

Dieter M.M. Rohe, Max-Planck-Institut für Kohlenforschung, P.O. Box 01 13 25, D - 4330 Mülheim a.d. Ruhr

Bryant E. Rossiter, Hoffman LaRoche, Inc. Nutley, New Jersey

Wolfgang M.H. Sachtler, Department of Chemistry, Northwestern University, Evanston, Illinois.

W.E. Smith, General Electric Company, Corporate Research and Development, P.O. Box 8, Schenectady, New York

John F. Stenberg, Research Laboratories, Eastman Kodak Company, Rochester, New York

J.K. Stille, Department of Chemistry, Colorado State University, Fort Collins, Colorado

R.J. Uriarte, Department of Chemistry, Yale University, New Haven, Connecticut

Jan F. VanPeppen, Allied Corporation, Petersburg, Virginia

Current Affiliation:
*Monsanto Co., 800 N. Lindbergh Blvd., St. Louis, Missouri

PART I

HOMOGENEOUS CATALYSIS

1

Mechanism and Stereochemistry of Asymmetric Catalytic Hydrogenation

Jack Halpern
Department of Chemistry
The University of Chicago
Chicago, Illinois

ABSTRACT

The use of chiral catalysts to effect the asymmetric hydrogenation of prochiral olefinic substrates with high optical yields represents one of the most impressive achievements to date in catalytic selectivity. Notably high optical yields, approaching 100% enantiomeric excess, have been achieved in the hydrogenation of enamides to the corresponding amino acid derivatives, using cationic rhodium catalysts containing chiral phosphine (especially bis(tertiary) phosphines) ligands. The commercial synthesis of L-dopa by such a route constitutes an important practical application of this extraordinarily stereoselective catalysis.

The mechanisms of such asymmetric hydrogenation reactions and the origin of the enantioselection will be discussed. A remarkable conclusion of the studies to be described is that, contrary to the hitherto prevailing "lock and key" view concerning these and related stereoselective catalysts, the enantioselection in these systems is determined not by the preferred mode of initial binding of the prochiral substrate to the chiral catalyst but, rather, by the much higher reactivity of the minor diastereomer of the catalyst-substrate adduct corresponding to the less favored binding mode. This result serves to explain the unusual effects of temperature and hydrogen pressure on the optical yields obtained with such catalysts.

I. INTRODUCTION

The use of chiral catalysts to effect the asymmetric hydrogenation of prochiral olefinic substrates with high optical yields represents one of the most impressive achievements to date in catalytic selectivity, rivaling the corresponding stereoselectivity of enzymic catalysts.[1,2] Notably high optical yields, approaching 100% enantiomeric excess (e.e.), have been achieved in the hydrogenation of α-acylaminocinnamic acid derivatives such as **1** to the corresponding phenylalanine derivatives (Eq. 1) using cationic rhodium complexes containing chiral diphosphine ligands such as R,R-DIPAMP, **2**, or S,S-CHIRAPHOS, **3**, (generally designated as $\widehat{P*P}$) as catalysts, yielding predominantly S- and R-phenylalanine derivatives, respectively. The commercial synthesis of L-dopa (3,4-dihydroxyphenylalanine) by such a route constitutes an important practical application of this remarkably stereoselective catalysis.

$$(H)(C_6H_5)C{=}C(COR)(NHCCH_3{=}O) + H_2 \xrightarrow{[Rh(\widehat{P*P})]^+} C_6H_5CH_2\overset{*}{C}(COR)(H)(NHCR{=}O) \quad (1)$$

1a (MAC) $R{=}CH_3$ R or S

1b (EAC) $R{=}C_2H_5$

This paper discusses the mechanistic aspects of such reactions and the origin of the enantioselection.

II. MODEL STUDIES USING CATALYSTS CONTAINING THE ACHIRAL DIPHOSPHINE LIGAND, DIPHOS

Recognizing that several of the most effective and widely used chiral ligands for homogeneous catalytic hydrogenation, such as **2** and **3**, are simple derivatives of the familiar achiral ligand 1,2-bis(diphenylphosphino)ethane (abbreviated DIPHOS, **4**), our initial kinetic and mechanistic studies were

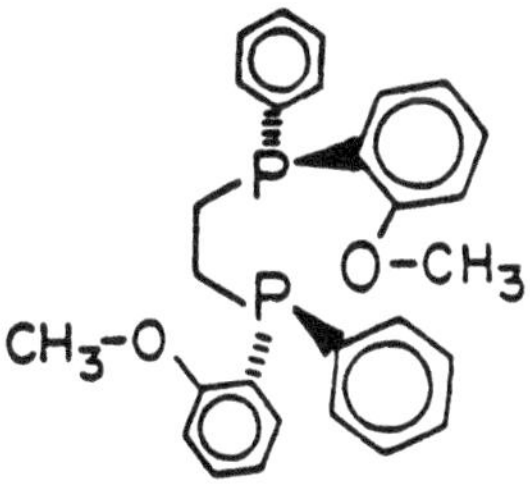

2 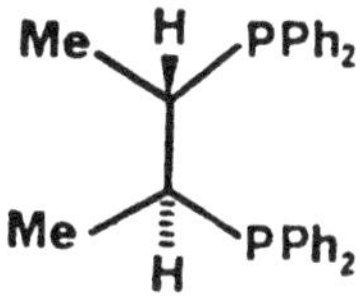 3 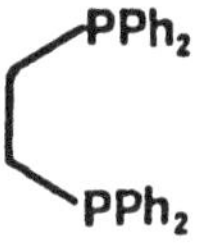 4

directed at catalytic systems containing the latter.[3-5] The mechanism of Eq. 1, catalyzed by $[Rh(DIPHOS)]^+$, as deduced from studies encompassing kinetic measurements as well as characterization of several intermediates by spectroscopic (notably NMR) and structural methods, is depicted in Fig. 1. The kinetic parameters are summarized in Table I.

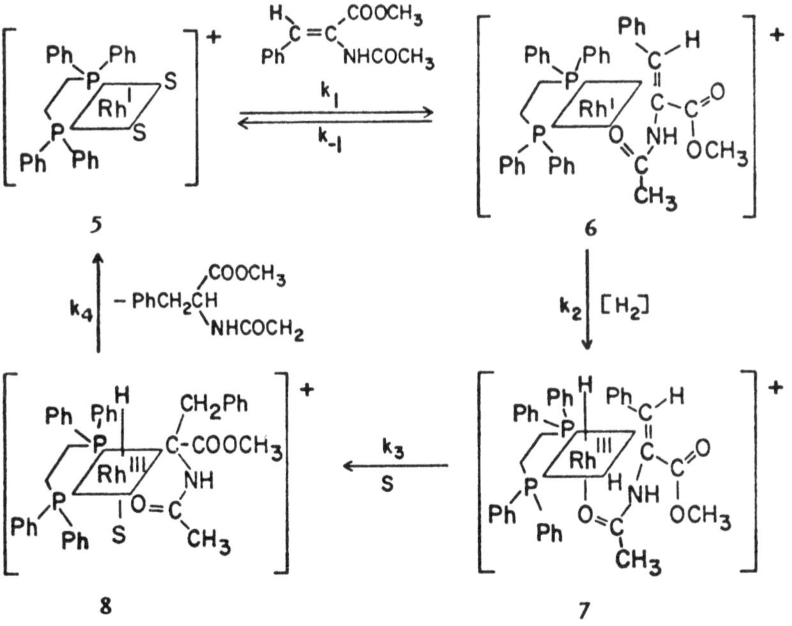

Fig. 1. Mechanism of the $[Rh(DIPHOS)]^+$-catalyzed hydrogenation of methyl-(Z)-α-acetamidocinnamate (MAC).

TABLE I

Kinetic Parameters for the [Rh(DIPHOS)]$^+$ Catalyzed Hydrogenation of Methyl-(Z)-α-Acetamidocinnamate in Methanol[5,7]

Rate constant	(units)	k(25°C)	ΔH≠ (kcal/mol)	ΔS≠ (cal/mol °K)
k_1	(M^{-1} sec^{-1})	1.4×10^4		
k_{-1}	(sec^{-1})	5.2×10^{-1}	18.3	+2
k_2	(M^{-1} sec^{-1})	1.0×10^2	6.3	-28
k_3	(sec^{-1})	1		
k_4	(sec^{-1})	23	17.0	+6

The formation of the [Rh(DIPHOS)(MAC)]$^+$ adduct, **6**, is rapid and essentially complete ($K^{eq}_1 = k_1/k_{-1}$ = [6]/[5][MAC] = 2×10^4 M^{-1} at 25°C) even at moderate MAC concentrations. The structure of **6** was established by NMR (^{31}P, ^{13}C and ^{1}H) spectroscopy and by single crystal X-ray analysis, revealing chelation of the MAC substrate through the carbonyl oxygen of the amide group as well as through the symmetrical (η^2) coordination of the C=C bond.[4]

Under ambient conditions, the second step of the catalytic cycle (corresponding to k_2), i.e., the reaction of [Rh(DIPHOS)(MAC)]$^+$ with H_2, was found to be rate-determining for the overall catalytic hydrogenation reaction and, thus, [Rh(DIPHOS)(MAC)]$^+$ is the principal species present under steady state conditions of the catalytic reaction.[5] However, the final product-forming reductive elimination step (corresponding to k_4) exhibited a sufficiently higher activation enthalpy compared with k_2 (17.0 vs 6.3 kcal/mol) that this step became rate-limiting below -40°C, permitting the hydridoalkyl intermediate **8** to be intercepted and characterized.

III. HYDROGENATION WITH CATALYSTS CONTAINING CHIRAL DISPHOSPHINE LIGANDS

A. Mechanism and Origin of Enantioselection

With the exception of the features identified below that are specifically related to the formation of diastereomeric forms of the adduct corresponding to **6** and subsequent reaction intermediates, the chemistry of cationic rhodium complexes of DIPAMP, **2**, and CHIRAPHOS, **3**, and the mechanisms of catalysis by these complexes of prochiral substrates such as MAC, **1**, parallel closely those of the corresponding DIPHOS complexes.[6-10] When the mechanistic scheme of Fig. 1 is extended to catalysts containing such chiral ligands, it must be modified in accord with Fig. 2 to accommodate the formation of diasteromeric forms of **6** (i.e., **6'** and **6"**) and of the subsequent reaction intermediates. The stereochemical features of the various steps of the catalytic cycle (i.e., **6** → **7**; **7** → **8** and **8** → **5** are known, making it possible to correlate the absolute configurations of the two enantiomeric products with the absolute configurations of the intermediate species of the two diastereomeric manifolds as depicted in Fig. 2.

The enantioselection in these reactions is determined by the first irreversible step, i.e., **6** → **7**. Two limiting interpretations may be accorded to the origin of this enantioselection, namely:

(a) The prevailing product chirality is determined by the preferred mode of initial binding of the substrate to the catalyst (i.e., thermodynamic preference for the formation of **6'** or **6"**), the <u>predominant</u> enantiomer of the product arising from the <u>predominant</u> diastereomer of the catalyst-substrate adduct.

(b) The <u>predominant</u> enantiomer of the product arises from the <u>minor</u> (thermodynamically unfavorable) diastereomer of the catalyst-substrate adduct by virtue of the much higher reactivity of the latter, compared with that of the predominant diastereomer, toward H_2.

Because the formation and dissociation of the catalyst-substrate adducts, and hence interconversion and equilibration of their diastereomeric forms, are rapid (compared with their reactions with H_2) at ambient

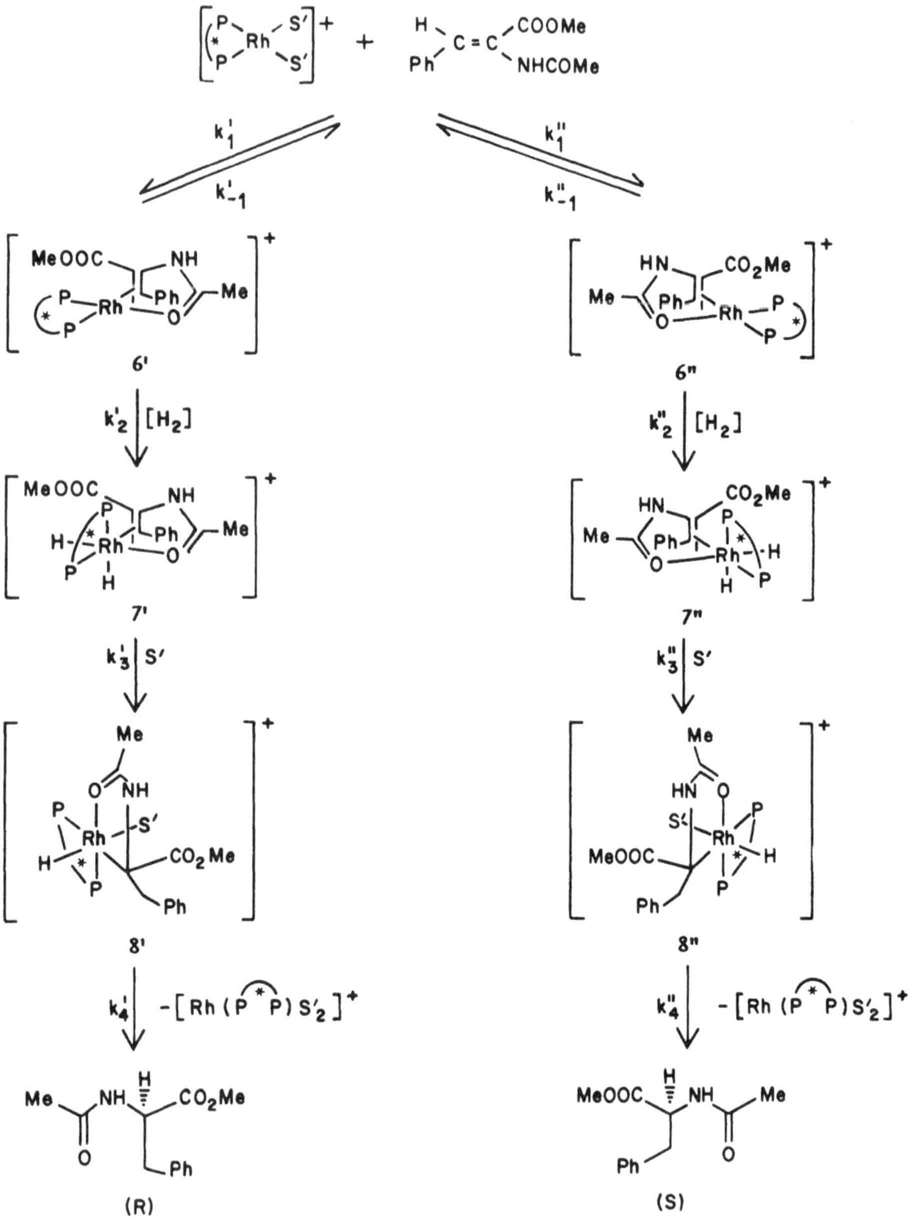

Fig. 2. Mechanistic scheme for the hydrogenation of a prochiral substrate (MAC) with a catalyst containing a chiral chelating diphosphine ligand ($\widehat{P*P}$), (S = methanol).

temperatures and H_2 pressures, these two alternatives are kinetically indistinguishable under the usual catalytic conditions. The first of the above interpretations, which corresponds most closely to the familiar "lock and key" concept that has been invoked so widely to explain the characteristically high selectivities of enzymic catalysts, initially appeared to be more attractive on conceptual grounds and was widely accepted and supported by various claims[9b,11] of indirect "evidence" which, however, proved to be unreliable and invalid.

B. [Rh(S,S-CHIRAPHOS)]$^+$-Catalyzed Hydrogenation

To distinguish between these alternative interpretations and establish the origin of the enantioselection, it was necessary to correlate the absolute configurations of the products with those of the intermediate catalyst-substrate adducts. This was first achieved for the hydrogenation of ethyl-(Z)-α-acetamidocinnamate (EAC, **1b** catalyzed by the rhodium complex of S,S-CHIRAPHOS, **3**, in accord with Eq. 2.[6,12] This reaction was shown to proceed with high enantioselectivcity yielding N-acetyl-(R)-phenylalanine

$$\mathbf{1b} + H_2 \xrightarrow{[Rh(S,S\text{-CHIRAPHOS})]^+} C_6H_5CH_2\text{-}C(COC_2H_5)(H)(NH\text{-}C(=O)\text{-}CH_3) \quad (2)$$

ethyl ester (>95 per cent enantiomeric excess).

The essential features of Eq. 2 were found to parallel those of the corresponding [Rh(DIPHOS)]$^+$-catalyzed reaction as depicted by Fig. 1. Formation of a [Rh(S,S-CHIRAPHOS)(EAC)]$^+$ adduct, **9**, analogous to **6** occurred with a similar equilibrium constant. The electronic spectrum and ^{31}P NMR spectrum of **9** also were virtually identical to those **6.** Only a single diastereomer of [Rh(S,S-CHIRAPHOS)(EAC)]$^+$ could be identified in solution by NMR and, hence, it could be concluded that the other diastereomer (which is expected to exhibit a distinguishable NMR spectrum) must be present to the extent of less than 5 per cent.[6]

The rate law for Eq. 2 was found to be similar to that for the [Rh(DIPHOS)]$^+$ -catalyzed reaction, namely $-[H_2]/dt = k_5[H_2][\mathbf{9}]$ with

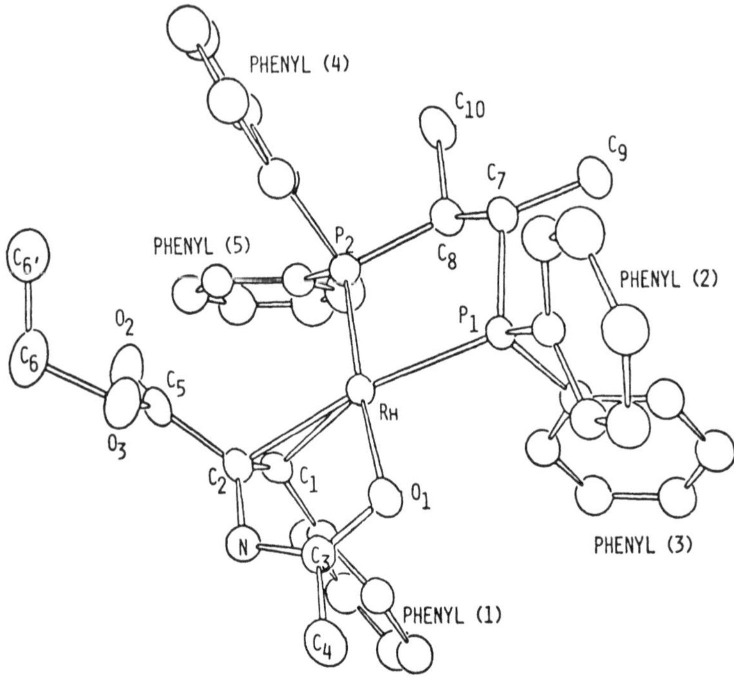

Fig. 3. Structure of the predominant diastereomer of [Rh(S,S-CHIRAPHOS)(EAC)]$^+$, **9.**

k_5 = 1.6 M^{-1} sec^{-1}, i.e., only about 1/60 the corresponding value (k_2) for the [Rh(DIPHOS)]$^+$ -catalyzed reaction. This rate difference represents the only significant disparity revealed by the various solution measurements on the [Rh(DIPHOS)]$^+$ and [Rh(S,S-CHIRAPHOS)]$^+$ systems. It provided the first indication that the major [Rh(CHIRAPHOS)(EAC)]$^+$ diastereomer might be unreactive, the low apparent value of k_5 reflecting the low concentration of the reactive minor diastereomer.[6]

The structure of the predominant diastereomer of the [Rh(S,S-CHIRAPHOS)(EAC)]$^+$ ion, determined by X-ray analysis of single crystals of the perchlorate salt, is depicted in Fig. 3 and is essentially identical to that previously determined for **6.** Of crucial significance in the present context is the finding that the $C_{\alpha\text{-}re}$ face of EAC is coordinated to the Rh atom. Addition of H_2 to this face, in accord with the mechanism of Fig. 1, would yield N-acetyl-(S)-phenylalanine ethyl ester. Instead, it was found that the predominant product of Reaction 2 (>95% e.e.) was the R isomer.

These results lead to the conclusion that, contrary to the prevalent earlier views, it is not the preferred mode of initial binding of the prochiral substrate to the catalyst but, rather, differences in the rates of the subsequent reactions of the diastereomeric catalyst-substrate adducts with H_2, that dictate the enantioselectivity of these catalyst systems. Apparently, the minor diastereomer is sufficiently more reactive than the major one that it determines the predominant chirality of the product. To account for the observed enantioselection this difference in reactivity (k_2'/k_2'') must be greater than 10^3.

C. $[Rh(R,R\text{-DIPAMP})]^+$-Catalyzed Hydrogenation

Reinforcement of this interpretation was provided by related studies on the hydrogenation of MAC, **1a**, using the catalyst containing the chiral diphosphine ligand, R,R-DIPAMP, **2**, which yields N-acetyl-(S)-phenylalanine methyl ester in high optical yield (>95% enantiomeric excess).[8,9c] In this case both diastereomers of the $[Rh(DIPAMP)(MAC)]^+$ adduct could be detected in solution by NMR. Equilibrium and kinetic measurements yielded the values listed in Table II for the constants defined by Eq. 3 and 4.[8]

$$[Rh(DIPAMP)]^+ + MAC \;\; \begin{array}{l} \underset{k_{-6}}{\overset{k_6}{\rightleftharpoons}} [Rh(DIPAMP)(MAC)]^+_{major} \\ \underset{k'_{-6}}{\overset{k'_6}{\rightleftharpoons}} [Rh(DIPAMP)(MAC)]^+_{minor} \end{array} \qquad (3)$$

$$K^6_{eq} = \frac{K^{Major}_{eq}}{K^{Minor}_{eq}} = \frac{k_6/k_{-6}}{k'_6/k'_{-6}} \qquad (4)$$

TABLE II

Kinetic Data for the [Rh(R,R-DIPAMP)]$^+$-catalyzed Hydrogenation of Methyl-(Z)- α-acetamidocinnamate in Methanol[8]

Constant (units)	Value at 25°C	$\Delta H^{\neq}$ or ΔH^o kcal/mol	$\Delta S^{\neq}$ or ΔS^o cal/mol °K
k_6 ($m^{-1}sec^{-1}$)	5.3×10^3	4.9	-25
k_{-6} (sec^{-1})	1.5×10^{-1}	13.3	-18
k_7 ($M^{-1}sec^{-1}$)	1.1	10.7	-23
k'_6 ($M^{-1}sec^{-1}$)	1.0×10^4	6.9	-17
k'_{-6} (sec^{-1})	3.2	13.0	-13
k'_7 ($M^{-1}sec^{-1}$)	6.3×10^2	7.5	-21
K_6^{eq} (M^{-1})	3.5×10^4	-8.4	-7
$K_6^{eq'}$ (M^{-1})	3.1×10^3	-6.1	-4

Under ambient conditions, interconversion and equilibration of the two diastereomeric adducts is rapid compared with their reactions with H_2; hence, they react together, essentially as a single manifold, and it is not possible to deduce which product enantiomer is derived from each of the diastereomeric adducts. However, cooling the solutions to ca -35°C resulted in freezing out of the interconversion of the diastereomeric adducts so that the reaction of each with H_2 could be separately and directly observed. Under these conditions it was found that the minor diastereomer reacted rapidly with H_2 to give the major (i.e., S) enantiomer of the product. The major diastereomer reacted much more slowly, principally by interconversion to the minor diastereomer followed by rapid reaction of the latter with H_2.[8] Thus, these observations provide direct evidence for the conclusions that the major enantiomer of the product is derived from the

$$\frac{-d[\mathrm{Rh(DIPAMP)(MAC)}]^+_{major}}{dt} = \frac{d[\mathrm{Rh(DIPAMP)(MAC)}]^+_{minor}}{dt} = \frac{k_{-6}k'_6[\mathrm{Rh(DIPAMP)(MAC)}]_{major}}{k_6 + k'_6} \quad (5)$$

minor catalyst-substrate adduct by virtue of the much higher reactivity of the latter.

The results of these measurements, as well as of kinetic measurements on the catalytic reaction, are summarized by Scheme 1 and by the data in Table II.

Application of the steady state treatment to Scheme 1 yields the rate laws corresponding to Eq. 6 and 7 for the formation of the two enantiomeric products, D- and L-N-acetylphenylalanine methyl ester, respectively. These

$$\frac{d[D]}{dt} = k_7[H_2][Rh]_{total} \quad (6)$$

$$\frac{d[L]}{dt} = \frac{k'_6k'_7[H_2][Rh]_{total}}{(k_6/k_{-6})(k'_{-6} + k'_7[H_2])} \quad (7)$$

are in good agreement with the observed kinetics.[8]

The enantioselectivity of this catalyst system results from the much higher reactivity of $[\mathrm{Rh(DIPAMP)(MAC)}]^+_{minor}$, compared with $[\mathrm{Rh(DIPAMP)(MAC)}]^+_{major}$, toward H_2. This reactivity ratio, k'_7/k_7 = 630/1.1 = 573, is offset by the relative concentrations of $[\mathrm{Rh(DIPAMP)(MAC)}]^+_{minor}$ and $[\mathrm{Rh(DIPAMP)(MAC)}]^+_{major}$, whose equilibrium ratio, $(K^6_{eq})'/(K^6_{eq})$, is 0.09 at 25°. These values yield a limiting product ratio at low H_2 pressures (*vide infra*), [L]/[D] = 52:1, corresponding to the enantiomeric excess of about 96% that is obtained under these conditions.

D. Effect of Temperature and H_2 Pressure

According to the rate laws of Eq. 6 and 7 the rate of formation of the N-acetyl-D-phenylalanine methyl ester increases linearly with the H_2

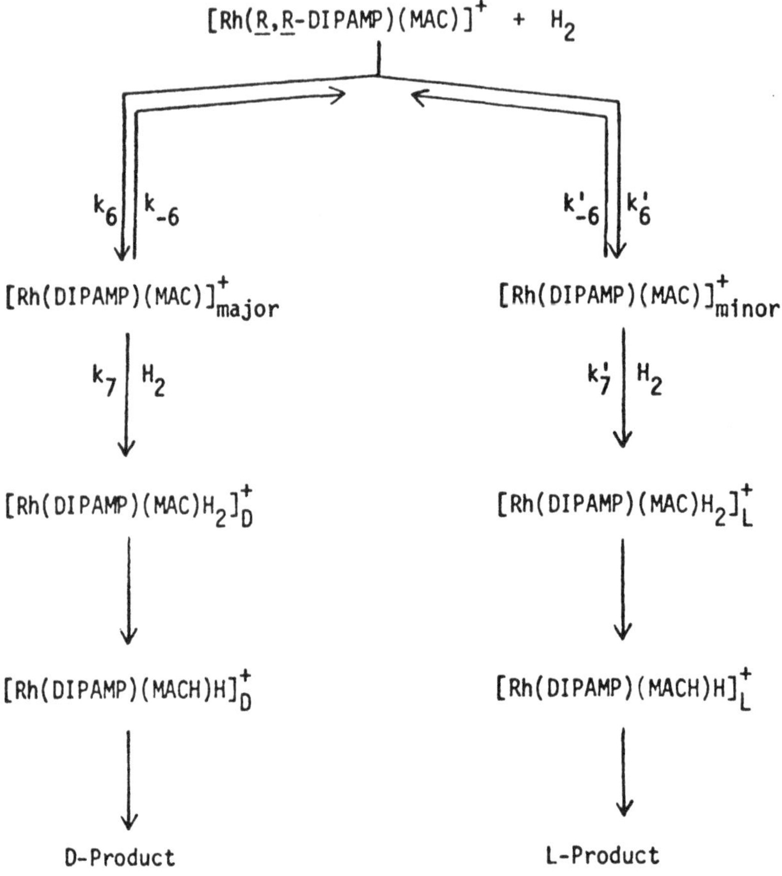

Scheme 1

concentration or partial pressure, while the rate of formation of the major L-enantiomer exhibits saturation behavior. Accordingly, the enantiomeric excess falls off with increasing H_2 pressure as depicted in Fig. 4.[8]

The high enantioselectivity that this catalyst system is capable of achieving depends on rapid interconversion of the two $[Rh(DIPAMP)(MAC)]^+$ diastereomers, compared with their reactions with H_2, since only through such rapid interconversion can the major diastereomer, which constitutes the predominant form of the rhodium under steady state catalytic conditions, cross over into the minor diastereomer manifold leading to the formation of the major (L) product. This condition

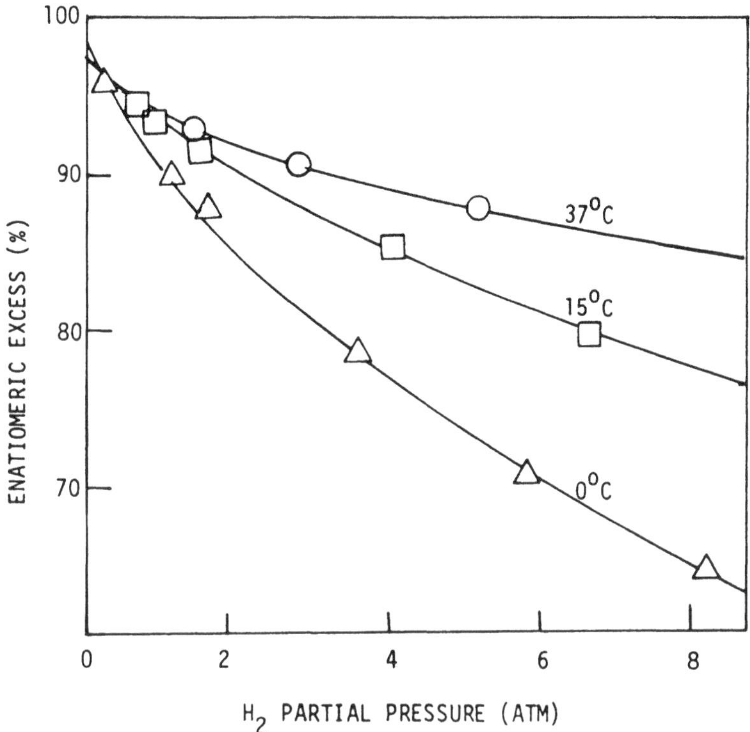

Fig. 4. Dependence of the enantioselection of the $[Rh(\underline{R},\underline{R}\text{-DIPAMP})]^+$-catalyzed hydrogenation of MAC on the temperature of H_2 partial pressure.[8]

is attained at low H_2 pressures (corresponding to slow reactions of the diastereomeric $[Rh(DIPAMP)(MAC)]^+$ adducts with H_2) and/or high temperatures since the activation enthalpy of the dissociation of $[Rh(DIPAMP)(MAC)]^+_{major}$ ($\Delta H^{\neq}_{-6}$ = 13.3 kcal/mol), which is the rate-determining step of the transformation of the latter to $[Rh(DIPAMP)(MAC)]^+{}_{minor}$, is significantly higher than the activation enthalpies of the reactions of the adducts with H_2 ($\Delta H^{\neq}_7$ = 10.7 kcal/mol; $(\Delta H^{\neq}_7)'$ = 7.5 kcal/mol). Under limiting conditions of low H_2 pressure and/or high temperature the enantioselection is given by Eq. 8. In the other limit of high H_2 pressure and/or low temperature, trapping of the diastereomeric $[Rh(DIPAMP)(MAC)]^+$ adducts by reaction with H_2 is fast compared with

$$\frac{[D]}{[L]} = \frac{k_7\,(k_6/k_{-6})}{k'_7(k'_6/k'_{-6})} \qquad (= 52 \text{ at } 25°C) \qquad (8)$$

$$\frac{[D]}{[L]} = \frac{k_6}{k'_6} \qquad (= 2.0 \text{ at } 25°C) \qquad (9)$$

their rates of interconversion and the product ratio is determined by the relative rates of formation of the two diastereomeric adducts (Eq. 9).

The influence of temperature and H_2 pressure on the enantioselectivity of the $[Rh(DIPAMP)]^+$-catalyzed hydrogenation of MAC is depicted in Fig. 4. The results are in quantitative agreement with the predictions of Eq. 6 and 7, using the values of $\Delta H^{\neq}$ and $\Delta S^{\neq}$ listed in Table II. A remarkable result, that is explained by this mechanism, is the finding that the enantioselectivity increases with increasing temperature and, indeed, that increasing temperature is markedly effective in counteracting the detrimental effect of high H_2 pressures on the selectivity.[8]

The distinctive pattern of temperature and H_2 pressure effects, depicted by Fig. 4, can serve as a criterion of the origin of enantioselection in other catalyst systems where the direct criteria that have been employed for the $[Rh(CHIRAPHOS)]^+$ and $[Rh(DIPAMP)]^+$ systems cannot readily be invoked. Indeed, qualitatively similar patterns of temperature and H_2 pressure have been found for the hydrogenation of α-acylamino acid esters by virtually every other catalyst system (i.e., with different chiral diphosphine ligands) for which these variables have been examined.[13,14] The inverse dependence of enantioselectivity on the temperature also appears to be widespread.[14] In one particularly dramatic case, the optical yield increases from 0 to 60 per cent enantiomeric excess in going from 0° to 100°C.[15] In view of the generality of these trends, it may be concluded that the same interpretation of the origin of enantioselection extends to this whole class of catalysts and substrates.

E. Origin of the Difference of Stability and Reactivity of the Diastereomeric Catalyst-Substrate Adducts

Our conclusion concerning the origin of enantioselection in these systems implies very large differences in reactivity toward H_2 between the two diastereomeric forms of the catalyst-substrate adduct. With $[Rh(CHIRAPHOS)(EAC)]^+$ the minor diastereomer must be at least 10^3 times as reactive toward H_2 at 25°C as the major (more stable) diastereomer in order to account for the observed enantiomeric excess. In the case of $[Rh(DIPAMP)(MAC)]^+$, where both diastereomers are observed, the corresponding reactivity difference at 25°C has been directly measured to be 573. The origin of these marked differences in reactivity clearly constitutes an important aspect of the behavior of these catalytic systems.

A general feature of the conformations of the bis(diarylphosphine) ligands that are common to these catalysts is the arrangement of the four phenyl rings in a chiral alternating "edge-face" array depicted schematically for [Rh(R,R-DIPAMP)]$^+$ by structure **10** and also revealed by the structure

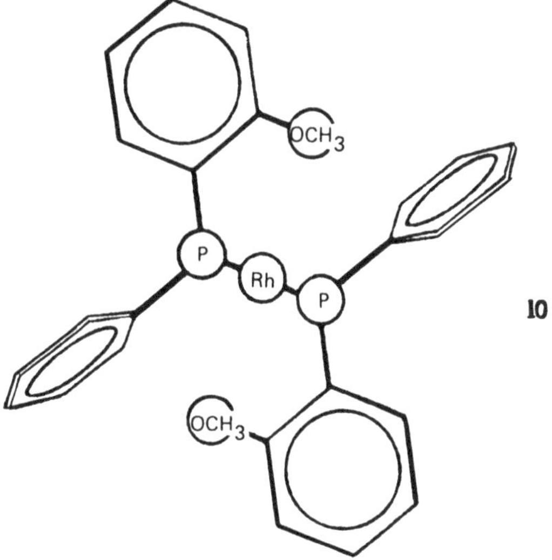

10

of Fig. 3. The enantiorecognition involved in the binding of prochiral substrates such as (Z)-α-acylaminocinnamic acid derivatives to such a chiral template, and the differences in stability between the diastereomers of the resulting adducts (for example, of [Rh(R,R-DIPAMP)(MAC)]$^+$ or [Rh(S,S-CHIRAPHOS)(EAC)]$^+$), have been attributed to "matching" of the two enantiofaces of the prochiral substrate to the chiral "template" corresponding to this array of phenyl rings.[11,16] The chirality and rigidity of this template, in turn, are determined by the nature of the chiral centers on the phosphine ligand and of the backbone connecting the two phosphorus atoms.[16] This model, as reflected in the structure depicted in Fig. 3, appears to provide a satisfactory basis for interpreting the stability difference between the diastereomers of the initial catalyst substrate adducts, e.g., ΔG^o = 1.4 kcal/mol, ΔH^o = 2.2 kcal/mol and ΔS^o = 2.7 kcal/(mol °K) for [Rh(R,R-DIPAMP)(MAC)]$^+$.

The origin of the striking differences between the reactivities of the diastereomeric adducts toward H_2 is less clear. It is, of course, not unexpected that the less stable of a pair of diastereomers will exhibit the higher reactivity by virtue of its higher initial free energy. However, to account for the enantioselectivity of the reaction, the difference in reactivity must be much greater than the difference in stability (i.e., than the difference in equilibrium concentrations) of the diastereomers, indeed at least 50 times greater to account, for an optical yield of 96 per cent enantiomeric excess. The 576-fold reactivity difference of the two diastereomers of [Rh(R,R-DIPAMP)(MAC)]$^+$ corresponds to $\Delta G^{\neq}$ of 3.7 kcal/mol ($\Delta H^{\neq}$ = 3.2 kcal/mol; $\Delta S^{\neq}$ = 2 cal/mol °K). This is made up of contributions of ΔG = 1.4 kcal/mol, representing the initial difference in free energy between the two diastereomeric adducts and ΔG = 2.3 kcal/mol representing the difference between the transition state free energies. The limiting enantioselectivity is determined solely by the latter difference.

At this stage, the origin of this marked reactivity difference is still a matter of speculation. A plausible interpretation is that the reactivity differences has its origin in the stability difference of the diastereomers of the initial product of the oxidative addition of H_2 (i.e., [RhH_2(DIPAMP)(MAC)]$^+$, etc.) the relative stabilities of the diasteromeric products being opposite to that of the parent catalyst-substrate adducts. Thus, the greater stability of the diastereomer of the dihydride [RhH_2(R,R-DIPHOS)(MAC)]$^+$, derived from the less stable diastereomer of [Rh(R,R-DIPHOS)(MAC)]$^+$, enhances the driving force and rate of reaction of the latter with H_2, compared with the rate of the more stable diastereomer. The reaction profiles corresponding to the behavior, which is characterized by crossing of the two profiles, are depicted in Fig. 5.

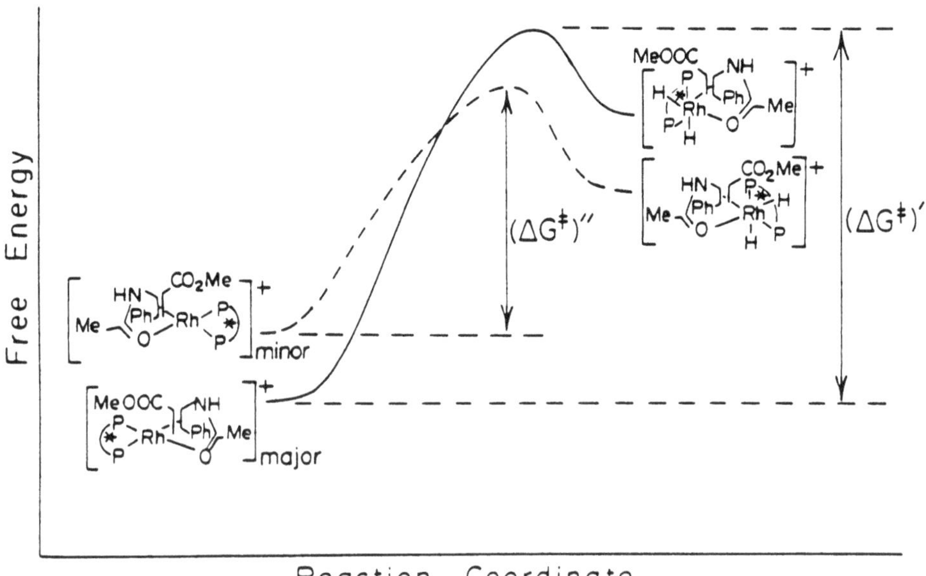

Fig. 5. Schematic reaction coordinate profiles for the enantio-determining reactions of the diastereomeric $[Rh(DIPAMP)(MAC)]^+$ catalyst-substrate adducts with H_2.

Since the behavior depicted in Fig. 5 appears to be quite general for this entire class of asymmetric hydrogenation reactions (i.e., involving a variety of different chiral phosphine ligands and substrates) its origin must reflect some systematic feature of the reactions. A plausible suggestion is that the reason for the inverted stabilities of the diastereomers of the initial catalyst-substrate adducts and the dihydrides derived from them, is the trans-disposition of the substrate and diphosphine chelate rings in the former case and the cis-disposition in the latter (Fig. 5). Unfortunately, space filling models (particularly of the dihydrides) are too crowded to afford a reliable test of this suggestion. Further elucidation of this important theme must await the interception and characterization of the dihydride intermediates.

E. Concluding Remarks

The most significant and unexpected conclusion yielded by the studies on the mechanism of asymmetric catalytic hydrogenation described in this chapter is that, contrary to the prevalent earlier view, it is not the preferred mode of initial binding of the prochiral substrate to the chiral catalyst, but rather an overcompensating difference in the rates of the subsequent reactions of the diastereomeric catalyst-substrate adducts that dictates the enantioselectivity of asymmetric catalytic hydrogenation. The predominant product arises from the minor (less stable) diastereomer of the adduct, which frequently does not acumulate in sufficient concentration to be detected. In contrast, the major diastereomer whose stability reflects the optimal fitting of the prochiral substrate to the chiral catalyst template, and which is the principal species present in solution under reaction conditions, is unreactive and corresponds to a "dead end" complex.

An important question that remains to be addressed is how widely our conclusion concerning the origin of enantioselection applies to other stereoselective catalysts. Indeed, for several other asymmetric catalytic reactions, e.g., hydroformylation[17] the opposite interpretation has been proposed, namely that the enantioselectivity is determined by the preferred mode of initial binding of the prochiral substrate to the chiral catalyst. However, the mechanisms of these reactions remain to be fully elucidated and, at this stage, the evidence for the proposed interpretations is indirect and not entirely convincing.

Particularly significant in this context is the question of the relevance of our conclusions concerning the origin of enantioselection to enzymic catalysts where a contrary interpretation (embodied in the familiar lock and key concept) often is assumed, namely that the stable adducts that are formed by optimal matching of the substrate to the enzyme, and that so frequently are the focus of direct observation and characterization, are intermediates in the enzymic reactions. Since the formation of such adducts typically is rapid compared with the catalytic reactions, direct support for such a conclusion often is lacking and the results of our studies on the mechanism of asymmetric hydrogenation raise serious questions about its widely accepted plausibility. The criteria for distinguishing between the alternative interpretations that need to be considered are not readily realized for enzymic reactions, but our studies on asymmetric catalytic hydrogenation do provide some leads as to how such criteria might be developed.

ACKNOWLEDGMENT

The research on which this article is based was supported by a grant from the National Science Foundation.

REFERENCES

1. For general reviews on asymmetric catalytic hydrogenation and related processes, see: (a) H.B. Kagan, in **Comprehensive Organometallic Chemistry** (G. Wilkinson and F.G.A. Stone, eds.), Pergamon, Oxford 1982, Vol. 8 pp. 463-498. (b) B. Bosnich and M. Fryzuk, Top. Stereochem., **12**, 119 (1981). (c) J. Halpern, Science, **217**, 401 (1982).
2. B.D. Vineyard, W.S. Knowles, and M.J. Sabacky, J. Mol. Catal., **19**, 159 (1983).
3. J. Halpern, D.P. Riley, A.S.C. Chan, and J.J. Pluth, J. Am. Chem. Soc., **99**, 8055 (1977).
4. A.S.C. Chan, J.J. Pluth, and J. Halpern, Inorg. Chim. Acta, **37**, 2477 (1979).
5. A.S.C.Chan and J. Halpern, J. Am. Chem. Soc., **102**, 838 (1980).
6. A.S.C.Chan, J.J. Pluth and J. Halpern, J. Am. Chem. Soc, **102**, 5952 (1980).
7. A.S.C.Chan, Ph.D. Dissertation, The University of Chicago, 1979.
8. C.R. Landis, Ph.D. Dissertation, The University of Chicago, 1983.
9. J.M. Brown and P.A. Chaloner, (a) J. Chem. Soc., Chem. Commun., 321 (1978). (b) Tetrahedron Lett., 1877 (1978). (c) J. Chem. Soc., Chem. Commun., 344 (1980). (d) J. Am. Chem. Soc., **102**, 3040 (1980).
10. J.M. Brown, L.R. Canning, A.J. Downs,, and A.M. Forster, J, Organomet. Chem., **255**, 103 (1980).

11. W.S. Knowles, B.D. Vineyard, M.J. Sabacky, and B.R. Stults, in **Fundamental Research in Homogeneous Catalysis,** (M. Tsutsui, ed.), Plenum, New York, 1979, **Vol. 3** p. 537.
12. P.S. Chua, N.K. Roberts, B. Bosnich, S.J. Okrasinski, and J. Halpern, J. Chem. Soc., Chem. Commun., 1278 (1981).
13. W.S. Knowles, M.J. Sabacky and B.D. Vineyard, Advan. Chem. Series, **132**, 274 (1974).
14. I. Ojima, T. Kogure, and N. Yoda, J. Org. Chem., **45**, 4728 (1980).
15. D. Sinou, Tetrahedron Lett., 2987 (1981).
16. B. Bosnich and N.K. Roberts, Advan. Chem. Series, **196**, 337 (1982).
17. G. Consiglio and P. Pino, (a) Topics Curr. Chem., **105**, 77 (1982). (b) Advan. Chem. Series, **196**, 371 (1982).

2

Asymmetric Hydroformylation via Polymer-Supported Chiral Catalysts

J.K. Stille
Department of Chemistry
Colorado State University
Fort Collins, Colorado 80523

ABSTRACT

The suspension copolymerization of styrene, divinylbenzene and SS-2-p-styryl)-4,5-bis[(tosyloxy)methyl]-1,3-dioxolane **(SS-1)** gave cross-linked beads containing 10% incorporation of **SS-1.** This polymer was converted to the polymer containing chiral 4,5-bis[(diphenylphosphino)methyl]-1,3-dioxolane (DIOP-type ligand) **SS-4a** or the corresponding dibenzophosphole (DIPHOL-type ligand) **SS-4b.** Exchange of a Rh(I)-carbonyl species gave polymer-supported catalysts which were used to hydroformylate olefins. Compared to the corresponding homogeneous catalysts, these polymer-supported catalysts gave unique selectivities and equivalent optical yields. A polymer obtained from the copolymerization of styrene, divinylbenzene and N-acryloyl-(2S,4S)-4-(diphenylphosphino)-2-[(diphenylphosphino)methyl] pyrrolidine **(SS-3)** gave polymer beads onto which platinum chloride and stannous chloride were exchanged. Hydroformylation of styrene with this catalyst gave relatively low branched to normal ratios ($\sim$0.5), but gave optical yields (70-75% ee) of hydrotropaldehyde that were comparable to those obtained from the analogous homogeneous catalyst (70-80% ee). In all of these hydroformylations, the catalyst could be separated from the product by filtration and used again without giving a decrease in the optical yield of hydrotropaldehyde.

I. INTRODUCTION

Homogeneous transition metal catalysis of organic reactions has the advantage of being more selective and more stereospecific than the heterogenously catalyzed reactions, or the analogous reactions carried out without the aid of catalysis. The greater selectivity and stereospecificity is due in part, to the mild reaction conditions and the ability of the transition metal to assemble the reactants at one site. However, a practical disadvantage of performing homogeneously catalyzed reactions in the liquid phase, is the difficulty in separating the product from the catalyst and recycling the catalyst or removing the product continuously. Recovery of the catalyst becomes an important concern when the transition metals such as rhodium, palladium, and platinum or the chiral ligands are expensive.

To overcome the difficulty of separating the product and catalyst, catalysts have been attached to a variety of insoluble supports, including crosslinked polymers.[1] These catalysts have the advantage that they are insoluble, may be readily separated, and retain the same reactivity as the homogeneous catalyst. Most of the synthetic polymer supports for catalysts containing polymer attached phosphine ligands are polystyrenes, and have been prepared by carrying out various reactions on the crosslinked polystyrene bead.

In designing a polymer support for a catalyst, a number of requirements should be met. It should be easily prepared from low cost materials. The support must be compatible with the solvent but be chemically and thermally stable under the reaction conditions. The polymer-supported catalyst should show the same rates as the homogeneous catalyst or at most show only minimal losses in rate, and should be able to be recycled many times without loss in activity (or loss of transition metal). Finally, the interaction between the catalytic site and the support must be either negligible or beneficial.

Asymmetric catalytic reactions are especially valuable for the synthesis of chiral organic compounds. Of the methods of asymmetric synthesis, the generation of an enantiomeric product from a prochiral reactant by the use of an optically active catalyst or enzyme has several advantages, the most important of which is that resolution, if necessary, is achieved with the catalyst instead of with the product. Thus, large amounts of product of high enantiomeric excess can be obtained from catalytic quantities of optically active catalyst. Amino acids and dipeptides, for example, can be synthesized in high enantiomeric excess by the asymmetric hydrogenation of achiral substrates, such as α-N-acylaminocinnamic acids, with optically active Wilkinson-type catalysts.[2] In these reactions, neither the rhodium nor the chiral phosphine ligand is readily recovered.

Although initial attempts to effect these same asymmetric hydrogenations with chiral catalysts supported on polymers were not particularly successful[3-8] (high enantiomeric excesses were not achieved), the careful design of the polymer allowed the same high enantiomeric excesses to be achieved as were attained in the analogous homogeneous reactions.[9-18] In the synthesis of a polymer-supported catalyst, the polymer support must be compatible with the reaction solvent. Also, it is desirable to introduce the chiral phosphine into the polymer as a chiral monomer in a copolymerization reaction.

The homogeneous hydroformylation of olefins by chiral rhodium catalysts has not provided the corresponding chiral aldehydes in high enantiomeric excess. Generally, the aldehydes are produced only in 25-30% enantiomeric excess.[18-21] Similarly, the hydroformylation of styrene by a rhodium catalyst attached to polystyrene containing the 2,3-O-isopropylidene-2,3-dihydroxy-1,4-bis(diphenylphosphino)butane (DIOP) ligand, gave mostly 2-phenylpropanal in an optical yield of only 2%.[3] Because the homogeneously catalyzed hydroformylation reactions[18-21] gave higher optical yields, initially, hydroformylation reactions with rhodium catalysts derived from styrene-divinyl benzene copolymers with 2-(p-styryl)-4,5-bis-[(tosyloxy)methyl]-1,3-dioxolane were undertaken.

II. RESULTS AND DISCUSSION

The most challenging problems in polymer-supported catalysis are encountered in the choice of the polymer matrix and the synthesis of the catalyst site in the matrix. Most of the reported methods introduce a reactive site into a crosslinked polystyrene bead via a copolymerization reaction, followed by the reaction of an optically active phosphine-containing ligand at that site. Our approach, the synthesis of a ligand-bearing monomer and its copolymerization with other monomers, has several advantages. First, the optical purity of the ligand on the monomer can be assured. Second, the concentration of the ligand-bearing monomer in the polymer can be controlled, and a polymer containing a wide range of ligand concentrations can be synthesized. Third, depending on the comonomer, and hence the reactivity ratios of the two monomers, isolation of the ligand-bearing monomer in the polymer chain can be assured. Fourth, the nature of the polymer backbone, polar or nonpolar, can be varied depending on the selection of the comonomer. Finally, varying degrees of crosslinking can be introduced.

A. Chiral Monomers

Several different types of chiral monomers have been synthesized for use in the preparation of chiral polymer supports. All of these monomers are chiral at a carbon adjacent to phosphorous or at one carbon atom removed from phosphorous (or the site to which phosphorous becomes attached). Although monomer **1** does not contain the phosphine ligands, the tosyl groups may be displaced by diphenylphosphide or the dibenzophosphole anion once the monomer has been copolymerized. However, the synthesis of a monomer such as **2** is a much more desirable approach to the introduction of the chiral phosphines into the polymer support. The **RR** enantiomers of monomers **1** and **3** have been synthesized as well as the **SS** enantiomers shown. All these monomers contain chelating groups, and therefore, should not be as susceptible as monophosphines to leaching of the transition metal from the support with repeated use.

Because there is no reliable way to predict which enantiomer will hydroformylate a given olefin to the desired enantiomeric aldehyde at this time, it is advantageous to have access to both enantiomeric phosphines. Further, it is a distinct advantage to utilize available chiral precursors for the synthesis of these phosphines rather than working with racemic mixtures

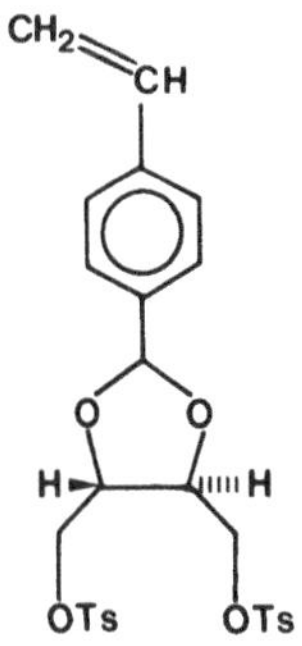

SS−1 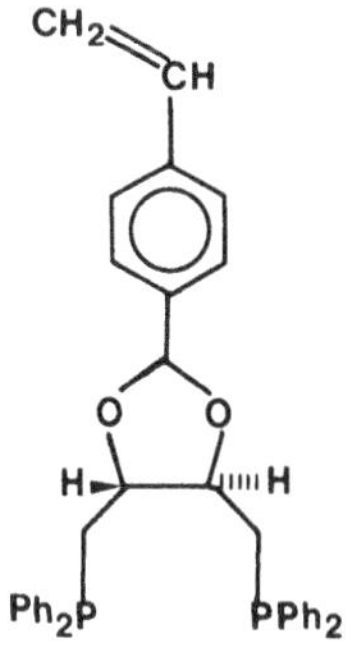 SS−2 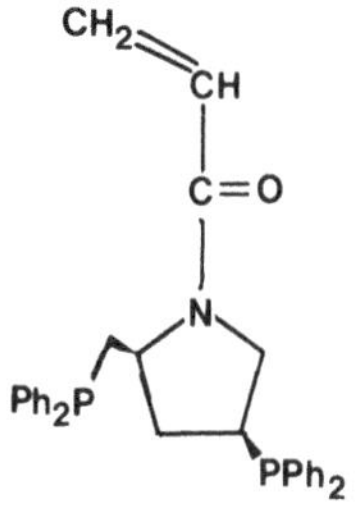 SS−3

that must be subjected to resolution at some stage during the synthesis. The enantiomers of tartaric acid, both of which are readily available, are starting compounds for the synthesis of **RR** and **SS-1**[9] and **2**, while L-hydroxyproline, the natural amino acid, is available for the synthesis of **SS-3** by a relatively direct pathway.[13] Its epimer, **RR-3** also can be obtained from L-hydroxyproline, although by a more tedious synthetic scheme.[13]

B. Catalytic Asymmetric Hydroformylations

Catalytic asymmetric homogeneous hydroformylation reactions with rhodium(I) catalysts generally are carried out in nonpolar solvents, so a crosslinked polymer catalyst containing a nonpolar backbone would be ideally suited for this reaction. The suspension copolymerization of **1** with styrene and varying amounts of divinylbenzene (0-20%) gave polymer beads 60µ in diameter.[22] Subsequent reaction with sodium diphenylphosphide, or sodium dibenzophosphole, and hydridocarbonyltris(triphenylphosphine) rhodium(I) gave the hydroformylation catalyst **SS-4**. Electron microprobe analysis of these beads showed uniform distribution of both phosphorous and rhodium atoms throughout the beads.

Hydroformylation of styrene with **SS-4a,b** (both 20% crosslinked) gave 2-phenylpropanal in lower optical yields than had been obtained with the homogeneous analogs[18-21] (Table I). The product aldehyde was readily racemized under the reaction conditions, so a comparison of optical yields was not especially meaningful.

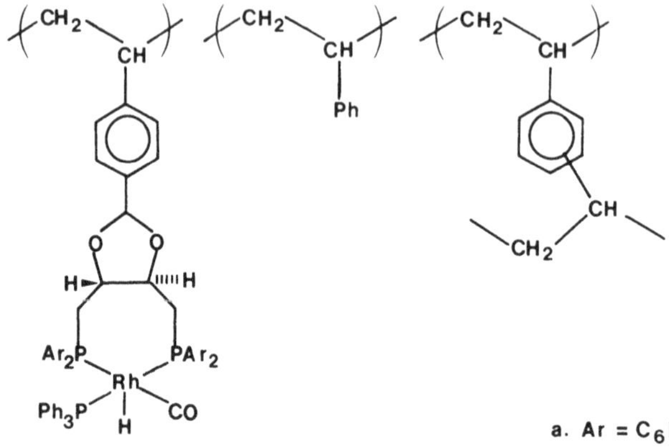

SS−4

a. Ar = C_6H_5

b. Ar_2 =

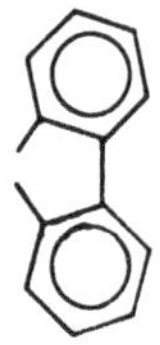

The reaction was highly selective, giving only aldehydes. In no case were alcohols or alkanes detected. However, the production of branched chain aldehydes was quite pronounced with the polymer-attached catalysts. The branched to normal ratio (b/n) of aldehydes obtained from the homogeneous hydroformylation of styrene with a rhodium-DIOP catalyst was about 2,[19] while the b/n ratios obtained with **SS-4a** and **SS-4b** were 5.2 and 21, respectively, under comparable conditons.

Hydroformylation of vinylcyclohexane with either homogeneous or polymer-attached catalysts gave predominately the linear aldehyde. However, even in this case, the polymer-attached catalyst produced a b/n ratio that was six times that of the homogeneous catalyst. Despite this, the amount of branched aldehyde was too low to determine the optical yield.

Cis-2-butene has been hydroformylated by a homogeneous rhodium-DIOP catalyst to give aldehyde in 27% enantiomeric excess.[18] In this case, the product aldehyde was not racemized under the reaction conditions. Hydroformylation of cis-2-butene with **SS-4a** gave 2-methylbutanal in 28% enantiomeric excess, demonstrating that the polymer-bound catalyst behaves identically to the homogeneous catalyst, at least with this olefin

TABLE I

Hydroformylation of Styrene and Vinyl Cyclohexane

Olefin	Catalyst	P(psi)	T(°C)	t(hr)	b/n	% ee
Styrene	Rh-DIOP[a]	16	40	192	2.2	25.2
	Rh-DIPHOL[a]	1500	80	28	8.4	30.9
	SS-4a	400	40	12	5.2	4.25
	SS-4a	45	25	72	2	11.4
	SS-4b	400	40	12	21	5.2
Vinyl Cyclohexane	Rh-DIOP[a]	16	50	168	0.025	—
	SS-4a	400	40	96	0.305	—

[a]DIOP or DIPHOL plus $Rh(H)(CO)(PPh_3)_3$

TABLE II

Hydroformylation of Aliphatic Olefins

Olefin	Catalyst	P(psi)	T(°C)	t(days)	b/n	% ee
Z-2-Butene	Rh-DIOP[a]	16	20	30	—	27
	SS-4a	45	25	26	—	28.5
E-2-Butene	**SS-4a**	45	25	21	—	7.2
1-Pentene	Rh-DIOP[a]	400	75	1	0.43	—
	Rh-DIOP[a]	16	25	1	0.074	—
	SS-4a	400	75	1	0.95	—
	Rh-DIPHOL[b]	400	75	1	0.38	—
	SS-4b	400	75	1	1.03	—

[a]DIOP plus $Rh(H)(CO)(PPh_3)_3$

[b]Rh(CO)(Cl) (DIPHOL)

where there is no regio-choice (Table II). Although optical yields in the hydroformylation of 1-pentene were not determined, it is significant that again higher branched to normal ratios of aldehydes were obtained with the polymer-attached catalyst than with the homogeneous analog.

None of these optical yields is high enough for the hydroformylation reaction to be useful in asymmetric syntheses in reactions such as those shown in Eq. 1 and 2.[23-25] Recently, high optical yields (as high as 95% ee

$$\text{CH}_3\text{CH=CHCH}_3 + H_2 + CO \longrightarrow \text{CH}_3\text{CH}_2\text{CH(CHO)CH}_3 \qquad (1)$$

$$PhCH{=}CH_2 + CO + H_2 \longrightarrow Ph{-}\overset{\displaystyle CH_3}{\underset{\displaystyle H}{C}}{-}CHO + PhCH_2CH_2CHO \qquad (2)$$

in the hydroformylation of styrene)[26] have been obtained utilizing a catalyst prepared from homogeneous dibenzophosphole (DIPHOL)-$PtCl_2SnCl_2$. When DIOP was used instead of DIPHOL, only 30-35% ee was realized. A polystyrene-supported DIOP-$PtCl_2$-$SnCl_2$ catalyst gave comparable optical yields, 29% ee.

The vinyl monomer, (2S,4S)-N-acryloyl-4-(diphenylphosphino)-2-[(diphenylphosphino)methyl] pyrrolidine (**SS-3**) was copolymerized with hydroxyethyl methacrylate and ethylene dimethacrylate by free radical initiation.[27] The resulting crosslinked, flocculent polymer (**SS-5**) containing 5 mol % phosphine monomer, swelled in both nonpolar solvents, such as benzene, and polar solvents, such as ethanol. A suspension polymerization of **SS-3** with styrene and divinylbenzene gave 60µ polymer beads of **SS-6** which contained 10 mol % of both the phosphine monomer and crosslinking agent. Both polymers, **SS-5** and **SS-6**, reacted with bis(benzonitrile)dichloroplatinum(II) to exchange platinum onto the phosphine sites, after which stannous chloride was added.[26]

Since the effectiveness of the BPPM phosphine in asymmetric hydroformylation with the $PtCl_2$-$SnCl_2$ catalyst had not been tested, homogeneous hydroformylations were carried out first with N-(t-butoxycarbonyl)-(2S,4S)-4-(diphenylphosphino)-2-[(diphenylphosphino)methyl]

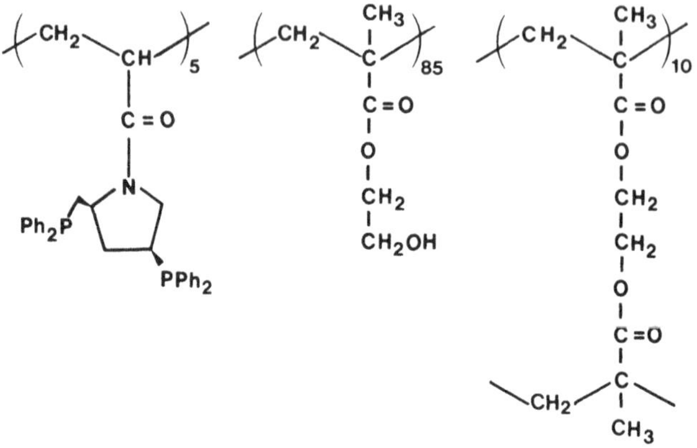

$$-(CH_2-CH)_5- \quad -(CH_2-C(CH_3))_{85}- \quad -(CH_2-C(CH_3))_{10}-$$

SS−5

$$-(CH_2-CH)_{10}- \quad -(CH_2-CH(Ph))_{80}- \quad -(CH_2-CH)_{10}-$$

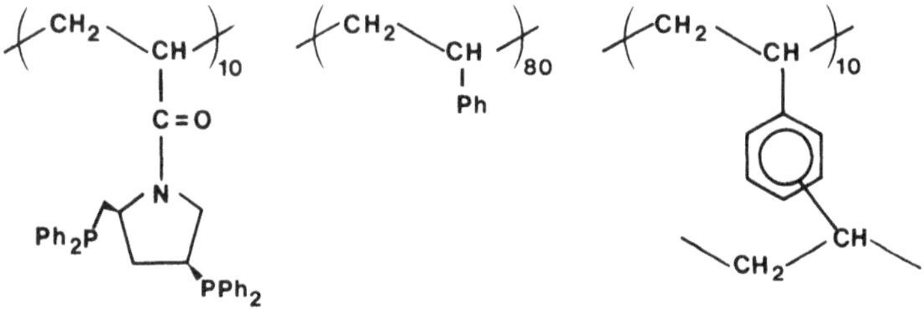

SS−6

pyrrolidine (7) (Eq. 3). The hydroformylation of styrene with this catalyst was carried out to yield hydrotropaldehyde in an 80% enantiomeric excess.

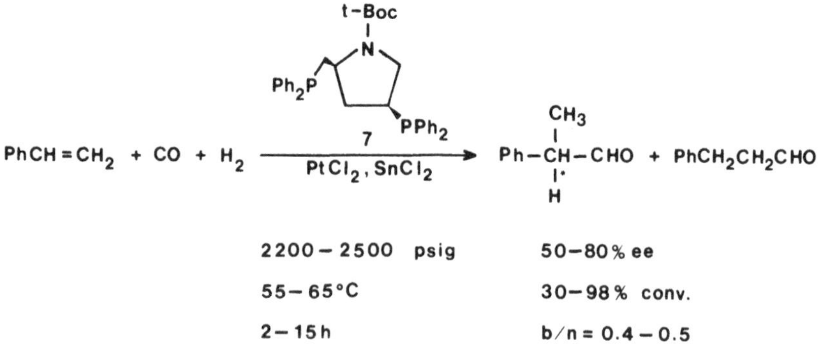

$$PhCH=CH_2 + CO + H_2 \xrightarrow[PtCl_2,\ SnCl_2]{7} Ph-\overset{CH_3}{\underset{H}{CH}}-CHO + PhCH_2CH_2CHO \quad (3)$$

2200 − 2500 psig	50−80% ee
55 − 65°C	30−98% conv.
2 − 15h	b/n = 0.4 − 0.5

TABLE III

Asymmetric Hydroformylation of Styrene Catalyzed by 5 and 6, $PtCl_2 \cdot SnCl_2$[a]

Catalyst	Styrene : Pt	t(h)	Conversion (%)	Selectivity (%)	b/n	Optical Yield %
1. 5, $PtCl_2 \cdot SnCl_2$	800	24	45	98.7	0.45	70
2. [b]	800	24	23.5	98.5	0.45	72
3. 6, $PtCl_2$, $SnCl_2$[c]	500	50	9	100	0.52	73
4. [c]		90	40	98.8	0.60	73
5. [c]		90	38	97.7	0.53	73

[a] 60°C, 2200 psi, $H_2/CO = 1$.

[b] Catalyst recycled from the first run.

[c] In each run, 3-5, the catalyst was recycled. Run 3 was the third time the catalyst had been used.

The b/n ratio was not high (b/n = 0.4-0.5 under various conditions) but no ethylbenzene was formed. The highest optical yields were obtained when the reaction times were short, and the temperature was low. Unfortunately, at low temperatures and short reaction times, the conversion to aldehyde was low. The lower optical yields at longer reaction times could be accounted for by racemization of the product, hydrotropaldehyde, under the reaction conditions.

Polymer-bound catalysts **SS-5** and **SS-6** containing exchanged platinum also catalyzed the hydroformylation of styrene (Table III). The rates of hydroformylation were appreciably lower than the homogeneous BPPM analog, but showed comparable b/n ratios and optical yields. This catalyst, from which a 70% optical yield, was obtained, could be recovered by filtration and reused. Recycled catalyst showed somewhat lower rates, but the optical yields were retained.

These optical yields obtained with the polymer-supported BPPM catalysts are the highest yet observed, and are beginning to reach the optical yields necessary for a useful asymmetric synthesis. The hydroformylation reaction of vinyl aromatics leads to chiral aldehydes which can be converted readily to the corresponding carboxylic acids. Such acids and their derivatives, which sometimes require the synthesis of one enantiomer, are powerful analgesic, antiinflamatory drugs.[28,29]

ACKNOWLEDGEMENT

This research was supported by a Grant DMR-8016503 from the National Science Foundation.

REFERENCES

1. C.U. Pittman, **Catalysis by Polymer supported Transition Metal Complexes,** (P. Hodge and D.C. Sherrington, eds.) Wiley, New York, 1980.
2. K.E. Koenig, M.H. Sabacky, G.L. Bachman, W.C. Christopfel, H.D. Burnstorff, R.B. Friedman, W.S. Knowles, B.R. Stults, B.D. Vineyard, and D.J. Weinkoff, Ann. N.Y. Acad. Sci., **333**, 16 (1980), and references therein.

3. W. Dumont, J.C. Poulin, T.P. Dang, and H.B. Kagan, J. Am. Chem. Soc., **95**, 8295 (1973).
4. G. Strukul, M. Bonivento, M. Graziani, E. Cernia, and N. Palladino, Inorg. Chim. Acta, **12**, 15 (1975.
5. K. Achiwa, Chem. Lett., 905 (1978).
6. H.W. Krause, React. Kinet. Catal. Lett., **10**, 243 (1979); German (East) Patent 133, 230 (1978).
7. K. Ohkubo, K. Fujimari, and K. Yoshinaga, Inorg. Nucl. Chem. Lett., **15**, 231 (1979).
8. K. Ohkubo, M. Huga, K. Yoshinaga, and Y. Motozato, Inorg. Nucl. Chem. Lett., **16** 155 (1980).
9. N. Takaishi, H. Imai, C.A. Bertelo, and J.K. Stille, J. Am. Chem. Soc., **98**, 5400 (1976); ibid., **100** 264 (1978).
10. T. Matsuda and J.K. Stille, J. Am. Chem. Soc., **100**, 268 (1978).
11. J.K. Stille, Colloques Internationaux du Centre National de la Recherche Scientifique, 281 (1977).
12. J.K. Stille, S.J. Fritschel, N. Takaishi, T. Matsuda, H. Imai, and C.A. Bertelo, Ann. N.Y. Acad. Sci., **333**, 35 (1980).
13. G.L. Baker, S.J. Fritschel and J.K. Stille, J. Org. Chem., **46**, 2960 (1981).
14. G.L. Baker, S.J. Fritschel, J.R. Stille, and J.K. Stille, J. Org. Chem., **46**, 2954 (1981).
15. P.D. Sybert, C. Bertelo, W.B. Bigelow, S. Varaprath, and J.K. Stille, Macromolecules, **14**, 502 (1981).
16. J.K. Stille, Pure Appl. Chem., **54**, 99 (1982).
17. G.L. Baker, S.J. Fritschel, and J.K. Stille, ACS Symposium Series, **212**, 137 (1983).
18. G. Consiglio, C. Botteghi, C. Salomon, and P. Pino, Angew. Chem. Int. Ed. Engl., **12**, 669 (1973).
19. C. Salomon, G. Consiglio, C. Botteghi, and P. Pino, Chimia, **27**, 215 (1973).
20. P. Pino, G. Consiglio, C. Botteghi, and C. Salomon, Adv. Chem. Ser., **132**, 295 (1974).
21. M. Tanaka, Y. Ikeda, and I. Ogata, Chem. Lett., 1115 (1975).

22. S.J. Fritschel, J.J.H. Ackerman, T. Keyser, and J.K. Stille, J. Org. Chem., **44**, 3152 (1979).
23. J.D. Morrison, W.F. Massler, and M.K. Neuberg, Adv. Catal., **25**, 81 (1976).
24. H.B. Kagan and J.C. Fiaud, Top. Stereochem., **10**, 175 (1978).
25. D. Valentine, Jr. and J.W. Scott, Synthesis, 329 (1978).
26. (a) C.U. Pittman, Jr., Y. Kawabata, and L.I. Flowers, J. Chem. Soc., Chem. Commun., 473 (1982). (b) G. Consiglio, P. Pino, L.I. Flowers, and C.U. Pittman, Jr., J. Chem. Soc., Chem. Commun., 612 (1983).
27. J.K. Stille and G. Parrinello, J. Mol. Catal., **21**, 203 (1983)
28. (a) J.S. Nicholson and S.S. Adams, U.S. 3,228,831 (1966). (b) S.S. Adams, J. Bernard, J.S. Nicholson, and A. Blancafort, U.S. 3,755,427 (1973). (c) J.S. Nicholson and J.G. Tantum, U.S. 4,209,638 (1980).
29. I.T. Harrison, B. Lewis, P. Nelson, W. Rooks, A. Roszkowski, A. Tomolonis, and J.H. Fried, J. Med. Chem., **13**, 203 1970.

3

Selective CH and CC Bond Activation in Alkanes by Transition Metal Complexes

R.H. Crabtree*, M.J. Burk, R.P. Dion, E.M. Holt,
M.E. Lavin, S.M. Morehouse, C.P. Parnell, and R.J. Uriarte
Department of Chemistry
Yale University
New Haven, Connecticut 06511

ABSTRACT

Alkanes can be dehydrogenated to alkenes and arenes stoichiometrically and in some cases catalytically by $[IrH_2(Me_2CO)_2(PPh_3)_2]SbF_6$. The structure of the related complex $[IrH_2(8\text{-methylquinoline})(PPh_3)_2]$ BF_4 shows a C-H-Ir bridging system. It is suggested that multiple coordinative unsaturation plays a role in promoting this chemistry. The same property also allows ligating groups on a substrate olefin to direct H_2 attact on the C=C group from one face of the substrate.

I. INTRODUCTION

This paper is intended to give a brief account of some of our most recent work on C-H[1] and C-C bond cleavage reactions in alkanes. The general problem of alkane C-H and C-C bond activation is of long standing in inorganic chemistry. Why does Reaction 1 go so easily and yet examples of Reaction 2 are so rare? Reaction 3 is unknown for unstrained alkanes.

$$M + H\text{-}H \longrightarrow M\begin{matrix} \diagup H \\ \diagdown H \end{matrix} \tag{1}$$

$$M + C\text{-}H \longrightarrow M\begin{matrix} \diagup C \\ \diagdown H \end{matrix} \tag{2}$$

$$M + C\text{-}C \longrightarrow M\begin{matrix} \diagup C \\ \diagdown C \end{matrix} \tag{3}$$

The answer seems to be that M-C bonds are on the order of 20 kcal/mole weaker than M-H bonds,[2] so that an alkane adduct tends to eliminate alkane spontaneously (i.e., Eq. 2 tends to go in the reverse direction). Direct C-C activation is probably even less favorable. Our plan has been to arrange for multiple C-H bond breaking steps to occur at the metal. The hydrogen removed the alkane in this dehydrogenation reaction is subsequently transferred to a hydrogen acceptor. This alters the thermodynamics of the overall process, and in principle allows catalytic alkane dehydrogenation (Eq. 4).

$$R^1CH_2\text{-}CH_3 + R^2CH{=}CH_2 \xrightarrow{\text{cat.}} R^1CH{=}CH_2 + R^2CH_2\text{-}CH_3 \tag{4}$$

Our initial experiments[3] (Eq. 5) showed that cyclopentane can indeed be dehydrogenated by $[IrH_2(Me_2CO)_2(PPh_3)_2]BF_4$ in 1,2-dichloroethane, $C_2H_4Cl_2$, at 80° in the presence of tert-buylethylene (tbe) as hydrogen acceptor. This system, the first of its kind, has a number of unusual features, which we have only recently begun to understand more recently.

$$IrH_2S_2L_2^{\ +} + C_5H_{10} + 3tbe \xrightarrow{C_2H_4Cl_2} Ir(C_5H_5)HL_2^{\ +} + tba + 2S$$

$$(S = Me_2CO;\ L = PPh_3;\ tbe = Bu^tCH{=}CH_2;\ tba = Bu^tEt) \tag{5}$$

A number of other systems[4] also operating on the basis of Eq. 2 have now been discovered. Alkane activation is now an accepted process in organometallic chemistry. In particular the groups of Bergman, Graham, and Jones[4] have managed to observe the initial alkyl hydride adduct of type M(R)H.

One unusual feature of our stystem is that a halocarbon solvent is used. This would normally be expected to destroy the iridium complex by oxidative addition (Eq. 6). More recently, however, we have shown that halocarbons bind as neutral ligands to our iridium system without C-Hal

$$\text{R-Hal} + \text{M} \longrightarrow \text{R-M-Hal} \qquad (6)$$

bond fission.[5] The first known complex of this type was $[Ir(\underline{o}\text{-}C_6H_4I_2)H_2(PPh_3)_2]BF_4$, for which the crystal structure, kindly performed for us by Professor J.W. Faller, is now published.[5] We now also have the structures of several other examples of this unusual class of complex; $[Ir(IMe)_2H_2(PPh_3)_2]BF_4$[6] and $[Ir(cod)(\eta^2PPh_2(\underline{o}\text{-}C_6H_4Br))]SbF_6$[7]. Interestingly, in all the complexes studied, the M-Hal-C unit is bent ($\sim 110^\circ$). Work described here shows other ways in which the halocarbon solvent can be involved in Eq. 5. We also report here on some work which helps show how the C-H bonds are broken in this reaction.

II. RESULTS AND DISCUSSION

The reaction shown in Eq. 5 does not work for cyclohexane, instead a mixture of complexes results in which the alkane is not incorporated. This initially defied separation and identification, but we have now characterized three of the complexes in the mixture which together amount to nearly 50% of the material. These compounds are $[HL_2Ir(\mu\text{-}Cl)_2(\mu\text{-}X)IrL_2H]BF_4$[8a] (X = Cl and X = H, 2 isomers). It is clear that the source of the halogen must be the solvent and that chlorocarbons are therefore not ideal solvents for this type of work. Unfortunately, none of the other usual organic solvents appeared to be suitable, either being coordinating (MeCN, THF, acetone) or being incapable of dissolving the complex (Et_2O, hydrocarbons). This prompted us to omit the halogenated solvent and carry out the reaction in neat alkane, containing tbe (tbe : M, 4 : 1 mol. ratio), in a resealable glass vessel at 85 - 150°C. We tend to use $[IrH_2S_2(P(\underline{p}\text{-}C_6H_4F)_3)_2]SbF_6$ **(1)** because of the SbF_6 salts give much better results (for reasons that are not clear) and the use of the fluorophenyl phosphine allows us to monitor P-C cleavage reactions (see below).

The dehydrogenation of cyclopentane now goes in much improved yield (82% *vs.* 32% by Eq. 5), but in particular methyl and ethylcyclopentane, previously unreactive, now dehydrogenate at 120°C in 78% and 36% yields, respectively, after 14 hours. The higher temperature required for these transformations simply leads to decomposition in chlorinated solvents under the conditions of Eq. 5.

$$C_6H_{12} \xrightarrow{IrH_2S_2L_2^+} (C_6H_7)IrHL_2^+ + (C_6H_6)IrL_2^+ + C_6H_6 + PhF$$

	$(C_6H_7)IrHL_2^+$	$(C_6H_6)IrL_2^+$	C_6H_6	PhF
85°C	35%	45%	32%	-
150°C	-	5%	60%	0.8*

(* in mol/mol Ir) (7)

Cyclohexane, also previously inert, now reacts as shown in Eq. 7. This was a very satisfying result for us because it was the transformation we initially hoped to bring about when we started our work on C-H bond breaking. We see the cyclohexadienyl hydride which is probably insolable because it is an 18 electron complex and requires loss of L or slippage of the ring to give the benzene complex. This in turn is relatively unstable and loses benzene to give the free arene. Unfortunately, the system has never given more than stoichiometric dehydrogenation of cyclohexane. Decomposition of the catalyst, possibly by P-C cleavage (see below) is probably responsible. Cyclohexene in contrast, appears to protect the catalyst from decomposition and this substrate can be dehydrogenated catalytically with 1 and excess tbe even in $C_2H_4Cl_2$[8b].

At 150°C, free benzene is the major product from cyclohexane, along with fluorobenzene[9,10] which must arise from P-C cleavage. This happens at temperatures above 100°C and most probably leads to deactivation of the complex. This may be a more general catalyst deactivation pathway than usually recognized.

The formation of a small amount of arene in a catalyst system must often escape detection.

C_6D_{12} gave essentially only C_6D_6 (gc-ms), as expected, and the active species was not metallic iridium, to judge from the fact that the reaction proceeds in the presence of metallic mercury.[11]

Methylcyclohexane reacts at 130°C to give a mixture of cyclohexadienyl complexes (25%) and the toluene complex (15%) and at 150°C free toluene (28%) is obtained, as well as fluorobenzene.

When the reaction is carried out with proton sponge (1·1 equivs/Ir) present, this amine deprotonates the metal,[12] the SbF_6 salt of the amine is precipitated, and the metal fraction now becomes soluble in the alkane. With cyclooctene at 130°C, 8 turnovers of cyclooctene are produced in the presence of 20 molar equivalent of tbe. This catalytic alkane dehydrogenation resembles the polyhydride systems studied by Felkin and coworkers.[4a] We assume that the neutral system releases the olefin rather than dehydrogenating it further. The base modified system, therefore, shows a different product from the unmodified system, which gives only stoichiometric amounts of $[Ir(cod)L_2]SbF_6$ under the same conditions.

In the first case of C-C bond cleavage in an unstrained alkane by a transition metal complex,[13] 1 reacts with tbe and 1,1-dimethylcyclopentane at 150°C after 8 hours to give a dimethylcyclopentadiene complex in 50% yield. After a further 12 hours of heating it is quantitatively converted to the methylcyclopentadienyl iridium methyl complex by a C-C cleavage reaction. The more difficult C-C bond cleavage does not occur directly, but only after C-H activation. This is reasonable because the aromatisation of the C_5 ring can drive the reaction.

1,1-Dimethylcyclohexane gives only the dimethylcyclohexadienyl hydride, which as an 18-electron compound canot give C-C cleavage.

The mechanism of these C-H and C-C bond cleavage reactions is probably a concerted oxidative addition as has been directly observed in alkanes by Bergman, _et al._[4d] Related cases of C-C activations in _activated_ substrates have been observed by Eilbracht,[14] Green, _et al_[15] and Suggs, _et al._[16]

A probable intermediate in the C-H cleavage reaction is a C-H...M bridged species. Species of this type have been observed by several groups and have been reviewed in detail by Brookhart and Green.[17] In order to see if our own system might involve an intermediate of this sort, we studied the

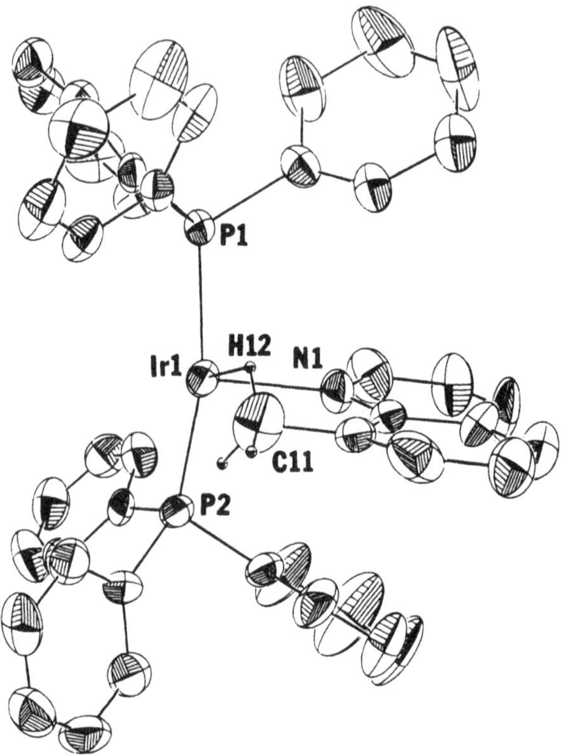

Fig. 1. The ORTEP diagram of $[IrH_2(8\text{-methylquinoline})(PPh_3)_2]^+$ in the SbF_6 salt. The Ir-H hydrogens were not detected in the X-ray experiment. The C-H-Ir bridge is shown in more detail in Fig. 2.

reaction of 8-methylquinoline (8-MeQ) and 1. This gives a complex, the structure of which has been determined crystallographically by Holt[18]

$$\text{C-H} + \text{M} \longrightarrow \overset{\text{H}}{\text{C}\diagup\;\diagdown\text{M}} \longrightarrow \text{M}\begin{smallmatrix}\diagup \text{H} \\ \diagdown \text{C}\end{smallmatrix}$$

(Fig. 1) with the bent C-H...M bridge shown in more detail in Fig. 2. The distances and angles in the bridging group are very similar to those seen in other cases.[17] The isotopic perturbation method of Saunders, et al.[19] was also successfully applied to this problem to confirm the presence of the C-H...Ir bridge in solution. Under D_2, the methyl group of the 8-MeQ is deuterated. Isotopic labelling experiments confirm that the methyl protons

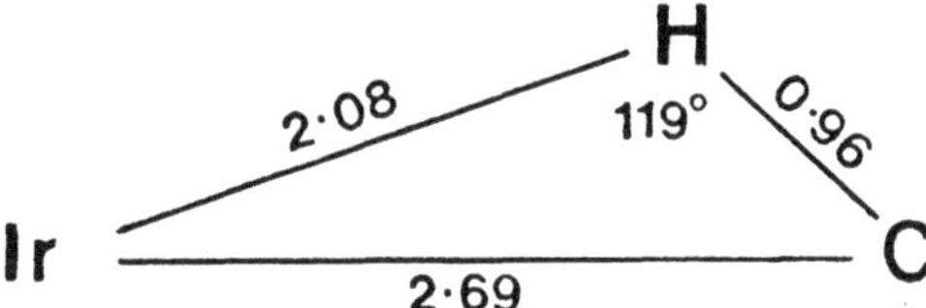

Fig. 2. A more detailed diagram of the C-H-M bridge in the complex shown in Fig. 1. The exact location of the hydrogen is subject to some uncertainty in this X-ray study.

exchange with the iridium hydride groups in the complex. In this compound, therefore, we see the C-H cleavage reaction taking place very readily at room temperature (albeit on an activated substrate, not an alkane).

III. CONCLUSIONS

We have shown that stoichiometric and catalytic dehydrogenation of alkanes can occur by C-H cleavage reactions. Selective C-C cleavage reactions can also occur subsequently. These results emphasize the importance of multiple coordinative unsaturation, obtained by H and Me_2CO dissociation from **1**, for alkane dehydrogenation and dealkylation.

C-H cleavage probably occurs via a C-H...M bridged intermediate, an example of which is shown in Fig. 1 and 2. The bent arrangement seen may indicate that apart from donation of C-H bonding electrons into an empty metal orbital, non-bonding metal d-electron density may also be donated into the σ^* orbital of the C-H bond. A complete 2-electron transfer into this orbital would cleave the C-H bond and give an alkyl hydride complex as product.

The C-C cleavage goes through the diene complex shown in Eq. 8, illustrating the more severe requirements placed on this reaction. It is

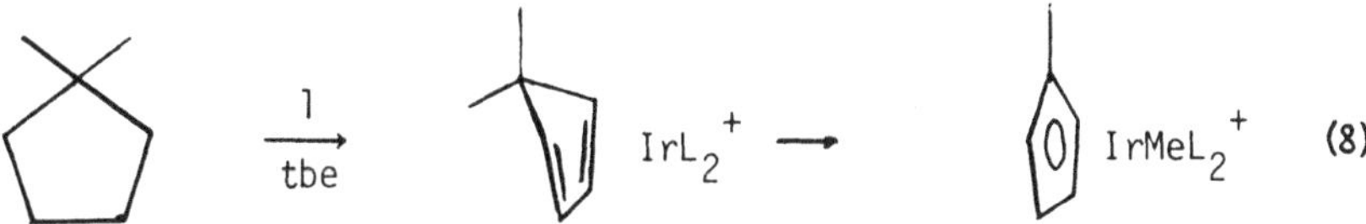

probable that for metal complexes, C-H cleavage may be necessary before C-C cleavage can take place.

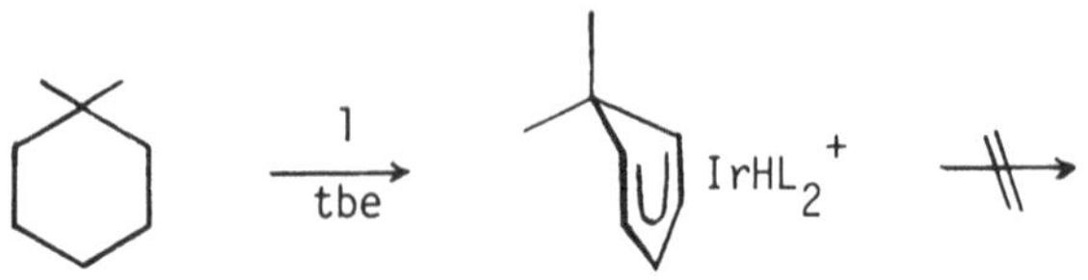

Far from being an exotic reaction, alkane C-H and C-C bond activation processes will probably turn out to be common reaction pathways for a wide variety of metal complexes. In order to observe this reaction, however, one must arrange for the suppression of the side reactions which would otherwise deactivate the system. In particular, cyclometallation and P-C cleavage of the tertiary phosphine are important. It is not yet clear why our system is so successful in this respect. Other factors are easier to understand. Reaction with the solvent is most easily avoided by making the alkane the "solvent", as described above. Reaction with the hydrogen acceptor is avoided by using tbe, a non-metallating compound.

III. EXPERIMENTAL

The details of these experiments will appear in several full papers on this work presently in preparation, but the methods used were essentially the same as those described previously,[3] except that a 5ml resealable pressure vessel was used for the alkane work. All products were detected and quantified by ^1H NMR. The authentic products were also made by independent routes and fully characterized. All the reactions also proceeded normally in the presence of metallic mercury. This shows that metallic iridium was not involved in any of these reactions.[11]

The 8-methylquinoline complex was prepared by the method[5] used previously for the synthesis of $[IrH_2(MeI)_2(PPh_3)_2]^+$, and was crystallized from CH_2Cl_2/Et_2O.

ACKNOWLEDGMENTS

We thank the National Science Foundation and The Petroleum Research Fund for financial support of this work and Johnson Matthey Co. for a loan of iridium.

REFERENCES

1. E.L. Muetterties, Chem. Soc. Rev., **12**, 283, (1983).
2. J. Halpern, Acc. Chem. Res., **15**, 238, (1982).
3. R.H. Crabtree, J.M. Mihelcic, and J.M. Quirk, J. Am. Chem. Soc., **101**, 7738, (1979), ibid, **104**, 107, (1982).
4. (a) D. Baudry, M. Ephritikine, and H. Felkin, J. Chem. Soc., Chem. Commun., 1243, (1980); ibid, 606 (1982); (b) A.H. Janowicz and R.G. Bergman, J. Am. Chem. Soc., **104**, 352, (1982) ; (c) J.K.L. Hoyano and W.A.G. Graham, ibid., **104**, 3722 (1982); (d) W.D. Jones and F.J. Feher, ibid, **104**, 4240 (1982); (e) P.L. Watson, ibid, **105**, 6491, (1983).
5. R.H. Crabtree, J.W. Faller, M.F. Mellea, and J.M. Quirk, Organometallics, **1**, 1361, (1982).
6. M.J. Burk, R.H. Crabtree, and B. Segmuller, J. Am. Chem. Soc., in preparation.
7. M.J. Burk, R.H. Crabtree, and E.M. Holt, Organometallics, **3**, 638, (1984).
8. (a) RH. Crabtree, H. Felkin and G.E. Morris, J. Organomet. Chem., **141**, 205, (1977); (b) R.H. Crabtree and C.P. Parnell, Organometallics, (In Press).
9. P-C cleavage[10] may be an important catalyst deactivation pathway in this and other systems.
10. J.V. Ortiz, Z. Havlas, and R. Hoffmann, Helv. Chim. Acta, **76**, 1, (1984), and references therein.
11. D.R. Anton and R.H. Crabtree, Organometallics, **2**, 855, (1983) and references therein.
12. The $HSbF_6$ adduct of proton sponge was also isolated among the products.
13. C-C cleavage has been observed in activated substrates.[14,15,16]
14. P. Eilbracht, Chem. Ber., **109**, 1429, (1976), ibid, **113**, 542, (1980).
15. M.L.H. Green and F.W.S. Benfield, J. Chem. Soc., Dalton, 1325, (1974).
16. J.W. Suggs and S.D. Cox, J. Organomet. Chem., **221**, 199, (1981).
17. M. Brookhart and M.L.H. Green, J. Organomet. Chem., **250**, 395, (1983) and references therein.
18. R.H. Crabtree, E.M. Holt, M.E. Lavin, and S.M. Morehouse, manuscript in preparation.

4

Catalytic Methods for the Synthesis of Cyclopropanes

Michael P. Doyle*
Department of Chemistry
Hope College
Holland, Michigan 49423

ABSTRACT

Cyclopropane formation is conveniently achieved by interaction of electrophilic metal carbenes with olefins. Numerous catalysts are employed for these transformations, but rhodium(II) carboxylates are generally the most effective. Their catalytic activities are dependent on their ability to react as electrophiles with carbenoid sources such as diazo compounds to produce transient metal-stabilized carbenes. Linear stereoselectivity and relative reactivity correlations between $Rh_2(OAc)_4$-catalyzed reactions of phenyldiazomethane and stoichiometric reactions of $(CO)_5WCHPh$ with the same set of alkenes define metal carbene involvement in $Rh_2(OAc)_4$-catalyzed reactions. Stereoselectivity and regioselectivity correlations for cyclopropane formation between $Rh_2(OAc)_4$-catalyzed reactions and those with a broad spectrum of copper, rhodium, and palladium catalysts provide their interlink to metal carbene intermediates. Stereoselectivity and regioselectivity in cyclopropane formation are functions of the electrophilic character of the metal carbene intermediate which, in turn, is dependent on

*Present Address: Department of Chemistry, Trinity University, San Antonio, Texas 78284

carbene substituents, including the metal. Olefin coordination at available sites on the metal catalyst depletes the reactivity of the coordinated olefin towards cyclopropanation with metal carbenes, and electronic factors influence carbene transfer from the metal in the transition state for cyclopropanation. α,β-Unsaturated carbonyl compounds and nitriles do not undergo direct cyclopropanation with metal carbenes, but certain metal complexes promote cyclopropane formation by dinitrogen extrusion from 1-pyrazolines which are produced by 1,3-dipolar addition of diazo compounds to these electron-deficient olefins. Electrophilic metal carbenes undergo ylide formation with nucleophilic substrates, and these ylide intermediates are capable of nucleophilic cyclopropanation reactions. Catalytic methods for cyclopropane formation will be addressed, and the development of stable metal carbenes as procatalysts for these reactions will be described.

I. INTRODUCTION

A wide spectrum of transition metal compounds are suitable catalysts for the production of cyclopropanes from olefins and diazo compounds.[1] Heterogeneous catalysts that have included copper bronze and cupric sulfate have been employed for more than 75 years[2] and, until recently, they have been the most effective and economical for carbenoid additions of diazo compounds to alkenes. Homogeneous copper catalysts that originated with Nozaki's chiral copper catalysts[3] and Moser's soluble $CuCl{\cdot}P(OR)_3$[4] were developed to provide a better understanding of the factors influencing selectivity and reactivity in catalytic cyclopropanation reactions. With the introduction of copper(I) triflate (CuOTf, Tf = SO_2CF_3) by Salomon and Kochi,[5] the importance of olefin coordination and the oxidation state of the metal could be evaluated. During the same time palladium catalysts ranging from π-allylpalladium(II) complexes[6] to $PdCl_2{\cdot}2PhCN$[7] were introduced, but these offered no apparent advantages over the traditional copper catalysts. Most recently, rhodium(II) acetate has been identified as an exceptionally effective catalyst[8] which is uniquely well suited to stoichiometric transformations.[9] The rhodium carbonyl cluster $Rh_6(CO)_{16}$[10] and iodorhodium(III)mesotetraphenylporphyrin (IRhTPP)[11] are also efficient cyclopropanation catalysts, but their unique advantages have not been fully developed.

The mechanism of the copper catalyzed decomposition of diazoketones was originally proposed by Yates[12] to proceed through a metal carbene complex. This model was uniformly adopted for cyclopropanation reactions

$$ML_n + R_2CN_2 \xrightarrow{-N_2} [L_nM{=}CR_2] \xrightarrow{=} \text{(cyclopropane with } R, R) + ML_n \qquad (1)$$

(Eq. 1) even though limited evidence supported the actual involvement of the metal during carbenoid addition to olefins, the most convincing of which was asymmetric induction in cyclopropane formation through the use of chiral catalysts.[3,13,14] During this same period, the development of stable metal carbene complexes by Fischer[15] offered selected examples of carbenoid additions to olefins, but heteroatom-stabilized metal carbenes in the chromium triad exhibited limited reactivity with alkenes. Not until Casey's report[16] that $(CO)_5WCHPh$ underwent efficient cyclopropanation of olefins did stoichiometric methods using metal carbene precursors, especially $[CpFe(CO)_2CHRZ]$ where R = H, CH_3, or Ph and Z = OMe[17] or SMe[18], advance to their current state of synthetic usefulness.

Catalytic methods have been most effective when diazocarbonyl compounds are employed, although examples of cyclopropanation with a limited variety of less stable diazo compounds have been reported.[19] Dimerization and polymerization of the carbenoid unit is the principal competing reaction, and the extent of this competition increases with decreasing stability of the diazo compound (eg., $N_2CHCOOEt \ll PhCHN_2 < CH_2N_2$). Consequently, diazocarbonyl compounds remain the most suitable carbenoid precursors for catalytic reactions, and stoichiometric methods are presently better adapted for aliphatic carbenoid addition reactions.

Olefins ranging from vinyl ethers to α,β-unsaturated esters have been reported to undergo catalytic cyclopropanation reactions with diazo compounds.[1] However, catalytic cyclopropanation of electron-deficient alkenes cannot be interpreted as involving metal carbene intermediates.[20] In these cases cyclopropane formation may be the direct result of 1,3-dipolar addition of the diazo compound to the electron-deficient olefin followed by

$$\text{CH}_2\text{=CHZ} + R_2CN_2 \longrightarrow \text{(1-pyrazoline: } R,R,Z; \text{N=N)} \longrightarrow \text{(cyclopropane: } R,R,Z) + N_2 \qquad (2)$$

extrusion of dinitrogen from the intermediate 1-pyrazoline (Eq. 2). Formal C-H insertion products often accompany the cyclopropane products, simple β-substituted compounds such as ethyl crotonate give drastically reduced product yields,[20,21] and reaction rates are independent of the "catalyst" employed. The principal role of the "catalyst" is to inhibit competitive tautomerization of the initially formed cycloaddition product, a reaction which appears to be autocatalytic in the absence of the transition metal promoter.[20] A considerable variety of transition metal compounds [Cp_2ZrCl_2, $V(CO)_6$, $Mo(CO)_6$, $Mo_2(OAc)_4$, $Fe(CO)_5$, $Os_3(CO)_{16}$, and $Pd(OAc]_2)]$ [20,21] are effective for these cyclopropanation reactions.

II. RESULTS AND DISCUSSION

A. Metal Carbenes in Catalytic Cyclopropanation Reactions

Evidence for the involvement of metal carbenes in catalytic cyclopropanation reactions has been inferred from asymmetric induction by the use of chiral copper catalysts[3,13] and, in one case, a chiral cobalt catlyst.[14] However, chirality transfer tells only that the chiral catalyst is involved in the product determining stage of cyclopropane formation and not that this transformation involves an actual metal carbene intermediate. Optimally, direct observation would demonstrate metal carbene formation from diazo compounds, but this has not yet been achieved. Alternately, reactivity and selectivity correlations for cyclopropane formation between catalytic reactions and those of known stable metal carbenes could provide the desired predictive relationships.

We have recently obtained linear stereoselectivity and relative reactivity correlations[22] between $Rh_2(OAc)_4$-catalyzed reactions of phenyldiazomethane and stoichiometric reactions of the analogous tungsten carbene $(CO)_5WCHPh$[16] with the same set of alkenes. Relative reactivity data (Fig. 1) correlate with substituent effects defined by linear free energy relationships. In the series of alkyl-monosubstituted ethylenes, log (relative reactivity) for cyclopropanation reactions of $(CO)_5WCHC_6H_5$ at -78°C

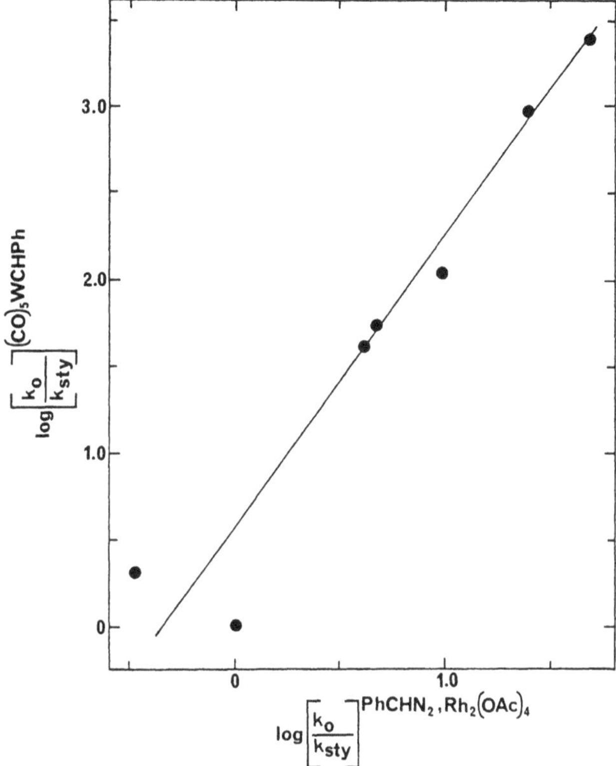

Fig. 1. Plot of log (relative reactivity) based on styrene for cyclopropanation phenyldiazomethane catalyzed by $Rh_2(OAc)_4$ at 25°C. Data points for styrene and for 2-methyl-2-butene fall below and above the line respectively.

provides a linear correspondence with the Taft σ^* values ($\rho^* = 3.7$), and for $Rh_2(OAc)_4$-catalyzed reactions a similar correspondence is evident ($\rho^* = 1.0$). Cyclopropanation stereoselectivities for $Rh_2(OAc)_4$-catalyzed reactions of phenylidiazomethane and stoichiometric reactions of $(CO)_5WCHPh$ with vinyl ethers and simple alkenes also show a distinct linear relationship (Fig. 2). This remarkable uniformity, whose only exception is 3,3-dimethyl-2-methoxy-1-butene, demands that those factors which govern the product-determining step in each of these cyclopropanation reactions are identical.

These correlations, which define metal carbene involvement in $Rh_2(OAc)_4$-catalyzed reactions, reflect the structural similarities of the

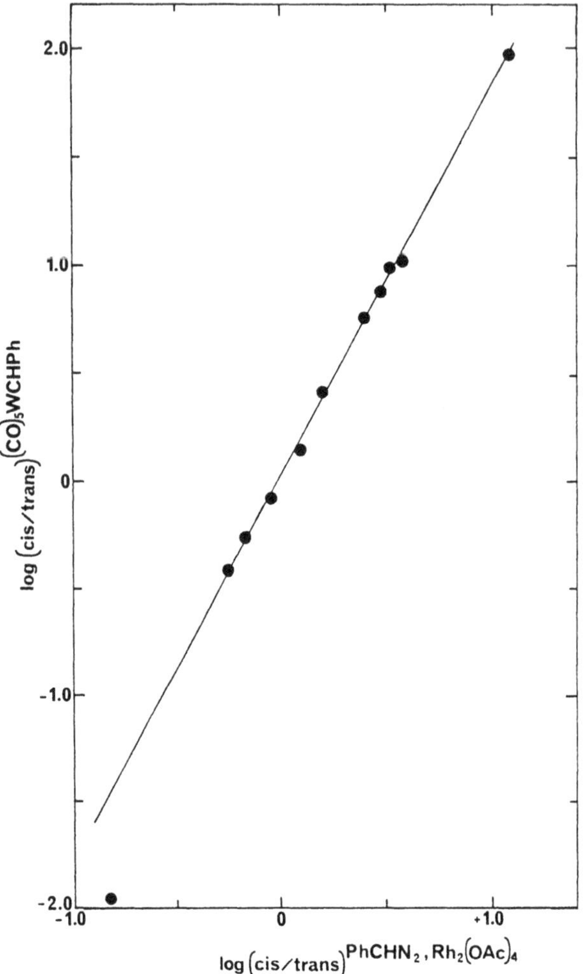

Fig. 2. Plot of log (cis/trans) or log (Z/E) for cyclopropanation ofalkenes with $(CO)_5WCHPh$ at -78°C vs. those with phenyldiazomethane catalyzed by $Rh_2(OAc)_4$ at 25°C.

metal environments of carbenes associated with $Rh_2(OAc)_4$ (**1**) and $W(CO)_5$ (**2**) toward interaction of olefins with the electrophilic carbon center. Stereoselectivity (Fig. 3)[23] and regioselectivity (Fig. 4)[24] correlations for cyclopropane formation between $Rh_2(OAc)_4$-catalyzed reactions of ethyl diazoacetate and those with a broad spectrum of copper, rhodium, and

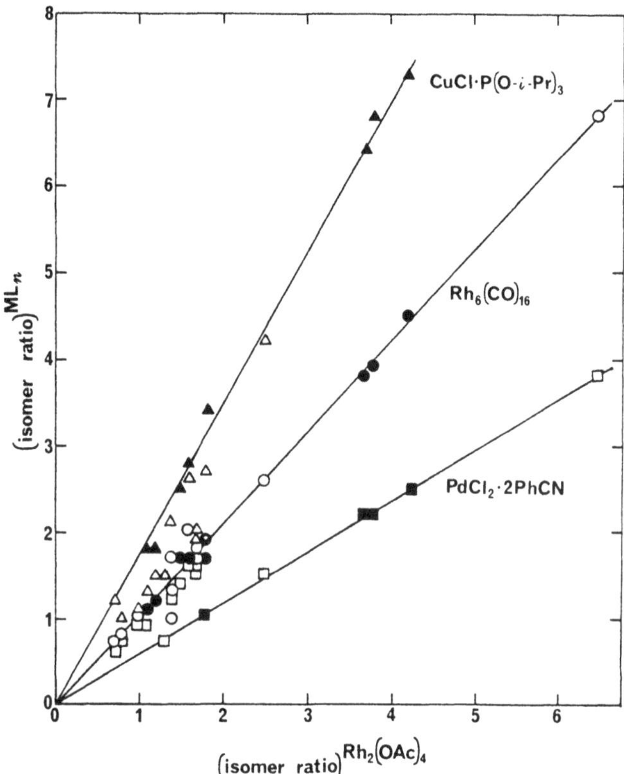

Fig. 3. Plot of observed stereoselectivities from $Rh_2(OAc)_4$-catalyzed reactions of ethyl diazoacetate with simple olefins (▲, ●, ■) or vinyl ethers and dienes (△, ○, □) vs. those from $CuCl{\cdot}P(O\text{-}\underline{i}\text{-}Pr)_3$ (▲, △), $Rh_6(CO)_{16}$ (●, ○), and $PdCl_2{\cdot}2PhCN$ (■, □) catalyzed cyclopropanation reactions.

1

2

palladium catalysts provide their link to metal carbene intermediates. The plot of stereoselectivities for cyclopropanation reactions catalyzed by $CuCl{\cdot}P(O\text{-}\underline{i}\text{-}Pr)_3$, $Rh_6(CO)_{16}$, and $PdCl_2{\cdot}2PhCN$ describes a linear correlation

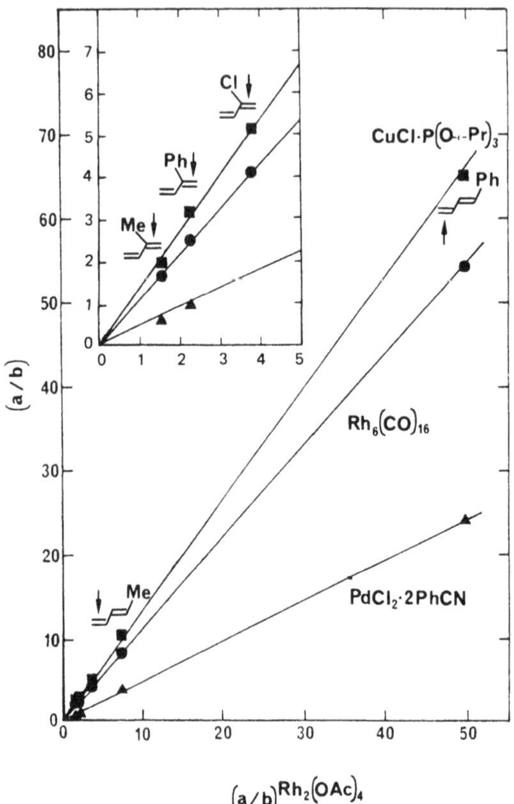

Fig. 4. Plot of observed regioselectivities (a/b) from $Rh_2(OAc)_4$-catalyzed reactions of ethyl diazoacetate with monosubstituted dienes vs. those from CuCl•P(O-i-Pr)$_3$ (■), $Rh_6(CO)_{16}$ (●), and $PdCl_2$•2PhCN (▲) catalyzed cyclopropanation reactions. Arrows define site of a-addition.

with stereoselectivities from $Rh_2(OAc)_4$-catalyzed reactions. Among the twenty-two olefins whose selectivities are recorded in Fig. 3, deviations from the linear relationship occur with vinyl ethers in CuCl•P(O-i-Pr)$_3$-catalyzed reactions and with both vinyl ethers and dienes, except for isoprene, with $PdCl_2$•2PhCN. In all other cases the selectivity correlations are remarkable. Vinyl ethers are the most reactive of the olefins employed (Table I), and they are also the most sensitive to ylide transformations[25,26] or to oxocyclopropane isomerization.[27] Similarly, vinyl cyclopropanes are

TABLE I

Relative Reactivities and Stereeoselectivities for $Rh_2(OAc)_4$ Catalyzed Reactions of Ethyl Diazoacetate with Representative Alkenes at 25°C

Alkene	Relative Reactivity	Cyclopropane trans/cis	Alkene	Relative Reactivity	Cyclopropane trans/cis
2-Methoxypropene	14.2	1.0	Cyclohexene	3.7	3.8
n-Butyl vinyl ether	12.9	1.7	2,5-Dimethyl-2,4-hexadiene	3.2	1.8
Ethyl vinyl ether	12.5	1.7	2-Methyl-2-butene	2.2	1.5
Dihydropyran	11.2	6.5	Vinyl acetate	1.7	1.6
Phenyl vinyl ether	5.8	1.4	1-Hexene	1.5	1.5
Styrene	5.3	1.6	3-Methyl-1-butene	1.3	2.0
Cyclopentene	4.2	2.1	3,3-Dimethyl-1-butene	1.00	4.2

subject to selective chloropalladation[28] so that these deviations are not surprising. Values for the slopes of the lines in Fig. 3, which provide indices of relative stereoselectivity, are 1.74 for CuCl•P(O-i-Pr)$_3$, 1.04 for $Rh_6(CO)_{16}$, and 0.59 for $PdCl_2$•2PhCN.[23] Similar indices for regioselectivity, 1.44 for CuCl•P(O-i-Pr)$_3$, 1.10 for $Rh_6(CO)_{16}$, and 0.48 for $PdCl_2$•2PhCN, relative to $Rh_2(OAc)_4$, characterize reactions with a series of monosubstituted dienes (Fig. 4).[24] There is substantial uniformity in cyclopropanation selectivity among catalysts derived from the same transition metal (eg., CuCl•P(O-i-Pr)$_3$, CuCl•P(OPh)$_3$, Cu(OTf)$_2$, Cu(acac)$_2$, and Cu bronze), despite significant differences in their initial structure and oxidation state.[23] Thus, although metal carbene involvement in catalytic reactions of diazo compounds has not yet been demonstrated by direct observation, the correlations that we have obtained disclose these species as the reactive intermediates of catalytic cyclopropanation reactions and, by inference, of catalytic ylide transformations.[25,26]

The catalytic activity of transition metal compounds in reactions of diazo compounds with olefins is dependent, at least in part, on their ability to react as electrophiles with diazo compounds, resulting in the loss of

$$L_nMB \underset{-B:}{\overset{+B:}{\rightleftharpoons}} L_nM \xrightarrow{RCHN_2} L_n\bar{M}\text{-}CHR(N_2^+) \longrightarrow L_nM{=}CHR + N_2 \quad (3)$$

dinitrogen and in the production of a metal-stabilized carbene (Eq. 3). Coordination of alkenes or Lewis bases (B:) at the coordinatively unsaturated center inhibits metal carbene formation. Rhodium(II) acetate is a superior catalyst for most cyclopropanation reactions because olefin coordination or association with alkyl halides, ethers, and esters does not occur with this transition metal catalyst.

B. Stable Metal Carbenes as Procatalysts for Cyclopropanation

Fischer carbenes, $(CO)_5WC(OMe)R$, have been reported to produce vinyl ethers and, presumably, $W(CO)_5$ by reactions with diazomethane (Eq. 4).[29]

$$(CO)_5W{=}C(OMe)R + CH_2N_2 \xrightarrow{0°C} H_2C{=}C(OMe)R + N_2 + W(CO)_5 \quad (4)$$

This transformation provides a remarkably convenient way to generate the highly electrophilic $W(CO)_5$ which, under mild conditions, could be expected to convert diazo compounds to the corresponding metal carbenes.

$$R_2CN_2 + \text{=}\!\!\diagdown_{R'} \xrightarrow{(CO)_5WC(OMe)Ph} \text{R,R-cyclopropane-R'} + N_2 \quad (5)$$

Application to reactions of diazo compounds with olefins (Eq. 5) has demonstrated the feasibility of this "procatalyst" methodology. The turnover of the procatalyst with ethyl diazoacetate is greater than 15 in reactions with vinyl ethers and, as had been anticipated from previous correlations (Figs. 2 and 3), stereoselectivities in these cyclopropanation reactions are linearly related to those from $Rh_2(OAc)_4$-catalyzed reactions (slope of 0.93 relative to $Rh_2(OAc)_4$). The procatalyst methodology offers an alternative to catalyst development that may be advantageous for enhanced selectivity in catalytic cyclopropanation reactions.

C. Stoichiometric Catalytic Cyclopropanation

The formation of cyclopropane compounds in metal-catalyzed reactions of ethyl diazoacetate is limited by competitive dimerization of the carbenoid equivalent of the diazo compound,[1] and dimerization is even more important with less stable diazo compounds.[22,30] If competition between olefin and

$$EtOOCCH{=}CHCOOEt + ML_n \xleftarrow[-N_2]{N_2CHCOOEt} L_nM{=}CHCOOEt \xrightarrow{||} \triangleright\!\!\sim COOEt + ML_n \quad (6)$$

ethyl diazoacetate for the metal carbene (Eq. 6) is the cause of limited cyclopropane production then merely minimizing the available concentration of ethyl diazoacetate should afford optimal yields of cyclopropane products. As a consequence, exact control of the rate of addition of the diazo compound to the reaction medium should afford superior cyclopropane yields even when only equivalent amounts of olefin and ethyl diazoacetate are employed. Such procedural control should also offer economical benefits by limiting the amount of catalyst required for effective cyclopropanations. These advantages have been realized for reactions catalyzed by $Rh_2(OAc)_4$,

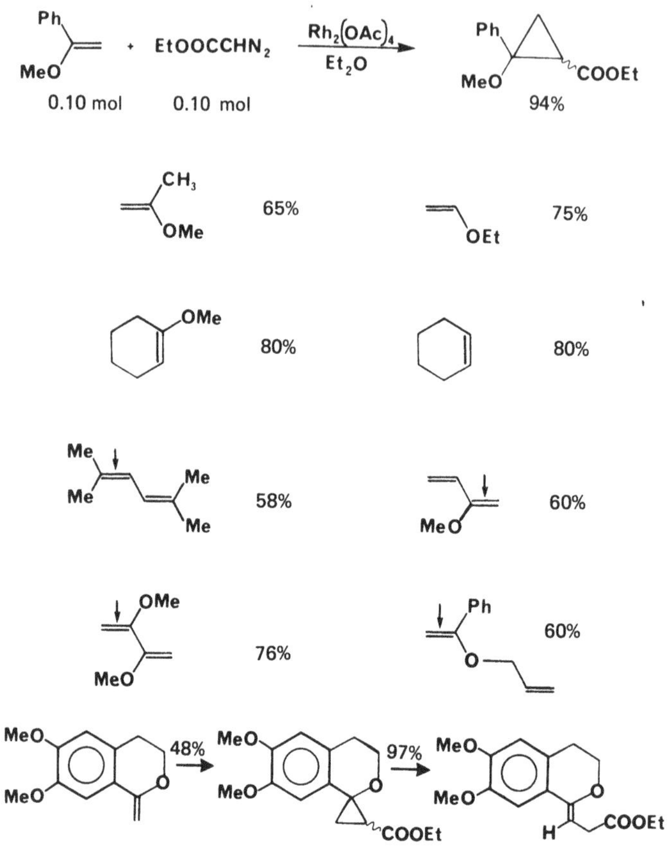

Ph, MeO (alkene) + $EtOOCCHN_2$ —$Rh_2(OAc)_4$, Et_2O→ Ph, MeO cyclopropane COOEt

0.10 mol 0.10 mol 94%

65% 75%

80% 80%

58% 60%

76% 60%

48% 97%

Fig. 5. Catalytic cyclopropanation of representative alkenes with a stoichiometric amount of ethyl diazoacetate using $Rh_2(OAc)_4$. Arrows identify site of exclusive addition to dienes.

$Rh_6(CO)_{16}$, and $CuCl{\cdot}P(O\text{-}\underline{i}\text{-}Pr)_3$ (Fig. 5),[9] but control of the rate of addition of diazo compound has only a marginal influence on cyclopropane product yield in reactions catalyzed by $PdCl_2{\cdot}2PhCN$ or $Cu(OTf)_2$.

The effect of this control on product yield is dramatically illustrated for reactions of 1-methoxycyclohexene with a stoichiometric amount of ethyl diazoacetate: at a $[EDA]/[Rh_2(OAc)_4]$ ratio of 200 and an addition rate of 5 mmol/hr, the corresponding cyclopropane product was obtained in 80% yield. With a $[EDA]/[Rh_2(OAc)_4]$ ratio of 2000 and an addition rate of 5 mmol/hr, the yield of this product was only 60%. However, by decreasing the rate of addition of ethyl diazoacetate to 0.5 mmol/hr, the cyclopropane product could again be isolated in 80% yield.

Although the use of $Rh_2(OAc)_4$ appears to be optimally suited to intermolecular addition of diazo compounds to olefins, the same is not necessarily true for intramolecular transformations. The diazoketone 3 was

$$\mathbf{3} \xrightarrow[CH_3NO_2]{Cu(OTf)_2} \mathbf{4} \quad (7)$$

$$\mathbf{5} \xrightarrow[CH_3NO_2]{Cu(OTf)_2} \mathbf{6} \quad (8)$$

converted to the tricyclic ketone 4 (Eq. 7) in 90% isolated by $Cu(OTf)_2$ in nitromethane.[31] In contrast, $Rh_2(OAc)_4$ generated 4 in only 36% yield, and yields ranging from 31-67% have been reported from $CuSO_4$ catalysis in refluxing cyclohexane.[32] Similarly, treatment of the diazoketone 5 with $Cu(OTf)_2$ in nitromethane resulted in the formation of Δ^3-hydrinden-2-one (Eq. 8) in 59% yield, which should be compared to the 50% yield of this product obtained from 5 with the use of $BF_3 \cdot Et_2O$ under similar reaction conditions.[33] The use of nitromethane as the solvent for these reactions offered distinctive advantages in product yield that could not be approached with alternative ether, halocarbon, or hydrocarbon solvents.

D. Stereoselectivity and the Mechanism of Catalytic Cyclopropanation

Metal catalyzed reactions of diazocarbonyl compounds with alkenes consistently produce greater yields of the more stable cyclopropane isomer.[23] In contrast, the less stable cyclopropane isomer is usually predominant in stoichiometric reactions of aryl- or methyl-substituted metal carbenes.[16,17] Since the stereoselectivities in $Rh_2(OAc)_4$- catalyzed reactions of phenyldiazomethane parallel those of $(CO)_5WCHPh$ and a similar correspondence exists for $Rh_2(OAc)_4$- and $W(CO)_5$-catalyzed reactions of ethyl diazoacetate, the divergence in stereoselectivity for

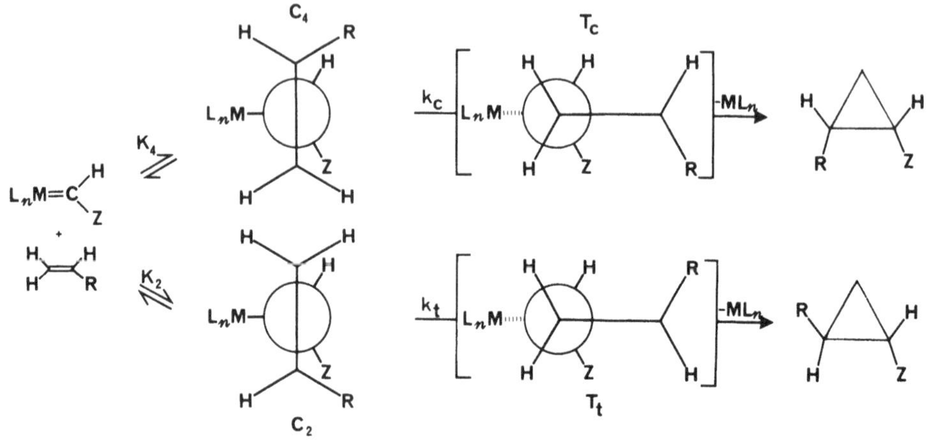

Scheme I

carbonylcarbenoid and phenylcarbenoid addition to olefins must originate with the carbenoid entity. High cis stereoselectivities in benzylidene transfer from $(CO)_5WCHPh$ and $CpFe(CO)_2CHC_6H_5^+$ have been previously attributed to a stabilizing interaction between the developing electrophilic center of the reacting alkene and the aryl group.[16,17a] However, ethylidene transfer reactions of $CpFe(CO)_2CH(OMe)CH_3$, for which such a stabilizing interaction is not possible, also exhibit a pronounced preference for formation of the cis-cyclopropane isomer.[17c] The predominant trans stereoselectivity observed in catalytic reactions of diazocarbonyl compounds has been attributed to the intervention of a metallocyclobutane-like transition state.[4]

We have recently provided a predictive model for stereoselectivity in cyclopropane formation that is universally applicable to stoichiometric reactions of stable metal carbenes and catalytic transformations of diazo compounds with alkenes.[22] According to this model, cyclopropane formation occurs by initial association of the olefin π bond with the electrophilic center of the metal carbene followed by σ-bond formation with backside displacement of the catalyst that results in cyclopropane formation (Scheme I). Stabilization of the developing electrophilic center by the electron-donating substituent R determines the motion of the alkene

TABLE II

Effect of Alkene Substituents on Cyclopropanation Stereoselectivity

	cis:trans Cyclopropane Isomer Ratios		
R =	$Rh_2(OAc)_4$, $PhCHN_2$ 25°C[22]	$(CO)_5WCHPh$ -78°C[16,22]	$Cp(CO)_2FeCHPh^+$ -78°C[17a]
C_6H_5-	3.3	9.7	≥100
EtO-	3.0	7.6	
CH_3CH_2-		0.9	6.5
$CH_3(CH_2)_3$-	0.90	0.84	
$(CH_3)_2CH$-	0.56	0.36	4.6

relative to the electrophilic carbon center of the reacting carbene. In proceeding from C_4 to the transition state T_c, the alkene carbon structure moves up relative to the carbenic carbon and concurrently rotates to the T_c configuration in which the original C-C bond axis is parallel to the original metal-carbon bond axis. Similarly, the alkene carbon structure in C_4 moves down relative to the carbenic carbon with rotation to the parallel allignment of T_t. Backside displacement of ML_n by the developing carbenium ion results in cyclopropane formation. Cyclopropane stereochemistry is determined by a combination of steric and/or electronic interactions at the π complex (C_2 and C_4) and cyclopropane-forming (T_c and T_t) states. In the absence of prominant steric interactions between R and Z in the cyclopropane-forming stage, predominant cis stereoselectivity is predicted ($k_t \approxeq k_c$, but C2 is less stable than C_4) and observed for reactions of monosubstituted alkenes possessing alkyl[17b,c] or aryl[16,17a,22] substituents. Increasing the size of R decreases k_c relative to k_t and leads to increased relative percentage of the <u>trans</u>-cyclopropane isomer (Table II).

Stereochemical preference for the <u>trans</u>-cyclopropane isomer in reactions involving transition-metal-catalyzed reactions of ethyl diazoacetate and related diazocarbonyl compounds appears to be the exception to this model. However, examination of the specific influence of

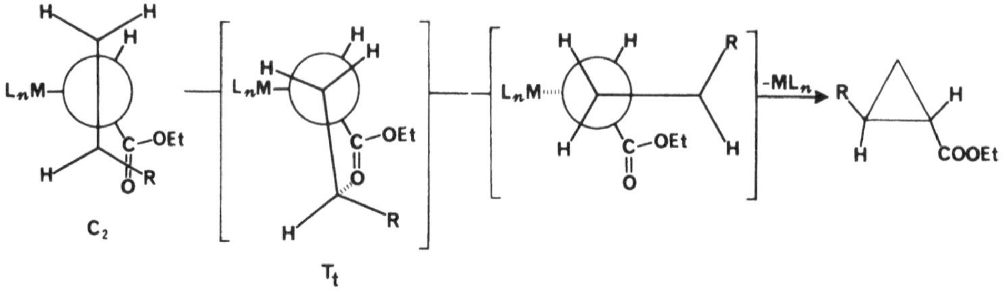

Scheme 2

the carbonyl group on the conversion of the intermediate π-complex to cyclopropane products explains this apparent discrepancy. Electronic stabilization of the transition state leading to trans-cyclopropane products by interaction of the developing electrophilic carbon of the original alkene with the α-carbonyl substituent of the metal carbene (T_t in Scheme II), which is only possible in the transition state leading to the trans cyclopropane, accounts for the predominant trans stereoselectivity in these reactions. Furthermore, increasing the nucleophilicity of the carbonyl oxygen and/or the electrophilicity of the developing electropositive center of the reacting alkene can be expected to alter the reaction course by forming dihydrofuran (Eq. 9) and/or apparent allylic C-H insertion (Eq. 10)

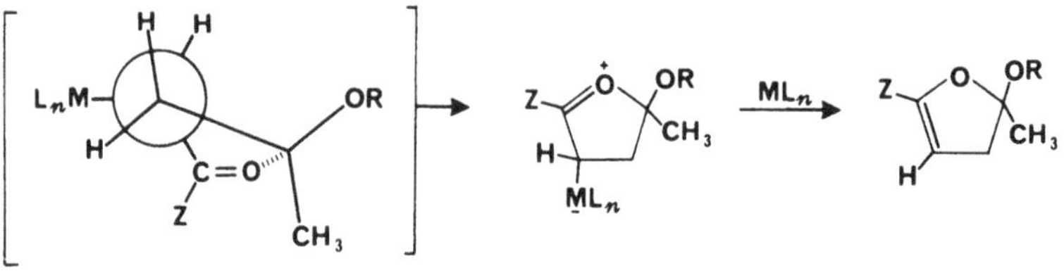

(9)

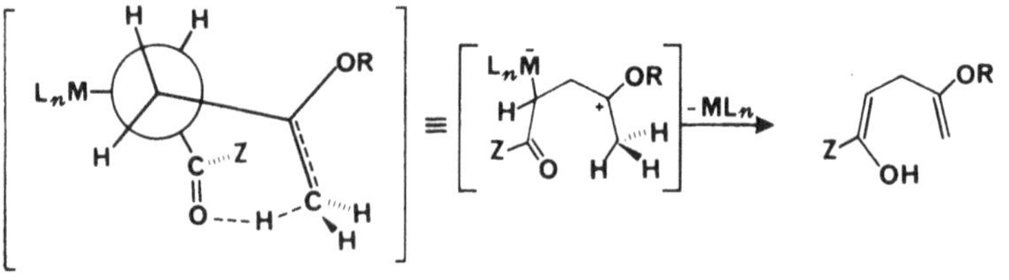

(10)

products. These transformations are, in fact, well documented in transition-metal-catalyzed reactions of diazomalonates and α-diazoketones with vinyl ethers[22,34,35] and certain simple alkenes[36] (Eq. 11-13).

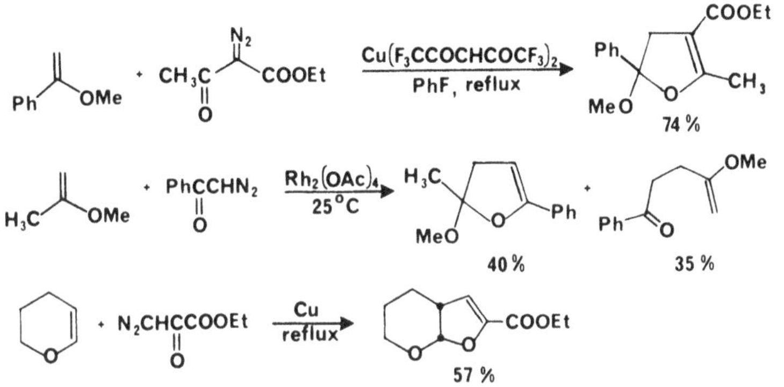

(11)

(12)

(13)

The model that we have presented differs considerably from that employed by Brookhart[17] to explain the predominant cis(syn) stereoselectivity observed in stoichiometric reactions of alkenes with metal carbenes possessing alkyl or aryl substituents. The Brookhart model requires the involvement of a metallocyclobutane intermediate to explain trans stereoselectivity and is apparently restricted to stoichiometric transformations of metal carbenes. Our model provides a unified explanation for all stereochemical features of catalytic and stoichiometric cyclopropanation reactions as well as dihydrofuran production and allylic C-H insertion observed in catalytic reactions of alkenes with certain diazocarbonyl compounds.

E. Influence of Olefin Coordination on Cyclopropanation Selectivity

The first reference to differential selectivity in transition metal catalyzed carbenoid transformations of diazo compounds was reported by Salomon and Kochi.[5] Comparing relative reactivities for cyclopropanation of tetramethylethylene and 1-hexene, they found that copper complexes capable of olefin coordination (CuOTf, $CuBF_4$) favored reaction with 1-hexene whereas those not capable of olefin coordination ($Cu(acac)_2$, $CuCl{\cdot}P(OR)_3$, $CuSO_4$) favored reaction with tetramethylethylene, and analogous differences were obtained for regioselective cyclopropanation of

7-methyl-1,6-octadiene. These observations were attributed to the differential influences of olefin coordination with the transition metal, and CuOTf was inferred to form cyclopropane products by reaction of the coordinated olefin with the metal-associated carbene. Intramolecular collapse of the coordinated olefin onto the carbene has been offered as one of the two fundamental mechanisms for olefin cyclopropanation.[37] However, there is one substantial problem with this "coordination mechanism" for cyclopropanation: the coordinated olefin is activated towards nucleophilic attack[38] but the proposed collapse involves electrophilic carbene addition.

The stable palladium(II) chloride complexes of 5-methylenebicyclo[2.2.1]-hept-ene (**7**) and <u>endo</u>-5-vinylbicyclo[2.2.1]hept-2-ene (**8**)[39] were subjected to ethyl diazoacetate in chloroform. Neither underwent any observable reaction with this diazo compound and remained unchanged even when either $Rh_2(OAc)_4$ or $PdCl_2 \cdot 2PhCN$ was added to the combination of ethyl diazoacetate and palladium complex. However,

7 Pd Cl_2

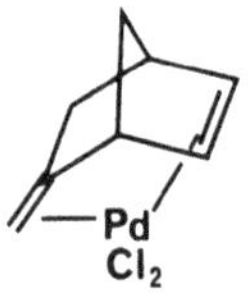

8 Cl H Pd—Cl

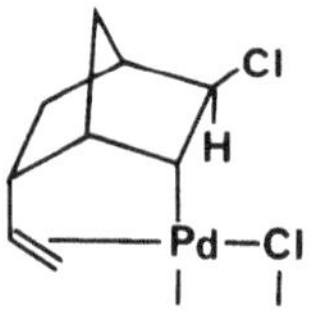

addition of excess diene to **7** or **8**, even in equivalent amounts, followed by ethyl diazoacetate resulted in rapid cyclopropanation. These data clearly exclude electrophilic addition of the coordinated olefin to the diazo compound, but they suggest that selective cyclopropanation of an uncoordinated carbon-carbon double bond of a diene could occur from an intramolecularly coordinated metal carbene.

Limonene and 5-methylenebicyclo[2.2.1]hept-2-ene are representative of dienes that we have examined for selectivity in catalytic cyclopropanation reactions. Electrophilic addition to one of the double bonds of 5-methylenebicyclo[2.2.1]hept-2-ene can occur with homoallylic stabilization of the transition state,[40,41] whereas no such stabilization is possible with limonene. The bicyclic diene is also suitably constructed for intramolecular carbene addition (Eq. 14). Catalysts with two available

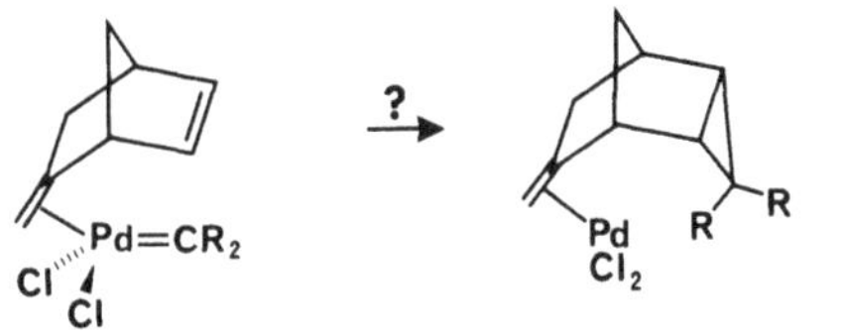

(14)

coordination sites were employed and $Rh_2(OAc)_4$, a catalyst that does not coordinate with alkenes, served as the reference. Treatment of excess olefin with ethyl diazoacetate in the presence of $Rh_2(OAc)_4$, $Cu(OTf)_2$ or $PdCl_2{\cdot}2PhCN$ gave the results described in Table III.[42] For reactions with limonene, regioselectivities are relatively invarient with the catalyst, and cyclopropanation of the disubstituted double bond is favored over addition to the trisubstituted double bond. In contrast, $Rh_2(OAc)_4$-catalyzed reactions of ethyl diazoacetate with 5-methylenebicyclo[2.2.1]hept-2-ene resulted in the favored production of cyclopropane products from electrophilic attack at the exocyclic double bond, whereas $PdCl_2{\cdot}2PhCN$ effects highly selective cyclopropanation of the exocyclic double bond. Furthermore, the use of diphenyldiazomethane (Eq. 15) gave **9** exclusively (> 100:1) with $Rh_2(OAc)_4$ and **10** in similar high selectivity (> 50:1) with $PdCl_2{\cdot}2PhCN$.

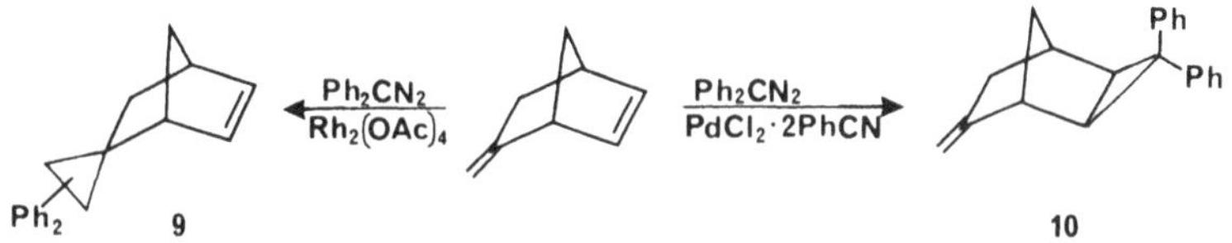

(15)

Competitive cyclopropanation of the two model compounds, bicyclo-[2.2.1]hept-2-ene and 2-methylenebicyclo[2.2.1]heptane, showed that the relative reactivity for cyclopropanation of the endocyclic double bond was 5 in $PdCl_2{\cdot}2PhCN$-catalyzed reactions of ethyl diazoacetate but only 0.45 with $Rh_2(OAc)_4$. Thus the relative reactivities of the two model compounds reflect the regioselectivities in cyclopropanation of 5-methylenebicyclo[2.2.1]hept-2-ene by ethyl diazoacetate. Furthermore, the stereoisomeric exo/endo ratios for cyclopropanation of bicyclo-[2.2.1]hept-2-ene by ethyl diazoacetate (1.6 for $PdCl_2{\cdot}2PhCN$ and 2.3 for $Rh_2(OAc)_4$) were very close to those observed for cyclopropanation of 5-methylenebicyclo[2.2.1]hept-2-ene (Table III). Based on the composite

TABLE III

Catalytic Cyclopropanation of Limonene and 5-exo-Methylene-2-norbornene by Ethyl Diazoacetate

Catalyst	Yield,%	A/B	Yield/%	A/B	B:exo/endo
$Rh_2(OAc)_4$	98	3.4	93	2.3	2.6
$Cu(OTf)_2$	59	3.9	53	1.0	1.3
$PdCl_2 \cdot 2PhCN$	32	3.6	82	0.16	1.2

results, intramolecular addition such as is depicted by Eq. 14 can be discounted, and the observed selectivity differences can be attributed to electronic influences on the transition state for cyclopropanation. An early transition state with little charge development at the olefinic carbons favors cyclopropanation at the endocyclic double bond (**11**), whereas a late transition state with considerable charge development (**12**) favors

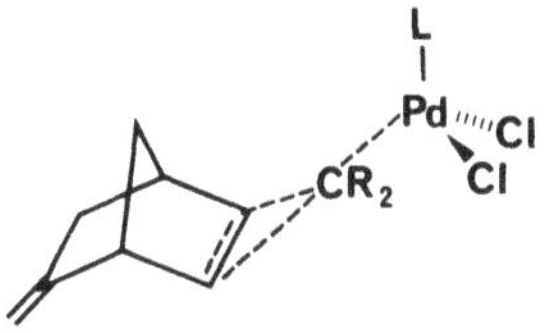

L
Pd
Cl
Cl
CR_2
11

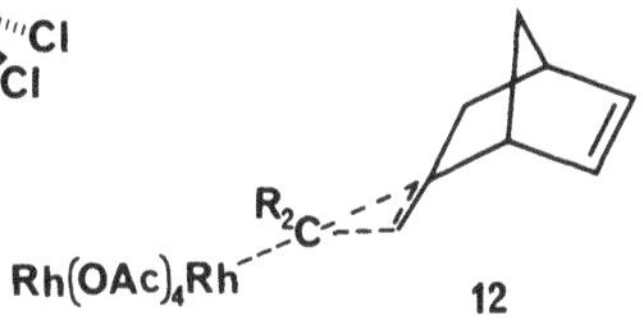

R_2C
$Rh(OAc)_4Rh$
12

cyclopropanation at the exocyclic double bond. The relative reactivity of 2-methylenebicyclo[2.2.1]heptane in $Rh_2(OAc)_4$-catalyzed reactions with ethyl diazoacetate, which closely approximates the regioselectivity for cyclopropanation of 5-methylenebicyclo[2.2.1]hept-2-ene, suggests that there is little, if any, homoallylic stabilization in **12**. As can now be predicted from the mechanism outlined in Schemes I and II for catalytic cyclopropanation, competition between 2,3-dimethyl-2-butene and 1-hexene should favor 2,3-dimethyl-2-butene in a late transition state. On the other hand, 1-hexene will be more reactive in an early transition state where charge development at the olefinic centers is minimal and steric influences in π-complex formation should be important.[43]

F. Steric Enhancement of Cyclopropanation Selectivity

The mechanism for catalytic cyclopropanation outlined in Scheme I projects increased trans(anti) stereoselectivity for cyclopropanation based on steric interactions between olefin and carbene substituents (R and Z) in transition state T_c. However, the relatively large distance between R and Z suggests that only very bulky substituents will provide steric interaction sufficient to lower the energy of T_t relative to T_c, particularly for catalytic transformations that occur at or above 25°C. The specific influences of olefin substituents on stereoselectivity for catalytic reactions of ethyl diazoacetate have already been described (Fig. 3) and, although weak, they follow the expected order.[23] The alternate use of tert-butyl diazoacetate has not indicated ay substantial improvement in trans(anti) stereoselectivity,[11] but this ester may not be sufficiently bulky to interact with the olefin substituent. We have recently employed 2,3,4-trimethyl-3-pentyl diazoacetate for catalytic cyclopropanation reactions, and the results that we have obtained suggest that this mode of ester modification is in the proper direction for highly selective cyclopropanation reactions.

2,3,4-Trimethyl-3-pentyl diazoacetate (13) is conveniently prepared from the corresponding alcohol and diketene followed by diazo transfer[44]

$$\text{R-OH} \xrightarrow{\text{diketene}} \text{R-OC(=O)CH}_2\text{C(=O)CH}_3 \xrightarrow[\text{2. KOH}]{\text{1. TsN}_3} \text{R-OC(=O)CHN}_2 \;(\mathbf{13}) \qquad (16)$$

and acyl cleavage (Eq. 16). Rhodium(II) acetate catalyzed reactions of 13 with representative olefins have resulted in substantial increases in trans (anti) selectivity (trans/cis (anti/syn): 2.6, ethyl vinyl ether; 3.8, styrene; 10, cyclohexene) over the comparative use of ethyl diazoacetate (1.7, ethyl vinyl ether; 1.6, styrene; 3.8, cyclohexene). However, trisubstituted olefins such as 2,5-dimethyl-2,4-hexadiene are not as responsive to this steric influence (t/c = 2.0 vs. 1.8 with ethyl diazoacetate). Reactions of 13 with isoprene exhibit decreased regioselectivity for addition to the 1,2-double bond (1.27 vs. 1.58 with ethyl diazoacetate). Further efforts are being directed to the achievement of exceptional selectivities in these catalytic reactions.

III. CONCLUSION

Catalytic cyclopropanation reactions with diazo compounds involve the intermediate production of electrophilic metal carbenes. Addition to nucleophilic olefins resulting in the formation of cyclopropane compounds occurs by initial association of the olefin π bond with the electrophilic center of the metal carbene followed by σ-bond formation with backside displacement of the catalyst. Both the metal (Cu > Rh > Pd) and carbene substituents (COOR < H < Ph) influence diene regioselectivity and olefin stereoselectivity. Transition metal catalysts with two available coordination sites, whose abnormal cyclopropanation selectivities have been previously explained by intramolecular collapse of a coordinated olefin onto the metal-associated carbene, do not undergo cyclopropanation through this "coordination" mechanism. Instead, their abnormal selectivities have been explained in terms of early and late transition states for cyclopropanation.

Rhodium(II) acetate is generally the most suitable catalyst for intermolecular cyclopropanation reactions. Relatively high product yields are obtained even with moderately reactive olefins. Intramolecular cyclopropanation of unsaturated diazoketones, however, is better suited to catalysis by $Cu(OTf)_2$ in nitromethane. Vinyl ethers are considerably more reactive than dienes or simple olefins towards cyclopropanation by diazo compounds, but reactions with diazoketones generally result in the formation of dihydrofurans and allylic C-H insertion products when vinyl ethers are employed. Each of these transformations is intimately linked to metal carbene reactivity, and predictions regarding the outcome of specific catalytic reactions can now be made.

IV. ACKNOWLEDGMENT

The research has been generously supported by the National Science Foundation. Recent results, not acknowledged by reference to the literature, are due to the efforts of Sandra Bellifeuille, Mitchell Chinn, Jose Conceicao, John Griffin, Randy Rodenhouse, Linda Wang, and Dr. Kuo-Liang Loh.

REFERENCES

1. Several authoritative reviews of selected aspects of catalytic cyclopropanation reactions are available: (a) V. Dave and E. Warnhoff, Org. React. (N.Y.), **18**, 217 (1970); (b) A.P. Marchand and N. MacBrockway, Chem. Rev., **74**, 431 (1974); (c) S.D. Burke and P.A. Grieco, Org. React. (N.Y.), **26**, 361 (1979).
2. (a) O. Silberrad and C.S. Roy, J. Chem. Soc., **89**, 179 (1906); (b) A. Loose, J. Prakt. Chem., [2] **79**, 505 (1909).

3. (a) H. Nozaki, S. Moriuti, H. Takaya, and R. Noyori, Tetrahedron Lett., 5239 (1966); (b) H. Nozaki, H. Takaya, S. Moriuti, and R. Noyori, Tetrahedron, **24**, 3655 (1968); (c) R. Noyori, H. Takaya, Y. Nakanishi, and H. Nozaki, Can. J. Chem., **47** 1242 (1969).
4. W. R. Moser, J. Am. Chem. Soc., **91**, 1135, 1141 (1969).
5 R.G. Salomon and J.K. Kochi, J. Am. Chem. Soc., **95**, 3300 (1973).
6. R.K. Armstrong, J. Org. Chem., **31**, 618 (1966).
7. A. Nakamura, T. Koyama, and S. Otsuka, Bull. Chem. Soc. Jpn., **51**, 593 (1978).
8. A.J. Hubert, A.F. Noels, A.J. Anciaux, and Ph. Teyssié, Synthesis, 600 (1976).
9. M.P. Doyle, D. van Leusen, and W.H. Tamblyn, Synthesis, 787 (1981).
10. M.P. Doyle, W.H. Tamblyn, W.E. Buhro, and R.L. Dorow, Tetrahedron Lett., **22**, 1783 (1981).
11. H.J. Callot and C. Piechocki, Tetrahedron Lett., **21**, 3489 (1980).
12. P. Yates, J. Am. Chem. Soc., **74**, 5376 (1952).
13. T. Aratani, Y. Yoneyoshi, and T. Nagase, Tetrahedron Lett., 1707 (1975) and 2599 (1977).
14. (a) A. Nakamura, A. Konishi, R. Tsujitani, M. Kudo, and S. Otsuka, J. Am. Chem. Soc., **100**, 3449 (1978); (b) A. Nakamura, A. Konishi, Y. Tatsuno, and S. Otsuka, ibid., **100** 3443, 6544 (1978).
15. (a) E.O. Fischer and K.H. Dötz, Chem. Ber., **105**, 3966 (1972); (b) K.H. Dötz and E.O. Fischer, ibid., **105**, 1356 (1972).
16. C.P. Casey, S.W. Polichnowski, A.J. Shusterman, and C.R. Jones, J. Am. Chem. Soc., **101**, 7282 (1979).
17. (a) M. Brookhart, B.H. Humphrey, H.J. Kratzner, and G.O. Nelson, J. Am. Chem. Soc., **102**, 7802 (1980); (b) M. Brookhart, J.R. Tucker, and G.R. Husk, ibid., **103**, 979 (1981); (c) M. Brookhart, J.R. Tucker, and G.R. Husk, ibid., **105**, 258 (1983); (d) M. Brookhart, D. Timmers, J.R. Tucker, and G.D. Williams, ibid., **105**, 6721 (1983).
18. (a) K.A.M. Kremer, P. Helquist, and R.C. Kerber, J. Am. Chem. Soc., **103**, 1862 (1981); (b) K.A.M. Kremer, G.-H. Kuo, E.J. O'Connor, P. Helquist, and R.C. Kerber, ibid., **104**, 6119 (1982).

19. (a) M.P. Doyle and D. van Leusen, J. Org. Chem., **47**, 5326 (1982); (b) M. Suda, Synthesis, 714 (1981); (c) H. Nozaki, S. Moriuti, M. Yamabe, and R. Noyori, Tetrahedron Lett., 59 (1966).
20. (a) M.P. Doyle, R.L. Dorow, and W.H. Tamblyn, J. Org. Chem., **47**, 4059 (1982); (b) M.P. Doyle and J.G. Davidson, ibid., **45**, 1538 (1980).
21. M.W. Majchrzak, A. Kotelko, and J.B. Lambert, Synthesis, 469 (1983).
22. M.P. Doyle, J.H. Griffin, V. Bagheri, and R.L. Dorow, Organometal., **3**, 53 (1984).
23. M.P. Doyle, R.L. Dorow, W.E. Buhro, J.H. Griffin, W.H. Tamblyn, and M.L. Trudell, Organometal. **3**, 44 (1984).
24. M.P. Doyle, R.L. Dorow, W.H. Tamblyn, and W.E. Buhro, Tetrahedron Lett., **23**, 2261 (1982).
25. M.P. Doyle, W.H. Tamblyn, and V. Bagheri, J. Org. Chem., **46**, 5094 (1981).
26. M.P. Doyle, J.H. Griffin, M.S. Chinn, and D. van Leusen, J. Org. Chem. **49**, 1917 (1984).
27. M.P. Doyle and D. van Leusen, J. Org. Chem., **47**, 5326 (1982); J. Am. Chem. Soc., **103**, 5917 (1981).
28. J.L. Williams and M.F. Rettig, Tetrahedron Lett., **22**, 385 (1981).
29. C.P. Casey, S.H. Bertz, and T.J. Burkhardt, Tetrahedron Lett., 1421 (1973).
30. B.K.R. Shankar and H. Shechter, Tetrahedron Lett., **23**, 2277 (1982).
31. M.P. Doyle and M.L. Trudell, J. Org. Chem., **49**, 1196 (1984).
32. (a) S.A. Monti, D.J. Bucheck, and J.C. Shepard, J. Org. Chem., **34**, 3080 (1969); (b) G.A. Russell, J.J. McDonnell, P.R. Whittle, R.S. Givens, and R.C. Keske, J. Am. Chem. Soc., **93**, 1452 (1971).
33. A.B. Smith III, B.H. Toder, S.J. Branka, and R.K. Dieter, J. Am. Chem. Soc., **103**, 1996 (1981).
34. (a) E. Wenkert, Acc. Chem. Res., **13**, 27 (1980); (b) E. Wenkert, Heterocycles, **14**, 1703 (1980).
35. (a) E. Wenkert, M.E. Alonso, B.L. Buckwalter, and E.L. Sanchez, J. Am. Chem. Soc., **105**, 2021 (1983); (b) M.E. Alonso, A. Morales, and A.W. Chitty, J. Org. Chem., **47**, 3747 (1982); (c) E. Wenkert, M.E. Alonso, B.L. Buckwalter, and J.K. Chow, J. Am. Chem. Soc., **99**, 4778 (1977).

36. M.E. Alonso and M. Gomez, Tetrahedron Lett., 2763 (1979).

37. (a) A.J. Anciaux, A.J., Hubert, A.F. Noels, N. Petiniot, and Ph. Teyssié, J. Org. Chem., **45**, 695 (1980); (b) A.J. Anciaux, A. Domonceau, A.F. Noels, R. Waren, A.J. Hubert, and P. Teyssié, Tetrahedron, **39**, 2169 (1983).

38. O. Eisenstein and R. Hoffmann, J. Am. Chem. Soc., **102**, 6148 (1980).

39. W.T. Wipke and G.L. Goeke, J. Am. Chem. Soc., **96**, 4244 (1974).

40. D.G. Garratt, P.L. Beaulieu, and V.M. Morisset, Can. J. Chem., **58**, 1021 (1980).

41. (a) S.A. Lermontor, N.A. Belikova, T.G. Skornyakova, T. Pehk, E. Lippmaa, and A. Plate, Zh. Org. Khim., **16**, 2322 (1980); (b) A.A. Bobyleva, N.A. Belikova, A.V. Dzhigirkhanova, and A.F. Plate, ibid. **16**, 915 (1980).

42. M.P. Doyle, L.C. Wang, and K.-L. Loh, Tetrahedron Lett., **25**, 4087 (1984).

43. Freeman, F. Chem. Rev., **75**, 439 (1975).

44. Regitz, M. Angew. Chem. Int. Ed. Engl., **6**, 733 (1967).

5

Interactions of Quinoline and Tetrahydroquinoline with Catalysts

Modeling Heterogeneous Catalysts with Homogeneous Catalysts

C.M. Giandomenico,+ A. Eisenstadt,+ M.F. Fredericks,
A.S. Hirschon and R.M. Laine
Organometallic Chemistry Program
SRI International, Menlo Park, CA 94025
and
Nicolet XRD Co., Fremont, CA 94539

ABSTRACT

The literature on the catalytic hydrodenitrogenation (HDN) of petroleum, coal, shale oil, and tar sands is replete with examples of studies on the reactions of HDN catalysts with model compounds such as pyridine, indole, or quinoline. Despite these extensive and often informative studies, to date, the exact mechanism(s) by which nitrogen heterocycles interact with the catalyst surface to undergo first hydrogenation and then ring cleavage and loss of ammonia remains a mystery. The present study attempts to develop a reasonable picture of how nitrogen heterocycles interact with HDN catalysts and undergo HDN through the exploration of the stoichiometric and catalytic reactions of these heterocycles with homogeneous catalysts. We present here studies on the reactions of quinoline and tetrahydroquinoline with $Os_3(CO)_{12}$ that provide information concerning how these species bind to metals and undergo hydrogenation and dehydrogenation.

+SRI Postdoctoral Fellow

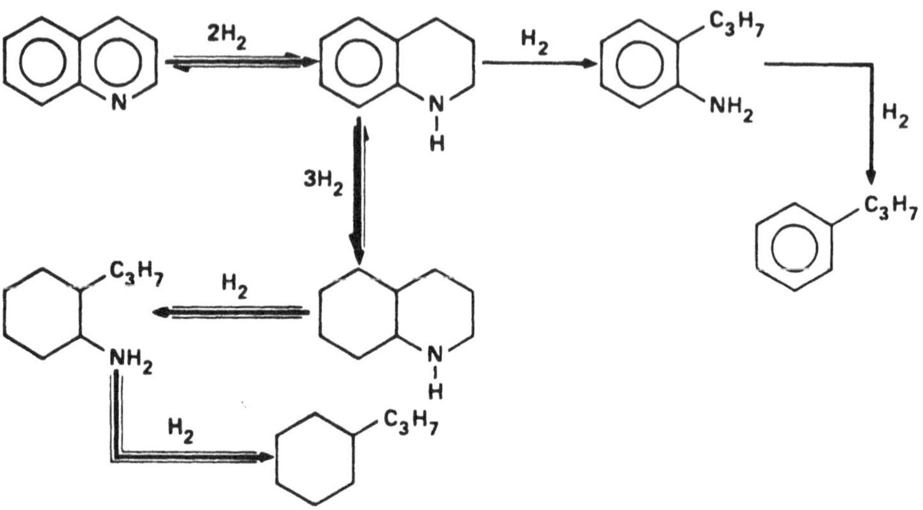

Fig. 1. Quinoline HDN reaction network.

I. INTRODUCTION

There is extensive literature devoted to the subject of catalytic hydrodenitrogenation (HDN) of petroleum, coal, shale oil, and tar sands.[3-10] This literature is replete with examples of studies on the reactions of HDN catalysts with model compounds such as pyridine,[4] indole,[5] or quinoline.[6-10] For example, studies of heterogeneously catalyzed HDN of the prototypical model compounds, quinoline and tetrahydroquinoline (THQ), provide the reaction network shown in Fig. 1.[6,7]

These careful studies have generated all the necessary rate constants and thermodynamic information to evaluate the HDN process, as it applies to quinoline, from a chemical engineering standpoint. Thus, the thermodynamically preferred reaction pathway for HDN of quinoline, as depicted in Fig. 1, is the heavily lined pathway.[6-10] The preferred pathway, in terms of hydrogen expenditure and quality of product, is the pathway that promotes HDN without hydrogenating the second aromatic ring; the unemphasized pathway in Fig. 1.

Currently, HDN costs account for ∿80% of the hydrotreating costs associated with refining petroleum. Consequently, there is a considerable economic driving force associated with reducing hydrogen consumption in

HDN through development of better catalysts; catalysts that promote HDN via the latter pathway.

Despite the extensive and informative HDN studies already in the literature, to date the exact mechanism(s) by which nitrogen heterocycles interact with the catalyst surface to undergo first hydrogenation and then C-N bond/ring cleavage and loss of ammonia remains a mystery. The motivating force for the work described below comes from the rationale that studies leading to the elucidation of these mechanisms may make it possible to design better HDN catalysts.

II. RESULTS AND DISCUSSION

Our initial approach to identifying the mechanisms operative in catalytic hydrogenation and in the C-N bond cleavage processes of heteroaromatic species was to explore the reactions of quinoline and THQ with a number of organometallic compounds. Our intent was to establish; (1) whether or not the reactions of quinoline/THQ with organometallic species can be compared with those observed for quinoline/THQ with a standard HDN catalyst; and, (2) if these comparisons will allow us to draw mechanistic analogies between the two systems. We have previously used such comparisons to propose a comprehensive picture of the mechanistic processes that occur in heterogeneously catalyzed HDN.[1,11,12] This proposal serves as a guideline for the research reported here.

Our preliminary work has centered on studies of the stoichiometric and catalytic reactions of $Os_3(CO)_{12}$ or $H_2Os_3(CO)_{10}$ with quinoline (Q) and THQ. We chose to study the stoichiometric rections of Q/THQ with osmium clusters predicated on our interest in preparing stable organometallic complexes of reactants (Q and H_2/D_2), products (THQ) and the desire to isolate reactive intermediates. Osmium clusters are well known to be extremely stable towards cluster degradation and to form stable, readily isolable, complexes under conditions where other metal complexes/catalysts will rapidly degrade (e.g., by comparison with similar reactions of $Ru_3(CO)_{12}$). An added incentive to pursue such studies is that the isolated intermediates may be identifiable in the catalytic reactions and hence permit the complete elucidation of the various steps in the catalytic cycle(s). Thus, concurrent with these stoichiometric studies, we are also studying the catalytic hydrogenation of quinoline using $Os_3(CO)_{12}$ and $H_2Os_3(CO)_{10}$.

Coincidental with our studies of the Q/THQ-osmium chemistry, we are exploring the catalytic reactions of Q/THQ with a sulfided cobalt-molybdenum catalyst (CoMo) in the presence of H_2 or D_2 in an attempt to establish the similarities and differences between the two catalyst systems.

A. Stoichiometric Reactions

In work that will be disclosed in greater detail elsewhere,[13] we have succeeded in isolating and characterizing the osmium complexes, 1-5.

$$Os_3(CO)_{12} + Q \xrightarrow{145^oC/16h/octane} H_2(\eta^2\text{-}C_9H_6N)_2Os_3(CO)_8 \quad (1)$$

$$Os_3(CO)_{12} + THQ \xrightarrow{145^oC/26h/octane} H_2(\eta^2\text{-}C_9H_6N)Os_3(CO)_{10} + H_2(\eta^2\text{-}C_9H_6N)(\eta^2\text{-}C_9H_8N)Os_3(CO)_8 \quad (2)$$

$$H_2Os_3(CO)_{10} + THQ \xrightarrow{125^oC/6h/octane} H_2(\eta^2\text{-}C_9H_6N(\eta^2\text{-}C_9H_8N)Os_3(CO)_8 + H(\eta^2\text{-}C_9H_8N)Os_3(CO)_{10} \quad (3)$$

$$Os_3(CO)_{12} + \text{piperidine} \xrightarrow{145^oC/6h/octane} H(\eta^2\text{-}C_5H_8N)Os_3(CO)_{10} \quad (4)$$

Compounds **1** and **2**, which were first prepared by Yin and Deeming,[14] have, for the first time, been characterized structurally as shown. Complexes **3-5** were characterized via mass spectroscopy, ^1H and ^{13}C NMR, elemental analysis, and IR. These characterizations, supported by the structures shown, allow us to propose the structures for **3-5**.

Visual examination of these compounds quickly leads to the pertinent observation that activation of a C-H bond to nitrogen in both the heteroaromatic ring in Q and the heterocyclic rings of THQ and piperidine has occurred. This is a common feature of cyclic amine osmium chemistry.

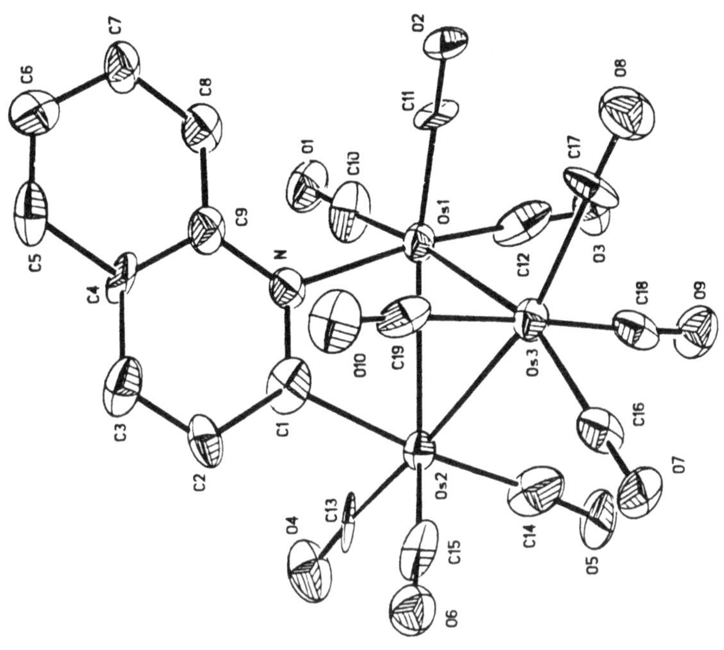
C6
C7
C8
C5
C9
C4
N
C3
C2
C1
O1
C10
C11
O2
Os1
C12
O3
C17
O8
O10
C19
Os3
C18
O9
C16
O7
Os2
C13
O4
C15
O6
C14
O5

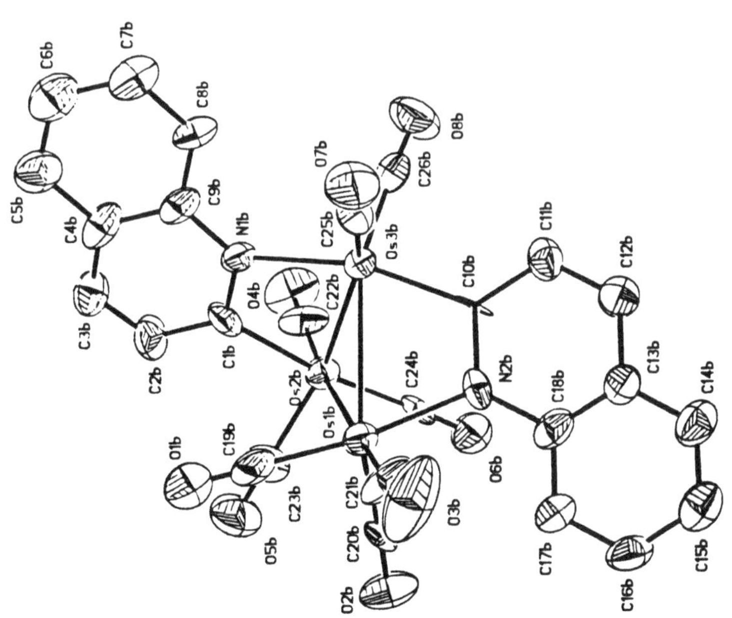
C6b
C7b
C5b
C8b
C4b
C9b
C3b
N1b
C2b
C1b
O4b
C22b
O7b
C25b
Os3b
C26b
O8b
Os2b
C24b
C10b
C11b
C12b
O1b
C19b
Os1b
N2b
C18b
C13b
C14b
O6b
C23b
C21b
O3b
O5b
C20b
O2b
C17b
C16b
C15b

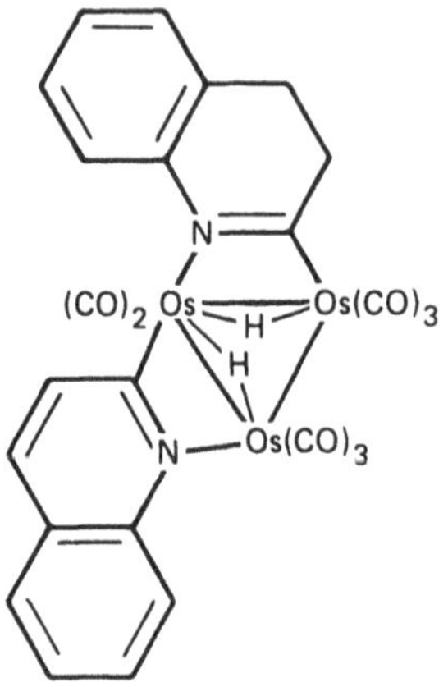

$(CO)_2Os$ — H — $Os(CO)_3$ (H) — $Os(CO)_3$ **3**

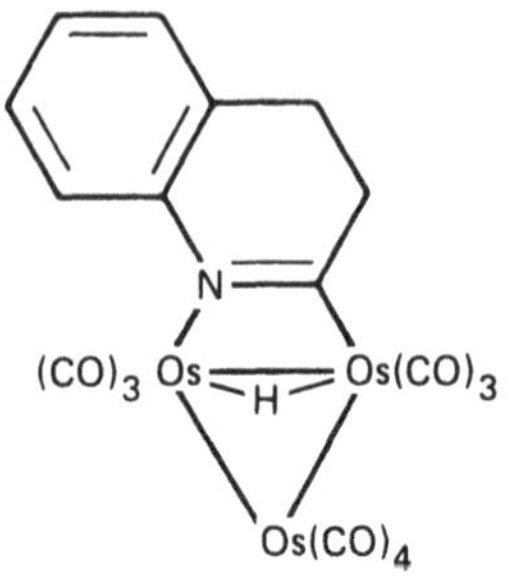

$(CO)_3Os$ — H — $Os(CO)_3$, $Os(CO)_4$ **4**

$(CO)_3Os$ — H — $Os(CO)_3$, $Os(CO)_4$ 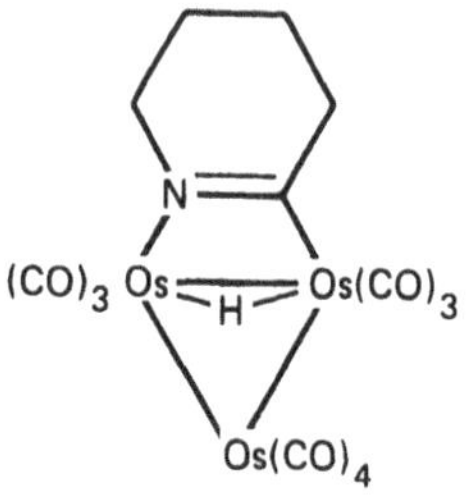 **5**

In fact, oxidative addition of C-H bonds to nitrogen in heteroaromatic rings is a general phenomena in both molecular organometallic chemistry and heterogeneous metal chemistry.[14-17] A second important observation is that we have finally isolated and characterized partially saturated nitrogen containing metallocycles, 4 and 5, resembling those we have previously proposed to form during catalytic C-N bond cleavage in the transalkylation reaction and in HDN.[1,11,18,19] In this instance, the metalloazocycles are four membered rings rather than the three membered rings we proposed; however, the intermediacy of three membered rings in the formation of the metalloazocyclobutenes cannot be ruled out. A third observation is that the osmium clusters catalyze the dehydrogenation of THQ (see below and Experimental) to form intermediates 3 and 4 and complexes of Q,[6-10] as shown by Eqs. 2 and 3. Because catalytic dehydrogenation of THQ occurs in CoMo catalyzed HDN of Q,[6-10] this observation provides an indication that the osmium system can be used to model some aspects of CoMo catalyzed

HDN. This is particularly important in view of the pioneering work of Fish,[20] and Kaesz.[21] In their HDN modeling studies on rhodium catalyzed hydrogenation of Q,[20] they were able to observe selective hydrogenation of Q to THQ. Unfortunately they were unable to obtain catalytic dehydrogenation of THQ, although they do observe dehydrogenation of the intermediate dehydroquinoline.

All of these observations permit us to draw conclusions concerning the mechanisms of Q hydrogenation under HDN conditions when considered in conjunction with the following hydrogenation and deuterium labeling experiments.

B. Catalytic Hydrogenation of Quinoline

We have briefly explored the kinetics of osmium and CoMo catalyzed hydrogenation of Q to THQ, Eq. 5, in order to develop an understanding of

$$\text{Q} \xrightarrow{\text{catalyst}} \text{THQ} \qquad (5)$$

Catalyst = $Os_3(CO)_{12}$, $H_2Os_3(CO)_{10}$, CoMo

the mechanisms involved and to be able to compare the two catalysts. The search for a suitable solvent for the kinetic studies led to the results shown in Table I. Of the three solvents tested, THF provides the highest reaction rates with the least complications. For the reactions carried out in heptane, after extended reaction times (24h), the catalysts were found to form an insoluble species tentatively identified (by IR) to be the higher cluster, $H_4Os_4(CO)_{12}$. For the reactions carried out in methanol, products resulting from methylation of the THQ were observed. Only in THF were the osmium catalysts found to be completely soluble and to produce THQ free of side products. Even in THF the catalyst activities diminished after significant conversions ($\sim$60%). A plausible explanation for this observation is that the catalyst is converted to a less active form such as 2 (vide infra). Consequently, we have relied on initial rates and deuteration studies to develop a mechanistic rationale for the osmium catalyzed hydrogenation. In

TABLE I

Rate of Hydrogenation of Quinoline by Various Catalysts[a]

Catalyst	Solvent	Temp. °C	Turnover Frequency hr^{-1}
CoMo	Heptane	145	6.5[b]
$Os_3(CO)_{12}$	Heptane	145	50
$H(C_9H_6N)Os_3(CO)_{10}$	Heptane	145	26
$H_2Os_3(CO)_{10}$	Heptane	145	70
$H_2Os_3(CO)_{10}$	MeOH	145	87
$H_2Os_3(CO)_{10}$	THF	90	11

[a] [Catalyst] = 0.7 - 3.0 x 10^{-3} M; [Quinoline] = 0.305 M; H_{2_1} = 600 psi.
[b] Based on total metal.

our preliminary rate studies, we evaluated the hydrogenation activity of $H_2Os_3(CO)_{12}$, Eq. 5, as a function of catalyst concentration. The results of this study, plotted in Fig. 2, indicate that the rate of Q hydrogenation exhibits a first order dependence on $H_2Os_3(CO)_{12}$ concentration which in turn signifies that the nuclearity of the catalyst is maintained under the reaction conditions. Therefore, the trinuclear cluster complexes, **1-4**, isolated from the stoichiometric reactions are reasonable intermediates in the Q hydrogenation/dehydrogenation equilibria.

We have begun studies of $H_2Os_3(CO)_{10}$ hydrogenation rates as a function of hydrogen pressure. Our results indicate that there is an apparent dependence on H_2 pressure; however, our data are incomplete at present, thus limiting what can be said concerning the involvement of H_2 in the hydrogenation reaction mechanisms. It is clear, though, that H_2 activation occurs prior to or as part of the slow step in the catalytic cycle.

In parallel with our hydrogenation studies, we have explored the use of D_2 to provide labelling information of value in identifying various steps in the hydrogenation/dehydrogenation cycles.[20] Through the use of deuterium NMR, we find that considerable deuterium becomes catalytically and

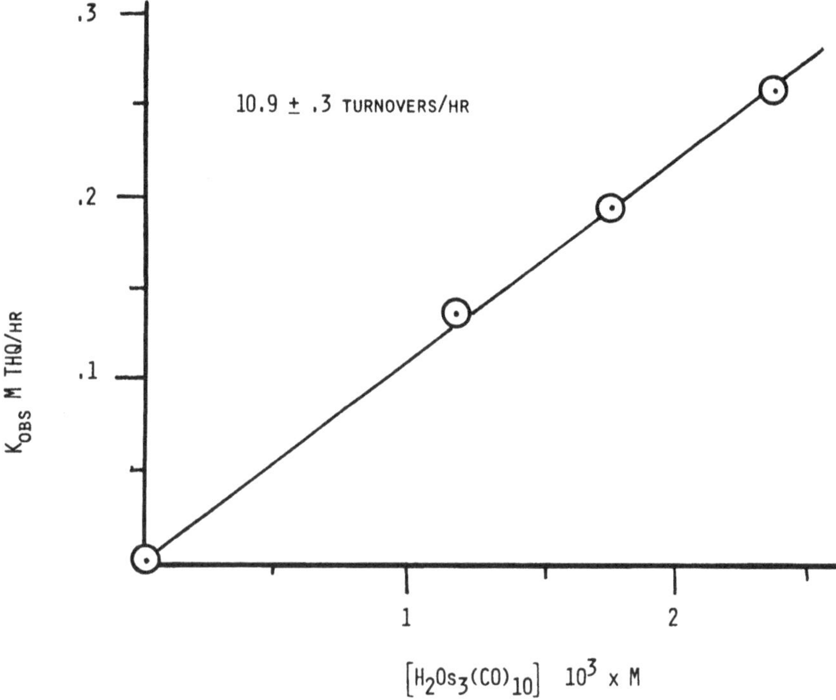

Fig. 2. Dependence of rate of appearance of THQ on $H_2Os_3(CO)_{10}$.

regioselectively incorporated in both Q and THQ. The incorporation in both Q and THQ are shown in Scheme 1. The deuteration results shown in Scheme 1 are a measure of the various catalyst-substrate interactions. These results suggest that there are quite a few similarities and some contrasts between the homogeneous osmium catalyst and the heterogeneous CoMo catalyst. The specific differences that are immediately noticeable are the relatively higher amounts of deuterium incorporated at the 2 position in THQ by osmium in comparison with CoMo and conversely the higher amounts of deuterium incorporated by CoMo at the 3 position. Reasonable mechanistic explanations for these results require an examination of the similarities.

The similarities are more numerous and must be separated into three areas: deuteration that preceeds hydrogenation, that which occurs during hydrogenation, and that which occurs in THQ. The last area was explored separately by studying the reactions of THQ with both catalysts in the

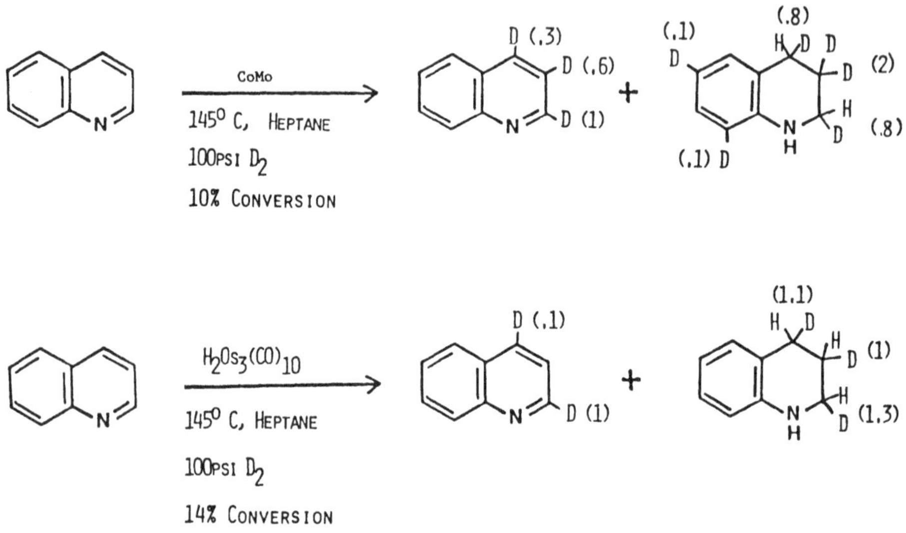
CoMo
145° C, Heptane
100psi D2
10% Conversion
D (.3)
D (.6)
D (1)
(.1)
D
(.8)
H D D
D (2)
H
D (.8)
N
H
(.1) D
H2Os3(CO)10
145° C, Heptane
100psi D2
14% Conversion
D (.1)
D (1)
(1.1)
H D
H
D (1)
H
D (1.3)

Scheme I

presence of D_2, and the absence of Q. Under these conditions deuterium incorporation is as shown in Scheme 2.

Both catalysts incorporate deuterium into the aromatic ring at positions ortho and para to the nitrogen as might be expected for C-H activation/D exchange in an activated aromatic ring (THQ is an aniline derivative). Aromatic C-H activation is an expected product in osmium rections,[14] but it's not clear what one would expect from a metal sulfide hydrogenation catlyst. In fact, it's not clear just which metal, cobalt or molybdenum, is responsible for which catalytic actions in the hydrogenation step. There is a continuing controversy about the physical interactions or lack of interactions between the two metals, although cobalt is generally considered to be the species responsible for hydrogenation. In view of Topsoe's recent suggestions that the two metals form strong chemical interactions at the active site, it is possible that a bimetallic catalyst site is responsible for the hydrogenation.[22,23]

In the osmium catalyzed reactions of THQ with D_2, the 2 and 3 positions on the ring are also equally deuterated. This observation is valuable for three reasons, first because it implicates compound 4 as the

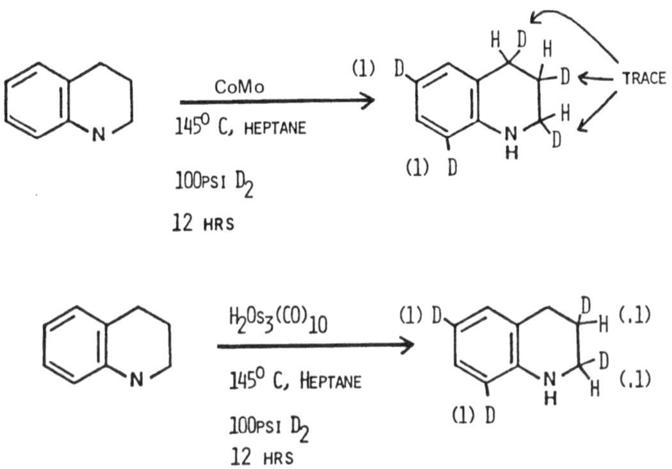

Scheme 2

first intermediate inTHQ dehydrogenation; second, based on the principle of microscopic reversibility it implicates compound **4** as the last step in the hydrogenation of Q; and third, because dehydrogenation was not complete, it suggests the existence of another intermediate.

This intermediate, which we have not been able to isolate, provides us with a clear idea of the complete, step by step catalytic cycle again based on the principle of microscopic reversibility (Scheme 3). Although, the compound was not isolated, the selectivity of the deuteration suggests that both the 2 and 3 positions are activated simultaneously. This result would readily obtain if compound **4** or a similar species were to react further to form a π-allyl complex, such as **6**.

In this light, analysis of the pattern of osmium catalyzed deuterium incorporation in the unreacted Q provides a picture of the mechanism of quinoline hydrogenation. Most of the deuterium is incorporated at the 2 position in quinoline, but some is incorporated at the 4 position. Incorporation at both positions could arise from competing 1,2 and 1,4 hydrogenation followed by ring inversion at the nitrogen and reductive elimination, Eq. 6 and 7.

However, the THQ deuteration experiments indicate that there is no significant difference between the amounts of deuterium incorporated at

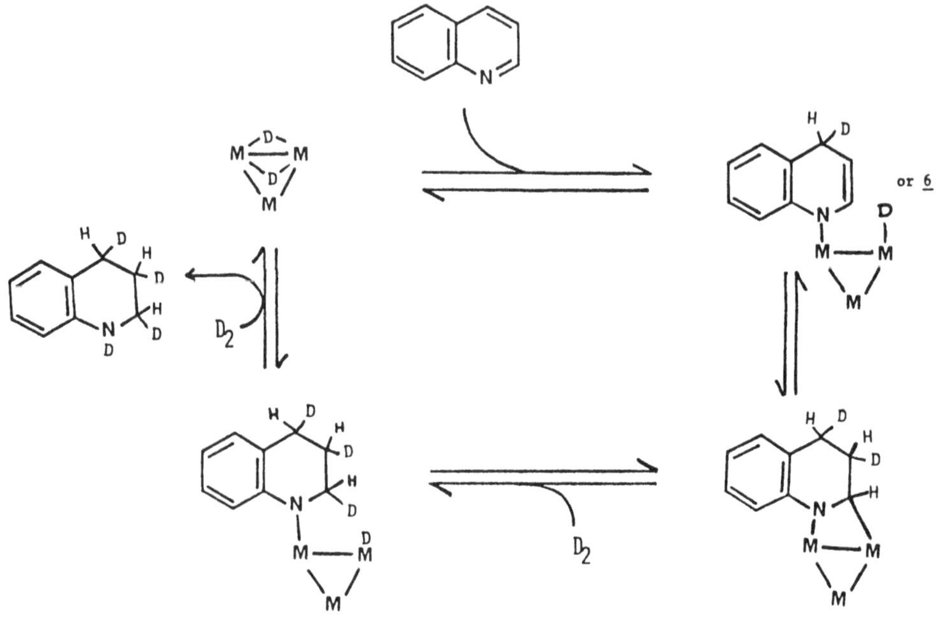

Scheme 3

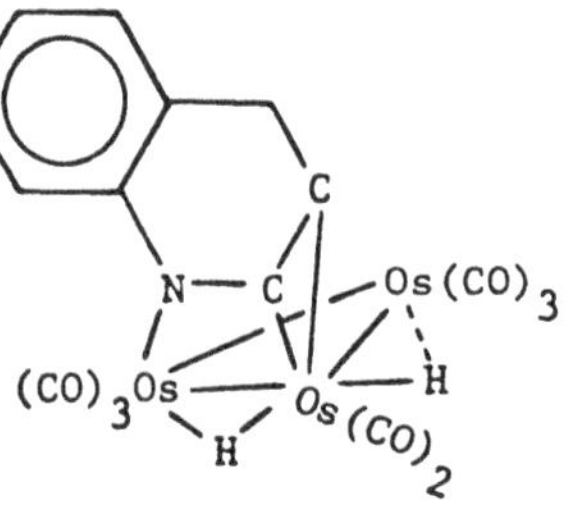

6

the 2 and 3 positions. If reductive elimination from the 1,2 hydrogenation intermediate (the reverse of Eq. 6) occurred readily then one would expect to see significantly more deuterium in the 2 position than in the 3 position in THQ. Furthermore, the small amounts of deuteration at the 4 position

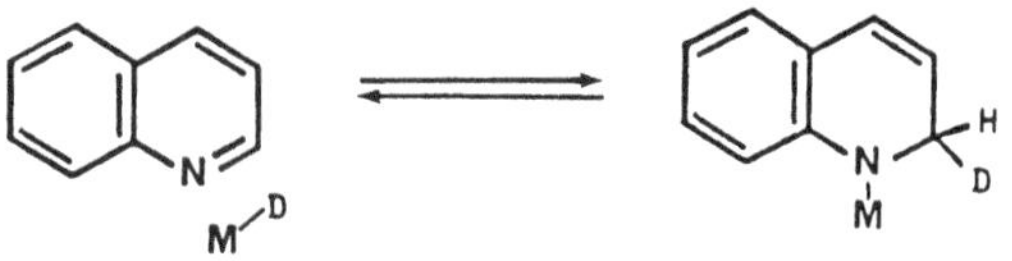

(6)

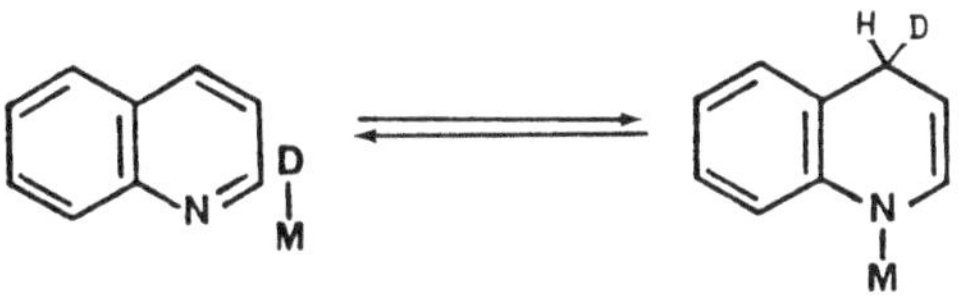

(7)

suggest (see below)that 1,4 addition is occurring. If this process proceeds via the same initial complex that gives 1,2 addition then the ratio of deuterium at the 4 position to that at the 2 position in both Q and THQ should remain constant over time. In fact, the ratio is not constant, increasing steadily as the reaction proceeds. Consequently, deuterium incorporation proceeds by intermediates such as **1** and **2** where C-H activation of the heteroaromatic ring occurs, Eq. 8 and 9.

$$H(\eta^2\text{-}C_9H_6N)Os_3(CO)_{10} + D_2 \rightleftharpoons H(\eta^2\text{-}C_9H_6N)(D)_2Os_3(CO)_9 + CO \quad (8)$$

$$H(\eta^2\text{-}C_9H_6N)(D)_2Os_3(CO)_9 + Q \rightleftharpoons QD + HD + H(\eta^2\text{-}C_9H_6N)Os_3(CO)_9 \quad (9)$$

This idea receives further support from the observation that **2** is not an intermediate in the hydrogenation catalytic cycle. If **2** is used as the catalyst species under hydrogenation conditions identical to those used with either $Os_3(CO)_{12}$ or $H_2Os(CO)_{10}$ (see Table I), then hydrogenation of Q to THQ is observed; however, the rate of reaction is one-third that observed with $H_2Os_3(CO)_{10}$. Thus, **2** cannot be an active intermediate in the catalytic cycle.[24] From these observations and the work of Muetterties and others,[14-17,25] we conclude that deuterium incorporation at the 2 position occurs via intermediates such as **1** and **2** rather than by 1,2 hydrogenation. Finally, we can conclude that the presence of more than one deuterium at the 2 position of THQ results from prior deuterium incorporation at the 2 position on Q.

The incorporation of small amounts of deuterium in the 4 position must proceed via a reversible 1,4 addition/elimination process; although this is not a particularly favorable process. This conclusion is based on the principle of microscopic reversibility. That is, if dehydrogenation proceeds via compounds **4** and **6**, then the next logical backward step must be the formation of compound 7 (Scheme 3). **7** must be the last step in the catalytic dehydrogenation cycle and therefore the first step in the hydrogenation pathway. We can then propose a complete picture of osmium catalyzed quinoline hydrogenation/THQ dehydrogenation as shown in Scheme 3.

In the case of CoMo catalyzed hydrogenation of quinoline, the analysis is much the same. The observation, in the THQ deuteration experiments, of only trace amounts of deuterium at all the positions in the heterocyclic ring does not permit significant discussion about the dehydrogenation mechanism; however, it does indicate that the process leading to extensive D exchange at the 3 position in THQ must occur during the hydrogenation of Q and not through THQ-catalyst interactions.

Although, only small amounts of deuterium are incorporated in Q during the CoMo catalyzed deuteration of Q, the incorporation is quite specific. The majority of the deuterium is found at the 3 position with lesser amounts at the 2 and 4 positions. We conclude that deuterium incorporation at the 2 position is likely to proceed by way of a quinoline C-H activation similar to that observed in the osmium case (as in compounds **1** and **2**) rather than via 1,2 hydrogenation, in order to readily account for deuteration at the 3 and 4 positions both in the Q and in THQ. The evidence implicates a 1,4 hydrogenation process as the pathway for incorporation at these positions.

The rationale for these statements comes from considering the possible steps in the catalytic cycle that will account for all of the observations in the deuteration experiments. For example, the C-H activation/D exchange that occurs at the 3 position leading to the 3,3-dideuterio THQ product, must occur during hydrogenation and must proceed by way of an intermediate or intermediates that allow changes in the stereochemistry of the catalyst vis a vis the face of the partially hydrogenated heterocycle. Furthermore, this same process must allow for reversible hydrogenation/dehydrogenation to provide pathways for the deuteration of the 3 and 4 positions in Q.

The apparently exclusive 1,4 hydrogenation process that occurs in the osmium catalyzed quinoline hydrogenation also appears to occur in the

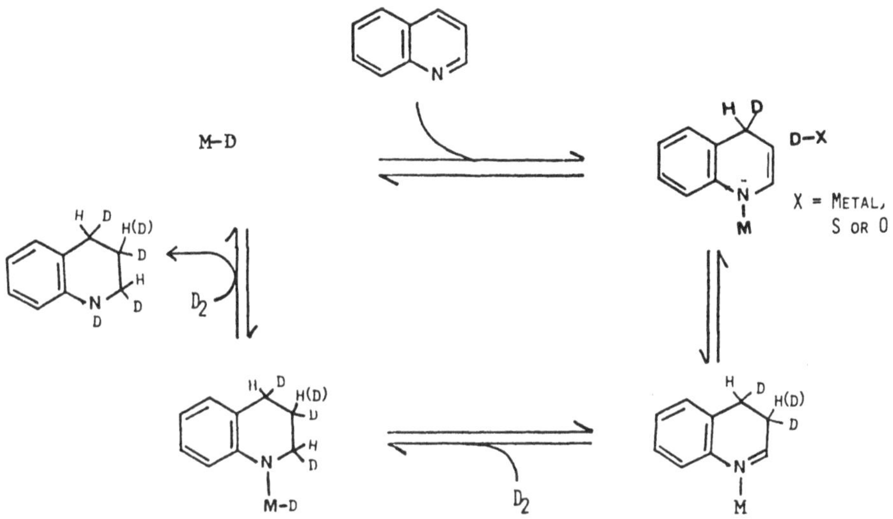

Scheme 4

CoMo process. This conclusion is supported by the observation that deuterium incorporates in the 4 position in Q. The deuterium experiments suggest that the difference between the two catalytic systems lies in reactivity of the resulting intermediates. In the osmium system, we propose the straightforward formation of a π-allyl-like complex that rapidly hydrogenates selectively to give the active, coordinately unsaturated catalyst and THQ. In the CoMo case, the formation of a true π-allyl complex may not be possible and it is at this juncture that we envision a reversible stepwise hydrogenation process wherein the stereochemistry of the catalyst-heterocycle interaction can be lost. The process that leads to multiple incorporation at the 3 position is unclear at present. Based on the known organic chemistry of enamines, however, we can suggest the reversible process shown in Scheme 4 where D-X is either metal hydride or a sulfhydryl group. This represents the complete mechanistic picture of CoMo catalyzed hydrogenation of Q as it is presently understood.

III. CONCLUSIONS

These mechanistic interpretations are far from definitive, they require further refinement; however, they do provide a preliminary picture of the catalyst-substrate interactions that occur during the HDN process. We will expand upon this picture in future publications.

IV. EXPERIMENTAL SECTION

A. General Methods

Both quinoline and tetrahydroquinoline were purchased from Aldrich and distilled from CaH_2 under N_2 prior to use. THF was dried and deoxygenated by distillation from sodium benzophenone ketyl under N_2 immediately before use. Methanol was used as received from Mallinkrodt. n-Heptane, n-octane and n-undecane of 99% purity were used as received from Aldrich. Catalyst precursors were either purchased from Strem Chemicals and used as received, or synthesized according to published methods. $Os_3(CO)_{12}$ was prepared from OsO_4 by the method of Johnson and Lewis,[26] and $H_2Os_3(CO)_{10}$ was prepared from $Os_3(CO)_{12}$ by the method of Knox, Koepke, Andrews, and Kaesz.[27]

B. Analytical Procedures

Product analyses for all the kinetic studies were performed on a Hewlett Packard 5710A GC equipped with FID and modified for use with 25 m dimethylsilicone capillary column. Chromatograms were collected and integrated on a Hewlett Packard 5880A GC terminal using n-undecane as an internal standard. Infrared spectra were obtained using a Perkin-Elmer 281 IR spectrophotometer. NMR spectra were recorded on a JOEL FX90Q equipped with a 5 mm broad band probe. GC-Mass Spectral analyses were performed using an LKB-9000 or Ribermag R 10 10 C mass spectrometer.

C. Catalyst Preparation

The commercial cobalt-molybdenum catalyst manufactured by Harshaw and used in these studies was HT-400 E 1/8 and consisted of 3% CoO and 15.1% MoO_3 by weight supported on 1/8 inch extruded Al_2O_3. This catalyst was ground to 60-200 mesh, and sufided by heating from room temperature to 400°C in 2 hours and then maintaining the temperature at 400°C for an additional 24 hours while blanketing the catalyst with a slow flow (ca 1-2 ml/min) of 10% H_2S - 90% H_2. Analysis provided the following stoichiometry $Co_4S_9(MoS_{2.1})_8$.

D. Kinetic Studies

All kinetic reactions were carried out in a 45 ml Parr screw cap bomb equipped with a gas inlet and pressure gauge. In a typical kinetic run, a 10 ml solution containing reactants ([cat] - 0.7-3 x 10^{-3} M, [quinoline] = 0.305 M for osmiun complexes; or 10-50 mg of CoMo catalyst, [quinoline] = 0.305 M for CoMo reactions) and a magnetic stirring bar were placed in a quartz liner which had been soaked for at least 12 hours in concentrated H_2SO_4-$(NH_4)_2S_2O_8$ to remove trace metals. The liner containing solution, was sealed in the bomb (total volume with liner, 34 ml) and flushed 5 times with at least 700 psi H_2 or N_2. After the final flush cycle the bomb was pressurized to the desired level, generally [H_2] = 600 psi, and placed in an oil bath maintained at the appropriate temperature $\pm$ 1°C, usually 90° or 145°. Periodically, aliquots were removed by cooling the bomb to room temperature, releasing the pressure and removing 0.1-0.3 mL of solution. The run was resumed by repeating the initial pressurization procedure. Progress of the reaction was monitored by GC as described above in the analytical procedures. In this manner 5 to 9 points were collected during the first 10-20% of the reaction. Initial reaction rates were calculated assuming pseudo zero-order reaction conditions. Mass balance was checked by comparing the area of the internal standard with sum of the areas of products and reactants and was found to agree within 5% even at > 90% conversion.

E. Deuterium Incorporation

Deuterium incorporation experiments were carried out using the same techniques and reaction conditions as the kinetic runs with the following modifications.

Each run was purged with N_2 prior to pressurization to 125 psi with D_2. In the studies of the reduction of quinoline, the reactions were quenched at selected times yielding conversions of quinoline to tetrahydroquinoline of 3%-85%. When tetrahydroquinoline was used as the substrate the solution was heated approximately 15 h. The contents of the bomb were concentrated under reduced pressure. Initially, the quinoline and tetrahydroquinoline were separated on a silica gel column eluting with 15:85

Et_2O:Petroleum Ether (30-60) and the ^2H-NMR spectrum of each product was collected in cyclohexane. Once the deuterium resonances were assigned, NMR spectra were collected without separating the products. Integration of the deuterium resonances gave the relative amounts of deuterium incorporation at each site on the deuterated Q and THQ.

F. Catalytic Dehydrogenation of THQ

A THF solution of $H_2Os_3(CO)_{10}$ (1.5×10^{-3} M) and THQ (.24 M) was placed in a bomb, purged 5 times with N_2 and pressurized with 250 psi of N_2. The bomb was placed in an oil bath at 145°C. Periodically, aliquots were removed from the bomb and the products were analyzed by GC as previously described. Initially, the solution contained 1.2% quinoline as an impurity in the starting THQ. However, dehydrogenation was observed as the amount of quinoline increased gradually with time. Hence, after 22 h the solution contained 3.0% quinoline, after 93 h 4.2%, and after 166 h 6.3%. The reaction mixture remained homogeneous throughout the run.

A solution containing 271 mg of $Os_3(CO)_{12}$ (.3 mmol), and .6 ml of THQ (4.8 mmol), dissolved in 50 mL of octane was heated to 145C in a sealed tube for 40 h. The solvent was removed under reduced pressure and most of the excess THQ was removed by drying in a vacuum (10^{-3} mm) for 8 h. The products were separated by preparative TLC on silica, eluting twice with hexane. The yields of $H_2(\eta^2\text{-}C_9H_6)_2Os_3(CO)_8$ and $H_2(\eta^2\text{-}C_9H_6N)(\eta^2\text{-}C_9H_8N)Os_3(CO)_8$ were 94 mg (30%) and 18 mg (6%) respectively.

ACKNOWLEDGEMENTS

We would like to thank Dr. Richard Fish for graciously sending us a copy of his manuscript prior to publication. We would like to thank Dr. David Thomas for his help in obtaining and interpreting the mass spectra. We would also like to thank Johnson-Matthey Inc. for a generous loan of the ruthenium. The major portion of this work was generously supported by the NSF Chem. Eng. Grant No. 82-19541. A portion of this work was supported through DOE grant No. DE-FG22-83PC60781.

REFERENCES

1. Previous paper in this series, R.M. Laine, J. Mol. Catal., **21**, 119 (1983).

2. Nicolet XRD Co.
3. H.G. McIlvried, Ind. Eng. Chem. Proc. Des. Dev., **10**, 125 (1971).
4. J. Sonnemans, G.H. van Den Berg, and P. Mars, J. Catal., **31**, 220 (1973).
5. E.W. Stern, J. Catal., **57**, 390 (1979).
6. C.N. Satterfield, and S. Gultekin, Ind. Eng. Chem. Proc. Des. Dev., **20**, 62 (1981).
7 M.V. Bhinde, S. Shih, R. Zawadski, J.R. Katzer, and H. Kwart, Third Internat. Conf. on the Chem. and Uses of Mo. 184 (1979).
8. J.F. Coccheto and C.N. Satterfield, Ind. Eng. Chem. Proc. Des. Dev. **20**, 49 (1981); ibid, **20** 53 (1981).
9. C.N. Satterfield and D.L. Carter, Ind. Eng. Chem. Proc. Des. Dev. **20**, 538 (1981).
10. H. Kwart, J. Katzer, and J. Horgan, J. Phys. Chem., **86**, 2641 (1982).
11. R.M. Laine, Catal. Rev., **25**, 459 (1983).
12. R.M. Laine, Ann. N.Y. Acad. Sci., **415**, 271 (1984).
13. A. Eisenstadt, C.M. Giandomenico, M.F. Fredericks, A.S. Hirschon, and R.M. Laine, Manuscript in preparation.
14. C.C. Yin and A.J. Deeming, J. Chem. Soc., Dalton, 2091 (1975).
15. E. Klei and J.H. Teuben, J. Organomet. Chem., **214**, 53 (1981).
16. P.O. Nubel, S.R. Wilson, and T.L. Brown, Organometallics, **2**, 515 (1983).
17. P.L. Watson, J. Am. Chem. Soc., **105**, 6491 (1983).
18. Y. Shvo, D.W. Thomas, and R.M. Laine, J. Am. Chem. Soc., **103**, 2461 (1981).
19. R.M. Laine, D.W. Thomas, and L.W. Cary, J. Am. Chem. Soc., **104**, 1763 (1982).
20. (a) R.H. Fish, A.D. Thormodsen, G.A. Cremer, J. Am. Chem. Soc., **104**, 5234 (1982); (b) R.H. Fish, Ann. N.Y. Acad. Sci., **415**, 292 (1983); (c) R.H. Fish and A.D. Thormodsen, J. Org. Chem. In press.
21. T.J. Lynch, M. Banah, H.D. Kaesz, and C.R. Porter, J. Org. Chem., **49**, 1266 (1984).
22. B.S. Clausen, B. Lengler, R. Candia, J. Als-Nielsen, and H. Topsoe, Bull. Soc. Chim. Belg., **90**, 1249 (1981).
23. N.-Y Topsoe and H. Topsoe, J. Catal., **85**, (1985) In press.

24. Note that while 1,2 addition does not appear reasonable in these processes it is proposed to occur in the mononuclear rhodium catalyzed hydrogenation of Q, see references 20 and 21.

25. R.M. Wexler, M.-C. Tasi, J. Stein, C.M. Friend, and E.L. Muetterties, J. Am. Chem. Soc., **104**, 2034 (1982).

26. B.F.G. Johnson and J. Lewis, Inorg. Synth., **13**, 92 (1972).

27. S.A.R. Knox, J.W. Koepke, M.A. Andrews, and H.D. Kaesz, J. Am. Chem. Soc., **97**, 3942 (1975).

PART II

H_2CO REACTIONS

6

Methanol Synthesis over Supported Palladium Catalysts

Jack H. Lunsford
Department of Chemistry
Texas A&M University
College Station, Texas 77843

ABSTRACT

Results from several laboratories have now established that on certain supports palladium is both active and selective for the synthesis of methanol. A review of current work reveals that the chemical composition of the support, the exact nature of the support, and the precursor of the surface palladium have a marked effect on the catalytic properties. For example, the specific activity for methanol synthesis of catalysts prepared from $Pd(\pi\text{-}C_3H_5)_2$ decreased in the order $Pd/La_2O_3 \gg Pd/ZrO_2 > Pd/ZnO \simeq Pd/MgO > Pd/TiO_2 > Pd/Al_2O_3 \simeq Pd/SiO_2$. Palladium on silica prepared from $PdCl_2$ is much more active than that prepared from $Pd(\pi\text{-}C_3H_5)_2$. Moreover, palladium supported on different types of silica yields completely different activities and selectivities. Incorporation of Li^+ or Na^+ into the palladium as M_2PdCl_4 (M = alkali metal) produced a large increase in methanol conversion, regardless of the support, but K^+, Rb^+ or Cs^+ had a negative effect. Co-impregnation of silica with Mg^{2+} or addition of lithium or sodium carbonate to a reduced Pd/SiO_2 catalyst also promoted the activity. Theories to explain the support and promoter effects include modification of the oxide ions, the presence of Pd^{n+} ions, metal support interaction, and structural requirements for the metal.

I. INTRODUCTION

In the conversion of synthesis gas to chemicals and gasoline-type fuels methanol may become a very important intermediate. Production of chemicals such as formaldehyde and acetic acid from methanol currently is being carried out on a commercial scale. In addition, the selective yields of gasoline-range hydrocarbons from methanol over ZSM-5 zeolites is well documented.[1] This discovery, in particular, has renewed the interest in the catlytic formation of methanol from synthesis gas.

Industrial low-pressure methanol synthesis catalysts are composed mainly of mixtures of copper and zinc oxides deposited on Cr_2O_3 or Al_2O_3. It has been demonstrated recently that supported palladium may also be active for methanol synthesis, although the activity and selectivity is highly dependent upon the type of support and, in certain cases, upon the presence of promoters. A study of Poutsma _et al._[2] showed that the palladium supported on Davison grade 57 silica was active under normal methanol synthesis conditions, and the work of Ichikawa,[3] which was carried out at atmospheric pressure, indicated the strong influence of the support. In addition, Tamaru and co-workers[4] were the first to show that certain alkali metal ions function as promoters for methanol synthesis. As will be subsequently shown, some of the most dramatic support/promoter effects that have been observed in heterogeneous catalysis influence this reaction. It is important to understand the origin of these effects not only to improve the commercial potential of these catalysts, but also to gain a broader understanding of metal-support and promoter interactions as they relate to catalysis by supported metals.

In this paper evidence for the support and promoter effects will be reviewed and several theories for the origin of the effect will be presented in light of recent experimental results. Before reviewing the experimental data it should be noted that the differences between support and promoter effects often are not easily distinguished because the palladium precursor (e.g. $PdCl_2$, $Pd(\pi\text{-}C_3H_5)_2$, etc.) frequently differs. The problem is accentuated because chloride ions appear to be promoters, and $PdCl_2$ in HCl is a favorite solution for impregnating catalysts. Thus, when chloride ions

TABLE I

Catalytic Properties of Supported and Unsupported Pd[5]

Catalyst	CO conv. (%)	Turnover freq. (X $10^3 s^{-1}$)			S (%)		
		CH_3OH	CH_3OCH_3	CH_4	CH_3OH	CH_3OCH_3	CH_4
Pd black	0.01	0.60	0	0.07	75.0	0	8.8
0.3% Pd/MgO	0.1	7.70	0.10	0.02	98.4	1.2	0.3
0.2% Pd/ZnO	0.4	8.40	0	0.01	99.8	0	0.1
1.5% Pd/Al_2O_3	0.1	2.61	4.92	0.26	33.2	62.7	3.3
0.2% Pd/La_2O_3	0.5	99.10	0	0.50	99.0	0	0.5
1.5% Pd/SiO_2(I)	0.1	2.33	0	0.08	91.6	0	3.4
7.9% Pd/SiO_2(II)	1.5	18.51	0	0.28	98.3	0	1.5
0.5% Pd/TiO_2	0.3	4.20	0.81	4.00	44.1	8.6	42.1
0.3% Pd/ZrO_2	0.7	11.40	0.07	3.40	74.7	0.5	22.3

Reaction conditions: T = 250°C; P = 10 atm; H_2/CO = 3; Q = 200 cm^3(STP)/min.

are introduced to a catalyst, it becomes difficult to separate the direct influence of the support on the metal and its indirect influence in retaining the promoter.

II. EVIDENCE FOR THE SUPPORT EFFECT

Following the preliminary work of Ichikawa,[3] Bell and co-workers[5] carried out a systematic study of the methanol synthesis reaction over the metal oxides listed in Table I. Except for the catalyst designated Pd/SiO_2(II) all of the samples were prepared by the addition of palladium as Pd(π-C_3H_5)$_2$. The Pd/SiO_2(II) catalyst, obtained from Union Carbide, was prepared from $PdCl_2$.[2] The catalysts were evaluated in a tubular micro-reactor operated under continuous flow conditions at 10 atmospheres. The specific activity (turnover frequency) for methanol synthesis decreased in the order Pd/La_2O_3 >> Pd/ZrO_2 > Pd/ZnO ≃ Pd/MgO > Pd/TiO_2 > Pd/Al_2O_3 ≃

Pd/SiO_2(I) >> Pd black. The supported catalysts differ in turnover frequency by a factor of 42. The selectivity for methanol synthesis was greatest for Pd/MgO, Pd/ZnO, Pd/La_2O_3 and Pd on the two silica catalysts, with each of these having selectivities >90%. It should be noted that Pd/Al_2O_3 exhibited considerable activity for dimethyl ether formation, which was derived from methanol. The Pd/SiO_2(II) was much more active than the Pd/SiO_2(I) catalyst, but here it is difficult to distinguish the role of the Cl^- ion from that of the different grades of silica. As an example of methanol synthesis on a non-oxide support Poels *et al.*[6] have reported that palladium on active carbon had a greater activity than their Pd/SiO_2 catalyst.

Other apparent examples of a support effect are found when one compares methanol synthesis activity over catalysts which have the same nominal chemical composition. An example of this support effect has been reported recently by Fajula *et al.*[7] for Pd (from $PdCl_2$) on various types of SiO_2, which is generally viewed as a weakly interacting material. Three types of silica were compared in this study: Davison grade 57 [SiO_2(57)], which was the same material used by Poutsma *et al.*[2] and by Ryndin *et al.*[5]; Davison grade (01) [SiO_2(01)]; and grade M5 Cab-O-Sil. The properties of these silicas are given in Table II. As indicated in Table III, palladium supported on these three types of silica exhibited remarkably different activities for the production of methanol, although within a factor of two the turnover frequency for the formation of methane was the same. Results obtained for palladium in zeolites are included. The experiments with silica gel have been repeated using another batch of SiO_2(01), and a moderate specific activity for methanol synthesis was observed which after 36 hours on stream amounted to about 20% of the activity for the SiO_2(57) catalyst. All of the above reactions were carried out at $P \geq 10$ atm; however, by using modern analytical techniques it is possible to follow methanol formation at about 1 atmosphere. Under these low-pressure conditions Tamaru and co-workers[4] have observed that a catalyst prepared by impregnating Aerosil SiO_2 with $Pd(NH_3)_4Cl_2$ was inactive for methanol formation, although some methane was produced.

The literature on Pd/Al_2O_3 suggests that similar variations in activity for CH_3OH may exist with this support. Poutsma *et al.*[2] reported that at

TABLE II

Physical Properties of SiO_2 Supports [7]

Designation	Purity SiO_2%	Surface Area m^2/g	Size of grains Mesh	Pore diameter Å	Density g/cc	pH 4 wt % in H_2O
Davison Grade 01[a] [SiO_2(01)]	99.5	750	3-8	22	0.6	3.6 - 3.9
Davison Grade 57[b] [SiO_2(57)]	99.5	300	6	150	0.4	6.3 - 6.7
Fumed Silica M5[c] [Cab-O-Sil]	99.8	200	∿300	140	0.2[d]	3.8 - 4.2

[a] Al, S, Ti, Ca < 0.02 wt%, Fe = 0.006 wt%.

[b] Al, S, Ti < 0.02 wt%, Ca = 0.06 wt%, Fe = 0.005 wt%

[c] Al, S, Ti, Ca < 0.02 wt%, Fe = 0.0003 wt%.

[d] Density of the impregnated catalyst.

TABLE III

Activity and Selectivity of Palladium Catalysts[7]

	Catalyst	Rate[a] of CO Consumption Moles h^{-1} kg^{-1} of Catalyst	Selectivity[a] (Mole %)			$N \times 10^3$ [b]	
			CH_4	CH_3OH	DME	N_{CO}	N_{CH_4}
I	2.4% PdNaY	0.08 ± 0.03	100			12 ± 4	12 ± 4
II	5.4% PdHY	1.15 ± 0.01	98	2	trace	42 ± 3	41 ± 3
III	1.7% Pd/SiO_2(01)	0.06 ± 0.03	100			1.7 ± 0.8	1.7 ± 0.8
IV	4.8% Pd/SiO_2(01)	0.07 ± 0.03	100			1.1 ± 0.4	1.1 ± 0.4
V	4.6% Pd/SiO_2(57)	1.56 ± 0.01	3.5	96.5		18.8 ± 0.1	0.65 ± 0.02
VI	3.8% Pd/Cab-O-Sil	0.935 ± 0.01	24	75	1	5.1 ± 0.1	1.2 ± 0.02

[a] 280 ± 5°C, 15 atm H_2/CO(2.8–2.4), SV = 1200 ± 200 h^{-1}.

[b] $N = \frac{\text{Molecules consumed (or produced)}}{\text{Metal Site X Sec.}}$

considerably higher pressures ($\sim$1,000 atm) Pd on γ-Al_2O_3 was about 60% as active for methanol synthesis as their active Pd/SiO_2 catalyst. They also noted the formation of small amounts of acetic acid, but they did not report the production of dimethyl ether. The work of Bell and co-workers[5] described above was also carried out over Pd on γ-Al_2O_3, and as expected for an acid catalyst, the methanol formed was extensively converted to dimethyl ether. By contrast, with Pd on η-Al_2O_3 prepared from $PdCl_2$, Vannice and Garten[8] observed good activity for methane synthesis, but at 300°C and 20 atmospheres no methanol or dimethyl ether were detected. Vannice[9] has repeated these experiments at lower temperatures and has found oxygenated products; thus, the variations for palladium on two types of alumina may not be as great as with palladium on the several types of silica.

III. MORPHOLOGICAL EFFECTS

Morphological effects have been considered at two levels: (1) the influence of dispersion or average particle size on activity and (2) the influence of specific crystal planes on activity. Bell and co-workers[10] have investigated the effect of dispersion on the specific activity for methanol and methane formation. For methanol synthesis the activity did not vary within experimental error over a range of dispersions from 10 to 30%; whereas, over this same range the activity for methane formation decreased by more than a factor of two. We also have observed that the specific activity for methanol synthesis was not a function of dispersion over the range from 16-27% dispersion with Pd/SiO_2(01) as the catalyst.[11]

Here it is interesting to note that extensive sintering occurs during the reaction, and the dispersions reported are those of the used catalysts. Dispersions determined by hydrogen chemisorption and from x-ray line broadening were generally in good agreement.

Although no single-crystal data is available to show the specific activity for methanol synthesis as a function of Pd surface planes, Bell and co-workers[10] have successfully used the infrared spectrum of adsorbed CO to infer the relative amounts of Pd(100) and (111) planes which exist when the metal is supported on La_2O_3. In addition to a linearly bonded CO which has

TABLE IV

Selectivity and Activity of Pd Catalysts;[a] the Effect of Alloying[6]

Catalysts[b]	CO(a) at room temp. (g-catalyst)$^{-1}$ cm^3 (s.t.p.)	Product formation rates (X 10^{-2})[c] MeOH	CH_4	C_2	CO_2	MeOH selectivity (%)
Pd/SiO_2-II	0.40	0.15	0.11	0.00	0.04	49.6
Pd-Li/SiO_2	1.06	6.93	0.06	0.01	0.31	94.7
Pd-Na/SiO_2	1.14	5.39	0.11	0.02	0.53	89.0
Pd-K/SiO_2	0.43	0.07	0.00	0.00	0.01	43.1
Pd-Rb/SiO_2	0.49	0.00	0.01	0.00	0.07	0
Pd-Cs/SiO_2	0.00	0.00	0.01	0.00	0.00	0

[a]Reaction temperature = 180°C, P(H_2) = 0.39 atm, P(CO) = 0.20 atm, in the closed circulating system (ca, 250 cm^3) with liquid-nitrogen cold traps.

[b]Catalysts were pretreated by hydrogen for a few hours at 500°C.

[c]cm^3 (s.t.p.) (g-catalyst h)$^{-1}$.

a band at 2055 cm^{-1}, there are bridge-bonded forms of CO with bands at 1965 and 1920 cm^{-1}. The latter band grows during the progress of the synthesis reaction. By comparison with the infrared spectra of CO adsorbed on Pd single crystals, these two bridge-bonded species are assigned to CO adsorbed on Pd(100) and Pd(111) surfaces, respectively. From a comparison of the intensities of these infrared bands and the specific activity for methanol synthesis these authors conclude that Pd(100) planes are 2.8 times more active than Pd(111) planes.

IV. ENSEMBLE EFFECTS

Closely related to morphological effects are ensemble effects. Poels et al.[6]

have prepared palladium-silver alloys on MgO. The activities of a Pd/MgO catalyst and two Pd-Ag alloys on MgO are given in Table IV. Upon comparing the percent surface composition of palladium with the conversion to methanol, it is evident that conversion decreases less rapidly than the surface concentration. This is in marked contrast to Fischer-Tropsch synthesis where the activity decreases much more rapidly than the surface concentration of the active component.

These results, which suggest that a palladium ensemble is not essential for methanol formation, are somewhat surprising in view of the morphological results presented in the previous section. It seems that a reaction which is sensitive to the particular plane exposed would also reflect the need for a particular ensemble, since the latter appears to be a less demanding requirement. It is possible, however, that the reaction requires only a pair of palladium atoms having a particular interatomic distance. Such a requirement could be satisfied either with an appropriate plane or a two-atom ensemble. Dilution of Pd with Ag may not break up such small two-atom ensembles, thus the requirement of the active site would be satisfied.

V. INFLUENCE OF PROMOTERS

The catalytic activity of palladium is not only affected by the nature of the support, but also by the addition of certain other elements. This was first noted by Tamaru and co-workers,[4] who added alkali metals in the form of complexes of the type M_2PdCl_4 (M = alkali metal); i.e., the alkali metal was introduced as a complex with the palladium. From the data of Table V it is evident that both Li^+ and Na^+ greatly enhanced the activity for methanol formation, although K^+, Rb^+ and Cs^+ had relatively little effect on activity. More generally the enhancement of activity is given by the sequence $Li^+ >$ $Na^+ \gg K^+ > Rb^+ > Cs^+$. The positive effect of Li^+ and Na^+ was also noted with other catalysts, and the activities of several catalysts prepared from Na_2PdCl_4 are given in Table VI. Alumina was clearly the most active catalyst for methanol synthesis, with very little dimethyl ether being formed. By contrast silica-alumina was about as active for total oxygenates, but the acidity of the catalyst prevails and most of the

TABLE V

Effects of Various Alkali Cations on the CO + H_2 Reactions[a] Over Silica Supported Pd (5 wt%) Catalysts[4]

Supports	Loaded Pd (wt %)	H_2 Reduction Temp. (°C)	H_2 Reduction Time (h)	Product Formation Rates[b] (X 10^{-1}) MeOH	DME[c]	CH_4	C_2	CO_2	MeOH Selectivity (%)
Silica	5.0	500	5	5.39	0.00	0.11	0.02	0.53	89.0
Silica-alumina Si/Al = 2.1	5.9	300	2	17.4	34.2	2.9	0.02	0.14	94.0
Alumina	5.0	400	2	45.2	0.60	2.4	0.14	0.56	94.0
NaY-Zeolite Si/Al = 2.4	5.9	450	5	9.1	1.00	0.20	0.00	0.00	98.0

[a] $P(H_2)$ = 0.39 atm, P(CO) = 0.20 atm, in the closed circulating system (*ca.* 250 cm^3) with liquid-nitrogen cold traps.

[b] cm^3 (s.t.p.) (g-catalyst h)$^{-1}$.

[c] Dimethyl ether.

TABLE VI

$CO + H_2$ Reactions[a] Over Na_2PdCl_4-Derived Supported Catalysts at 180°C[4]

Precursor/Support (+ promoter)	Rate[a] of Formation (moles $h^{-1}kg^{-1}$ of catalyst)		Selectivity to CH_3OH (mole %)
	CH_3OH	CO_2	
$Pd/SiO_2(57)$	0.73	0.06	92
$Li_2PdCl_4/SiO_2(57)$	3.51	0.05	99
$PdCl_2/SiO_2(57)$[b] + Li_2CO_3	1.78	0.13	93
$PdCl_2/SiO_2(57)$[c] + Li_2CO_3	4.58	0.07	98
$Pd/SiO_2(01)$	0.36	----	100
$Li_2PdCl_4/SiO_2(01)$	1.84	0.12	94
$PdCl_2/SiO_2(01)$[b] + Li_2CO_3	2.31	0.26	90
$PdCl_2/SiO_2(01)$[c] + Li_2CO_3	4.97	0.30	94
$PdCl_2/SiO_2(01)$ + Na_2CO_3	0.97	0.04	96

[a] 255 °C, 16 atm, H_2/CO = 2.2-2.4, SV = 5000 h^{-1}, 30-36 h of reaction time.

[b] Li_2CO_3 added to Pd/SiO_2 after reduction step.

[c] Li_2CO_3 added to SiO_2 before impregnation with $PdCl_2$.

methanol was converted to dimethyl ether. While operating at higher pressures we have confirmed the enhancement of activity upon forming the catalysts from M_2PdCl_4 complexes where M = Li or Na. After impregnation with these complexes both the $SiO_2(01)$ and $SiO_2(57)$ supports gave rise to active catalysts.[11]

TABLE VII

Activity and Selectivity of 5% Palladium on Silica[11]

Bulk Composition (% Pd)	Surface Composition Estimate (%)	Conversion[b] α (%)	Selectivity[c] S (%)
100	100	0.02	99
70	10	0.007	99
50	≈2	≈0.002	99

[a] T = 227 °C, carrier: MgO, loading 10 wt%.

[b] α = (moles CO converted/moles CO in feed) X 100.

[c] Selectivity is defined as % CH_3OH of all products.

Although Tamaru and co-workers found that impregnation of a Pd/SiO_2-II catalyst with NaOH or NaCl did not promote an enhancement in activity, we have observed that upon treating a prereduced Pd/SiO_2(01) catalyst with Li_2CO_3 or Na_2CO_3 the activity for methanol synthesis is greatly improved.[11] Moreover, the addition of Li_2CO_3 to SiO_2(01) before impregnation with $PdCl_2$ led to the most active catalyst of this series. In fact, as shown in Table VII, upon addition of Li_2CO_3 a Pd/SiO_2(01) catalyst can be made almost seven times as active as a Pd/SiO_2(57) catalyst which does not contain Li^+. As observed by Tamaru and co-workers,[4] the effect of Na^+ is less than that of Li^+.

Ponec and co-workers[12] have likewise noted that Mg^{+2} and La^{+3} are effective promoters for Pd/SiO_2 catalysts in that they increase the activity and selectivity for the formation of methanol, or in some cases dimethyl ether. The catalysts containing Mg^{+2} were prepared by impregnating SiO_2 (Kiesel gel) with solutions of $PdCl_2$ and $Mg(NO_3)_2$ or $MgCl_2$. Similarly, La_2O_3 was dissolved in HCl and mixed with $PdCl_2$. As depicted in Fig. 1, the influence of the promoter gives rise to a rather complex curve, but the overall effect is to increase activity except at the very lowest

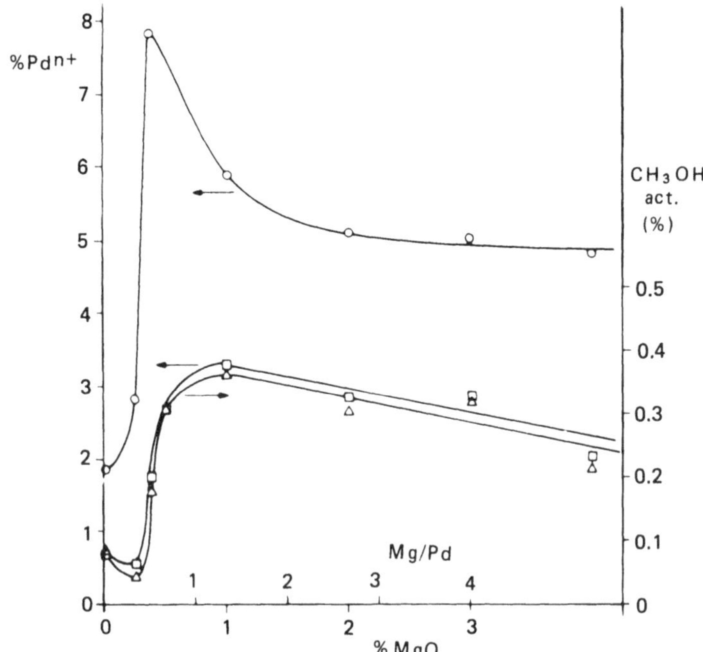

Fig. 1. Amounts of extractable Pd^{n+} and activity for CH_3OH synthesis as a function of Mg^{+2} added to a Pd/SiO_2catalyst: O , before reaction; □ , Δ , after reaction.[12]

concentration of promoter addition. (The amount of extractable palladium will be discussed below.) It was also observed that the formation of dimethyl ether generally increased more with addition of Mg^{+2} than did the formation of methanol, and on one catalyst prepared from $MgCl_2$ the conversion to dimethyl ether was complete.

The possible promotional effect of chloride ions has been noted previously. Catalysts prepared by impregnating SiO_2 with $Pd(\pi\text{-}C_3H_5)_2$ were much less active than those prepared with $PdCl_2$, but again the types of silica were different. Driessen et al.[12] prepared Mg^{+2}-promoted chloride-free catalysts from $Pd(NO_3)_2$ and $Mg(NO_3)_2$. These catalysts were about a factor of two less active than catalysts prepared from $PdCl_2$ and $Mg(NO_3)_2$. Tamaru and co-workers[4] observed that the activity of catalysts derived from $Na_2Pd(C_2O_4)_2$ had somewhat greater activity than catalysts derived from Na_2PdCl_4 provided the former were not reduced in H_2 at temperatures in excess of 200°C. At higher reduction temperatures the catalysts derived from the oxylate complex were essentially inactive. These authors suggest that the chloride ion plays "an important role in maintaining the surface structure for methanol formation". Unfortunately, in the latter two studies dispersion data were not complete so that it is impossible to establish whether the chloride ion simply stabilizes the Pd in a higher degree of dispersion.

Closely related to the possible role of chloride ions is the question of acidity or basicity. Earlier studies suggested that basic supports may favor alcohol formation;[3] however, comparison of the more extensive current results indicate that such is not the case. Acidic catalysts promote the conversion of methanol to dimethyl ether, but the primary step (i.e., the reaction of CO and H_2 to form CH_3OH) does not appear to depend upon the acidity of the catlyst. The results shown in Table I establish that La_2O_3, a basic catalyst, and Al_2O_3, an acidic catalyst, are both effective for producing oxygenates. Since support acidity is not of primary importance, one may conclude that the strong positive influence of Li^+ and Na^+ on catalyst activity is due to some phenomenon other than the poisoning of acid sites.

VI. SELECTIVITY

Although CO, CH_3COOH and C_2^+ hydrocarbons have been observed as products, the selectivity for CH_3OH or CH_4 is of primary concern at this point. Most of the current evidence suggests that these two products are formed on different sites. Evidence for this is found in the variation of product distribution with time on stream. For Pd/Cab-O-Sil it was observed that the rate of methanol formation remained constant over a 16 hour period while the rate of methane formation decreased by a factor of three.[7] Ryndin et al.[5] also noted a similar phenomenon with their Pd/SiO_2 catalysts for different run numbers. Several groups have found that activation energies for methanol and methane formation are approximately 14 and 25

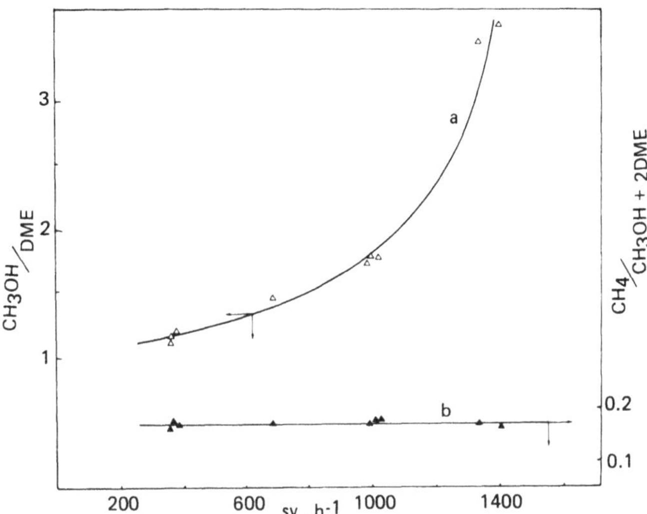

Fig. 2. Selectivity versus space velcity over a mixture of 4.6% Pd/SiO_2(57) and 5.4% PdHY at 277°C, PdHY at 277°C, 16 atm, H_2/CO = 2.25. (a) CH_3OH/DME, (b) $CH_4/(CH_3OH + 2DME)$.[7]

kcal/mole, respectively,[3-5] although the activation energy for methanol synthesis varies somewhat from one type of catalyst to another.

One must also be concerned with the possible conversion of methanol into methane. This hypothesis was tested over a physical mixture of Pd/SiO_2(57), which is a selective catalyst for the production of CH_3OH, and PdHY zeolite, which is a selective catalyst for the formation of CH_4.[7] If methanol were an intermediate in the production of methane, one would expect that as the space velocity was increased the ratio of methane to oxygenates would decrease, and such was not the case as shown in Fig. 2. Thus, methane must be derived by a separate reaction pathway. A possible exception to this may occur with Pd/TiO_2, as suggested by Ryndin et al.[5] Over rutile the decomposition of methanol yields coke and methane in the temperature range of interest. Production of methane from methanol over a TiO_2 catalyst needs to be tested by varying the space velocity as previously described.

VII. CURRENT THEORIES

Theories to explain the origin of activity for methanol synthesis and more recently the remarkable support effect are about as numerous as the number of research groups working on this problem. In this section an attempt will be made to summarize these theories, although one should realize that in most cases they were mainly given in the spirit of conjectures. Such is generally the case when one is attempting to understand new phenomena.

Poutsma et al.[2] observed that CO was not dissociated on supported Pd, in contrast to what occurs with supported Ni, and that the coordinated CO could therefore be hydrogenated to methanol. We now know that the ability to adsorb CO molecularly, rather than dissociatively, is a necessary but not sufficient condition for methanol activity. Palladium on SiO_2(01), for example, adsorbs CO molecularly but it is not a good catalyst for methanol synthesis.

Ponec and co-workers[12] favor a model which is similar to that proposed for the more traditional CuO-ZnO-Al_2O_3 catalyst in which one component is present for the coordination of CO and another for the activation of molecular H_2. It is suggested that Pd^{n+} ions are necessary for the former function and Pd metal for the latter. In support of this theory they have found that a certain form of the supported palladium is extractable with acetylacetone and that the amount of this extractable material correlates nicely with the catalytic activity as shown in Fig. 1. The extractable form of palladium is believed to be ionic. It will be interesting to see whether the correlation exists for palladium on the wide range of supports given in Table I. This view ultimately must be reconciled with the effect of reduction temperature on activity. In particular, it has been observed that the activities of samples prepared from Na_2PdCl_4 increased by a factor of five as the reduction temperature was increased from 200°C to 500°C.[4] Normally, the palladium would be more extensively reduced at the higher temperatures, which means that the ionic form of palladium would decrease as the activity increases. Another factor to consider in this theory is the role of CO_2 when added to the feed stream. To maintain activity over commercial copper oxide-zinc oxide catalysts, CO_2 is added to prevent complete reduction of copper to the metallic state. If Pd ions were important, one might expect CO_2 to enhance or at least prevent loss in activity. According to the observations, however, the addition of CO_2 in the feed had a slight deleterious effect on activity.[2,11]

Tamaru and co-workers[4] have presented infrared data which indicate that formate ions are intermediates in the formation of methanol. They propose that the alkali metal promoters, Li^+ and Na^+, increase the negative character of the surface oxygen, thereby stabilizing the oxygen-containing intermediates. The formate ion was not detected when potassium was incorporated into the catalyst.

In addition to pointing out the variations in activity with crystallographic planes,[10] Bell and co-workers[5] suggest that the very marked support effect may be the result of an electronic interaction between the metal and the support. Direct evidence for metal-support interaction between Pd and La_2O_3 comes from x-ray photoelectron spectroscopy which shows that the Pd $3d_{5/2}$ binding energy is shifted below the value for bulk Pd by as much as 0.7 eV.[10] This result indicates that Pd supported on La_2O_3 is more electronegative than zero valent Pd alone. A model has been proposed in which a thin layer of La_2O_3 partially covers the surface of the supported Pd crystallites.[13]

Fajula _et al._[7] noted that methanol was produced on Pd/SiO_2 catalysts exhibiting small size crystallites on which CO was weakly adsorbed. These observations suggested that structural effects may be important in determining activity for methanol synthesis. Although there is no evidence for the effect of different crystallographic planes, it appears that at least on a particular type of silica the distribution of planes is not related to dispersion.[10] It remains to be determined whether different grades of silica will give rise to different distributions of crystallographic planes. Differences in the ability of the silicas to retain chloride ions seems to be more likely explanation for the variations in activities which have been observed over the Pd/SiO_2 catalysts.

VIII. COMPARISON WITH A COMMERCIAL CATALYST

One factor in evaluating the commercial potential supported palladium as a methanol synthesis catalyst is its activity relative to an industrial catalyst. Such a comparison has been made with an ICI catalyst which is about 70% as active as a "state-of-the-art" catalyst. The commercial catalyst and an active 5% Pd/SiO_2(01) catalyst, promoted with Li_2CO_3, were tested in a Berty-type reactor operating at 16 atmospheres and 250°C.[11] The feed stream for the commercial catalyst was a mixture of H_2, CO, and CO_2 in the ratio 2.37 : 1 : 0.08. The commercial catalyst reached an activity of 10.2 mole $kg^{-1}h^{-1}$ after 33 hours of operation. On the basis of total catalyst mass this activity is about two times greater than that of our most active Li-promoted 5% Pd/SiO_2(01) catalyst. If comparisons of activities given in Table VI are valid here, one might expect that Pd/Al_2O_3 prepared from Na_2PdCl_4 would be about as active as the commercial catalyst.

In addition, at pressures of 15-40 atmospheres over Pd/SiO_2 the reaction was found to obey the rate law,[7]

$$\text{rate} \simeq k\, P_{H_2}^{2.08}\; P_{CO}^{-1.14}\,;$$

whereas, for a CuO-ZnO on alumina catalyst the rate law has the form[14]

$$\text{rate} = k\, P_{H_2}^{0.7}\; P_{CO}^{0.2 \text{ to } 0.6}\; \phi\, CO_2.$$

Here, $\phi\, CO_2$ represents the dependence of the rate on the partial pressure of CO_2. Upon comparison of the two equations it is apparent that an advantage could be gained from the second order dependence in H_2, and by operating at a greater H_2/CO ratios one could considerably improve the reaction rate over supported-palladium. These results suggest that at the present state of development the more active supported-palladium catalysts may be comparable to the commercial methanol synthesis catalysts.

ACKNOWLEDGMENT

The work carried out at Texas A&M University was supported by the Division of Basic Energy Sciences, Department of Energy. The author wishes to thank Professors Bell and Ponec for providing preprints of their manuscripts prior to publication.

REFERENCES

1. S.L. Meisel, J.P. McCullough, C.H. Lechthaler, and P.B. Weisz, Chemtech, 6, 86 (1976).

2. M.L. Poutsma, L.F. Elek, P.A. Ibarbia, A.P. Risch, and J.A. Rabo, J. Catal., **52**, 157 (1978).

3. M. Ichikawa, Shokubsi, **21**, 253 (1979); M. Ichikawa and K. Shikakura, Preprints; Seventh Intl. Congr. Catal., Tokyo, 1980.

4. Y. Kikuzono, S. Kagami, S. Naito, T. Onishi, and K. Tamaru, Faraday Disc. **72**, 135 (1981).

5. Yu. A. Ryndin, R.F. Hicks, A.T. Bell, and Yu. I. Yermakov, J. Catal., **70**, 287 (1981).

6. E.K. Poels, E.H. van Broekhoven, W.A.A. van Barneveld, and V. Ponec, React. Kinet. Catal. Lett. **18**, 223 (1981)

7. F. Fajula, R.G. Anthony, and J.H. Lunsford, J. Catal. **73**, 237 (1982).

8. M.A. Vannice and R.L. Garten, Ind. Chem. Prod. Res. Dev. **18**, 186 (1979).

9. M.A. Vannice, private communication.

10. R.F. Hicks, Q.-J. Yen, A.T. Bell, and T.H. Fleisch, submitted to J. Catal.

11. T. Tatsumi, K.P. Kelley, T. Uematsu, and J.H. Lunsford, unpublished results.

12. J.M. Driessen, E.K. Poels, J.P. Hindermann, and V. Ponec, J. Catal., **82**, 26 (1983).

13. T.H. Fleisch, R.F. Hicks, and A.T. Bell, submitted to J. Catal.

14. Reported in K. Klier, V. Chatikavanij, R.G. Herman, and G.W. Simmons, J. Catal., **74**, 343 (1982).

7

Supported Metal Cluster Catalysts for Synthesis Gas Conversion

Ronald Pierantozzi
Corporate Science Center, Air Products and Chemicals, Inc.
P.O. Box 538, Allentown, PA 18105

ABSTRACT

Supported metal clusters anchored to inorganic oxides such as SiO_2, Al_2O_3, MgO, and TiO_2 represent a potential new class of metal catalysts. Studies have been reported on the use of metal clusters as precursors to supported metal catalysts and also on the use of intact clusters as catalysts. This paper presents a review of some of the relevant literature in this field as well as results from our laboratories. New results from our laboratories on CeO_2 supported clusters show that the use of a cluster results in the formation of a catalyst that is selective for the synthesis of olefins from CO + H_2. Other precursors produce catalysts that give higher CH_4 yields as well as lower olefin selectivities.

I. INTRODUCTION

A major problem in the conversion of synthesis gas (CO + H_2) to hydrocarbon or oxygenated products has been catalyst selectivity. Only CH_4 and CH_3OH can be formed with 100% selectivity and the selectivity for higher hydrocarbon and alcohol formation is limited by the Anderson-Schulz-Flory distribution.[1] Recent reports in the literature suggest that the metal particle size of the catalyst could have a significant effect on the product distribution obtained from synthesis gas.[2] The problem associated with this is that supported metal catalysts generally do not give very narrow or well-defined particle size distributions. One approach to solving this problem of selectivity via the stabilization of metal particle sizes is the utilization of metal clusters as catalysts and as precursors in the formation of supported metals.

Supported metal cluster catalysts represent a class of metal catalysts that are intermediate between homogeneous metal catalysts (mono-nuclear or poly-nuclear) and supported metals. The use of supported clusters could provide flexibility in determining the metal particle size of the supported metal catalyst, but this requires an understanding of the mechanism for thermal decomposition and the surface organometallic chemstry of the supported cluster.

Two approaches have been followed in an attempt to develop novel catalytic materials from metal clusters; (1) maintaining the intact cluster framework with the ligands still bound to it thus generating a catalyst that resembles a homogeneous catalyst utilizing the support as a ligand, and (2) decomposition of the metal cluster to a metal particle that has a particle size distribution controlled by the cluster decomposition and perhaps the surface organometallic chemistry. The first approach allows one to gain a better understanding of the catalytic material because the techniques amenable to studying a metal complex are readily available. If the cluster remains intact on the surface, the nature of the catalytic site is well-defined. The stability of these types of catalysts, however, are suspect under most catalytic reaction conditions and only a few examples of intact, catalytically active clusters have been reported in the literature (see below).

The use of metal clusters as precursors to metal catalysts is an approach that is frought with the difficulties inherent in the study of solid catalysts, i.e., the definitive characterization of the active site as well as the problem of maintaining the metal particle size distribution initially obtained. This approach, however, has produced some interesting results for CO-hydrogenation suggesting that it could lead to some promising new materials.

This article is not intended to be a comprehensive review of the literature on supported metal clusters since several excellent reviews exist.[3] Rather it is an attempt to review some of the promising aspects of supported cluster catalysts. We will begin with a discussion of the reaction chemistry of clusters with surfaces, followed by a description of the surface orgnometallic chemistry and finish with an attempt to relate these two areas to the catalysis observed for synthesis gas reactions. We will draw on work that is in the literature as well as work that has been performed in our laboratories. We will cite examples of the use of intact clusters as well as the use of clusters as precursors to supported metals. We will not discuss the use of pendant ligands to support the organometallic species.

II. RESULTS AND DISCUSSION

A. Reaction of Clusters with Surfaces

Table I summarizes some of the reactions observed between metal clusters and inorganic oxides. Thermally stable clusters such as [$Ru_3(CO)_{12}$] and [$Os_3(CO)_{12}$] initially react with supports through the simple physical adsorption of the cluster onto the oxide. Typically the cluster geometry is retained and the infrared spectrum shows that the cluster is virtually identical to the molecular, unsupported complex. Kuznetsov et al.[4] have postulated that [$Ru_3(CO)_{12}$] interacts with the Lewis acid sites on Al_2O_3, a conclusion which is based on studies that show an effect due to the pretreatment of the support on the IR spectrum. Goodwin,[5] however, has reported that there are no low energy $\nu_{C\equiv O}$ bands observed in the infrared spectrum that would be characteristic of a Lewis acid adduct with a metal carbonyl. He concluded, therefore, that a Lewis acid interaction with the coordinated carbonyl probably does not occur. One other possibility is that the Lewis acid sites interact with the metal framework directly. This would result in an increase in the $\nu_{C\equiv O}$ band, which also is not observed. The major interaction of [$Ru_3(CO)_{12}$] with Al_2O_3, therefore, appears to be a very weak one, perhaps with the Lewis acid sites, although the interaction is not observable by infrared spectroscopy. The interaction, however does

TABLE I

Summary of Reactions of Metal Clusters with Inorganic Oxides

a Physical Adsorption

$$Ru_3(CO)_{12} + Al_2O_3 \rightarrow Ru_3(CO)_{12}/Al_2O_3$$

b OH Addition

$$H_2Os_3(CO)_{10} + Al\text{-}OH \rightarrow$$

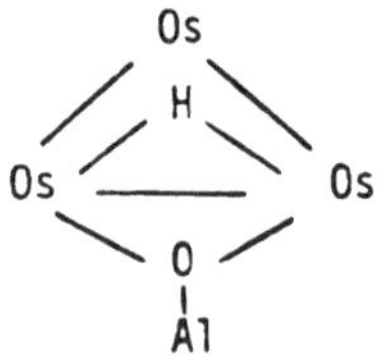

Os
H
Os Os
O
Al

c OH^- Attack

$$Ru_3(CO)_{12} + Mg\text{-}OH \rightarrow HRu_3(CO)_{11}^-$$

$$Fe_3(CO)_{12} + Mg\text{-}OH \rightarrow HFe_3(CO)_{11}^-$$

d Deprotonation

$$H_4Ru_4(CO)_{12} + Al_2O_3 \rightarrow H_3Ru_4(CO)_{12}^-$$

1

$$H_2RuOs_3(CO)_{13} + Al_2O_3 \rightarrow HRuOs_3(CO)_{13}^-$$

e Reactions with Zeolites

$$Fe_3(CO)_{12} + \text{NaY (hydrated)} \rightarrow HFe_3(CO)_{11}^-$$

$$Ru_3(CO)_{12} + \text{NaY} \rightarrow \text{encapsulation}$$

result in changes in the reactivity of the $[Ru_3(CO)_{12}]$ framework as discussed later.

$[Os_3(CO)_{12}]$ has been studied by numerous workers.[6] $[Os_3(CO)_{12}]$ and the unsaturated $[H_2Os_3(CO)_{10}]$ react with the hydroxyl groups on Al_2O_3, MgO, ZnO, and TiO_2 to give the structure shown in Eq. 1.[6c] This

$$Os_3(CO)_{12} + M\text{-}OH \rightarrow \qquad (1)$$

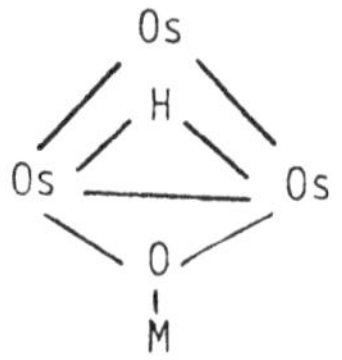

Os
H
Os Os
O
M

interaction, as well as the greater thermal stability of the Os_3 framework results in an intact, stable cluster on the surface. This interaction of $[Os_3(CO)_{12}]$ with the -OH groups on the surface is surprising in view of the lower reactivity of $[Os_3(CO)_{12}]$ relative to $[Ru_3(CO)_{12}]$, in general, and the lack of reactivity with Al_2O_3 exhibited by $[Ru_3(CO)_{12}]$. Basset et al.[7] have recently reported, however, that $[Ru_3(CO)_{12}]$ supported on carefully dried SiO_2 resulted in a grafting of the cluster onto the support that was similar to the reaction of $[Os_3(CO)_{12}]$. This reaction occurred on heating the physisorbed cluster to 80-100°C in an inert atmosphere. This reaction may not have been observed on Al_2O_3 due to the instability of the Ru_3 framework at these temperatures. On Al_2O_3 the decomposition of the cluster to oxidized Ru(II) occurred readily on heating.[4]

Basic supports such as MgO, ZnO, and in some cases Al_2O_3 result in cluster-surface reactions which are similar to the reactions of metal carbonyls with bases in solution. Both $[Ru_3(CO)_{12}]$[8] and $[Fe_3(CO)_{12}]$[9] have been reported to react with MgO by the reaction of an OH group with a coordinated CO to give a hydrido carbonyl (Table Ic). With $[Fe_3(CO)_{12}]$, the most reactive cluster in the Fe, Ru, Os series, the formation of the hydrido carbonyl was also formed on Al_2O_3. These reactions did not occur with the $[Os_3(CO)_{12}]$ cluster, once again reflecting the differences in reactivity between these species. The reaction of $[Ru_3(CO)_{12}]$ with MgO, also resulted in the formation of $[Ru_6(CO)_{18}]^{2-}$ by further reaction with the basic sites on the support or from decarbonylation of $[HRu_3(CO)_{11}]^{-1}$, the initial species formed on the surface.[8] The anionic clusters, especially the Ru derivatives, are typically more stable than their neutral analogs. For example, $[Ru_3(CO)_{12}]$/MgO is stable, intact, at temperatures as high as 215°C in the presence of synthesis gas. $[Fe_3(CO)_{12}]$/MgO, while not as stable as the Ru analog, decomposes in a more controlled manner than the neutral $[Fe_3(CO)_{12}]$.[9c]

Hydrido metal carbonyls such as $[H_4Ru_4(CO)_{12}]$[8] and $[H_2RuOs_3(CO)_{13}]$[10] were found to react with Al_2O_3 and MgO by deprotonation of the acidic cluster protons. $[H_3Ru_4(CO)_{12}]^-$ was formed on MgO and Al_2O_3 by the reaction of $[H_4Ru_4(CO)_{12}]$ with the supports. $[HRuOs_3(CO)_{13}]^-$ was also formed on Al_2O_3 from $[H_2RuOs_3(CO)_{13}]$.[10]

These reactions differed significantly from the reaction of $[Ru_3(CO)_{12}]$ with basic supports. No evidence of OH attack on the coordinated CO was observed for $[H_4Ru_4(CO)_{12}]/MgO$. These anionic clusters are stable at 200°C in the presence of CO and H_2.

Complete cluster decomposition occurs only when the reaction of the cluster with the support is run at elevated temperatures. Crawford et al.[11] reported the use of an extraction technique to prepare supported Ir, Os, and Ru carbonyls. This technique resulted in the formation of oxidized Ir and Os on Al_2O_3. These oxidized species gave highly dispersed Ir and Os catalysts upon reduction. These catalysts are significantly different from the catalysts prepared by dry grinding the organometallic and the support.

An interesting and widely studied method of preparing supported metal catalysts is the use of zeolite encapsulation. Metal clusters are potentially attractive for this approach since the chemistry and interaction with the support could be used to control the nature of the final catalyst.[12] Supported $[Fe_3(CO)_{12}]$ and $[Ru_3(CO)_{12}]$ have been prepared by vapor impregnation onto a zeolite support.[12] With dehydrated zeolites both $[Ru_3(CO)_{12}]$ and $[Fe_3(CO)_{12}]$ remain unchanged on reaction with the zeolite. On hydrated NaY, however, $[Fe_3(CO)_{12}]$ reacts to give $[HFe_3(CO)_{11}]^-$ which is more stable than the neutral analog.[13] $[Fe_3(CO)_{12}]$ supported on HY forms $[H_2Fe_3(CO)_{11}]$ in the zeolite cavity.[12a]

The variety of reactions taking place between metal carbonyl clusters and inorganic oxide supports results in a chemistry that is similar to that found in solution. These reactions also explain the pathways for some interesting organometallic chemistry of the supported complexes.

B. Surface Organometallic Chemistry

This area has been recently reviewed by Basset[14] and only selected examples will be discussed here. Although the interactions between the support and cluster have been studied in great detail, the organometallic chemistry of the supported cluster has not been investigated in much detail. Cluster rearrangements and decompositions on supports have been studied by Basset,[9] Kuznetsov,[4] Pierantozzi,[8] and Collier[15] for $[Ru_3(CO)_{12}]$, $[Fe_3(CO)_{12}]$, $[H_4Ru_4(CO)_{12}]$, and $[Os_3(CO)_{12}]$.

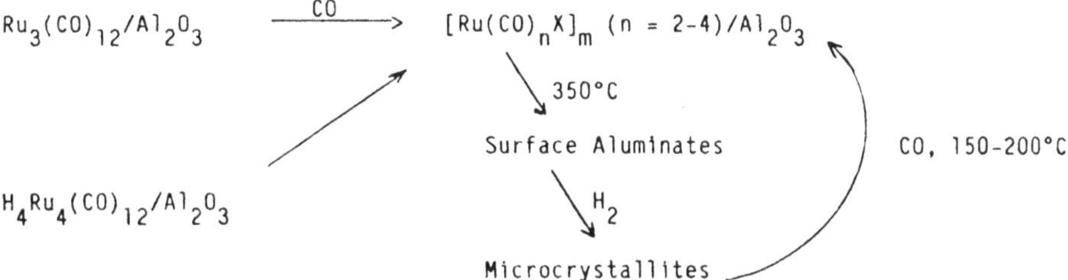

Scheme 1

Rearrangements of Supported Ru Clusters on Surfaces[4]

Kuznetsov et al.[4] reported that $[Ru_3(CO)_{12}]/Al_2O_3$ and $[H_4Ru_4(CO)_{12}]/Al_2O_3$ reacted on the support as shown in Scheme 1. The clusters were found to be thermally unstable, decomposing to mono-nuclear Ru compounds on the surface when heated in vacuum and exposed to air or CO at elevated temperatures. The reaction generally observed was the oxidation of the cluster by the support through the interaction with the OH groups.

$[H_4Ru_4(CO)_{12}]$ on Al_2O_3 and MgO and $[Ru_3(CO)_{12}]$ on MgO formed anionic clusters on the surface.[8] These anionic clusters, $[H_3Ru_4(CO)_{12}]^-$, $[HRu_3(CO)_{11}]^-$, and $[Ru_6(CO)_{18}]^{2-}$, rearranged on the surface in the presence of CO and H_2 to give the carbido-carbonyl cluster $[Ru_6C(CO)_{18}]^{2-}$ which remained intact at temperatures as high as 220°C. This rearrangement is summarized in Scheme 2. The rearrangement to the carbido cluster occurred more cleanly on the Al_2O_3 surface than on the MgO surface suggesting that the reaction involved the Lewis acid sites which are strong on Al_2O_3.[15] Indeed, infrared spectral evidence suggests that the Lewis acid sites on Al_2O_3 interact with the coordinated CO on the metal cluster and a $\nu_{C\equiv O}$ band at 1600 cm^{-1} is observed in the infrared spectrum which is characteristic of a Lewis acid interaction with coordinated CO groups.[16]

Collier et al.[17] studied the rearrangement of Os carbonyls on a variety of supports using temperature programmed decomposition, infrared spectroscopy, UV/VIS spectroscopy, and electron microscopy. They determined that $[Os_3(CO)_{12}]$ when supported on SiO_2, Al_2O_3, or TiO_2

$$H_4Ru_4(CO)_{12} + Al_2O_3 \rightarrow H_3Ru_4(CO)_{12}^-/Al_2O_3 \xrightarrow[200°C]{H_2/CO} Ru_6C(CO)_{16}^{2-}/Al_2O_3$$

$$Ru_3(CO)_{12} + MgO \rightarrow HRu_3(CO)_{11}^-/MgO + Ru_6(CO)_{18}^{2-}/MgO \xrightarrow[200°C]{H_2/CO} [Ru_6C(CO)_{16}]^{2-}/MgO$$

Scheme 2

Reactions of Ru Clusters with Basic Sites[8]

retained the Os-Os bond when heated to 250°C. The supports used were highly dehydroxylated so it was suggested that Os-Os bond rupture occurred when the degree of support hydroxylation was high, a factor which was found for $[Ru_3(CO)_{12}]$ as well. These workers further found that the Os clusters rearranged on heating to a carbido cluster with undetermined stoichiometry, $Os_n(CO)_mC_y$. This chemistry is similar to that exhibited by the Ru systems studied by Pierantozzi et al.[8]

The surface orgnometallic chemistry of carbonyl clusters differs significantly from the organometallic chemistry observed in solution when the supports are capable of oxidizing the cluster. The acid-base properties of the clusters on supports are similar to what has been found in solution with the exception of the Lewis acid site involvement in the chemistry which is virtually unexplored for the solution analogs. In solution the Lewis acid adducts generally reverse to give the original cluster.[16]

C. Catalysis by Supported Clusters

The catalytic reactions of supported metal clusters have been investigated in some detail with primary emphasis on the reactions of synthesis gas. Catalysts in which the cluster remained intact, decomposed to mono-nuclear systems, or decomposed to metal particles have all been observed with $[Os_3(CO)_{12}]$, $[Ru_3(CO)_{12}]$, $[H_4Ru_4(CO)_{12}]$, and $[Fe_3(CO)_{12}]$.

Basset[9] has found that $[Fe_3(CO)_{12}]$ supported on MgO or Al_2O_3 decomposed in the presence of synthesis gas at 180°C to produce supported iron particles that have a very narrow particle size distribution. Employing a wide variety of techniques, he found that the particle size distribution of the iron catalysts was centered around 14Å.[9] The conversion of synthesis gas over these materials resulted in a product distribution exhibiting a high selectivity to olefinic products with particularly high selectivity to C_3 and C_4 olefins. The catalyst, however, was fairly unstable, and with time on stream, the product distribution became less selective for the olefinic products and more characteristic of a supported iron catalyst such as that prepared from $Fe(NO_3)_3$ precursors.[9a] The results, however, do show that the use of the supported metal cluster can result in a narrow particle size distribution on the catalyst, and that this distribution does affect the product selectivity.

We have studied the conversion of synthesis gas over $[Ru_3(CO)_{12}]$ and $[H_4Ru_4(CO)_{12}]$ on Al_2O_3 and MgO.[8] Over catalysts in which the anionic ruthenium clusters are stable, namely MgO and $[H_4Ru_4(CO)_{12}]/Al_2O_3$, the product distributions listed in Table II were obtained. The product distributions, unlike those observed for Ru/Al_2O_3, consisted of a high yield of oxygenated compounds. With MgO the primary products were CH_3OH and C_2H_5OH, while on Al_2O_3 the primary product was CH_3OCH_3. It has been shown that under the reaction conditions studied $[Ru_6C(CO)_{16}]^{2-}$, formed on the surface, was stable. It has been hypothesized that this species is responsible for the catlytic behavior. It has also been shown that supporting the $[Ru_6C(CO)_{16}]^{2-}$ cluster on the support results in little or no catalytic activity at temperatures up to 300°. We have, therefore, postulated that the reaction requires both the presence of metal particles and the stable ruthenium cluster and that a synergism exists between the two such that the ruthenium metal supplies the hydrogen for the hydrogenation of the CO and the CO is coordinated to the ruthenium cluster as shown in Fig. 1. $[Ru_3(CO)_{12}]/Al_2O_3$, which decomposes to Ru/Al_2O_3 under reaction conditions, produces a product distribution that is typical of the Ru metal catalysts.[18]

Gates and coworkers[6c] have found that $[Os_3(CO)_{12}]$ remained intact on Al_2O_3 or MgO and was active for CO hydrogenation. In the presence of synthesis gas at low conversions, they found that low molecular weight

TABLE II

Product Distributions over Supported Ru Clusters[8]

Catalyst	T (°C)	CO conv (%)	Product Distribution C_1	C_2–C_4	C_5–C_{10}	Oxygenates	Olefins[f]
$Ru_3(CO)_{12}/Al_2O_3$	205[a]	90.7	63.4	24.3	4.9	7.3	19.7
	261[b]	39.8	54.8	32.2	7.4	0	22.0
	344[c]	87.4	86.8	11.0	2.0	0.3	3.3
$Ru_3(CO)_{12}/MgO$[d]	200	9.5	30.9	31.9	10.7	26.5	27.9
	205	15.8	36.9	25.4	4.6	33.0	19.1
	315	33.5	65.6	29.5	4.9	0	3.0
$H_4Ru_4(CO)_{12}/Al_2O_3$[e]	207	5.0	26.2	30.2	10.1	33.5	28.0
$H_4Ru_4(CO)_{12}/MgO$[e]	208	28.8	35.2	28.4	10.4	21.0	23.1

a $P = 2.1 \times 10^3$ kpa, GHSV = 300, $H_2/CO = 1$
b $P = 2.1 \times 10^3$ kpa, GHSV = 800, $H_2/CO = 1$
c $P = 2.1 \times 10^3$ kpa, GHSV = 1257, $H_2/CO = 1$
d $P = 2.1 \times 10^3$ kpa, GHSV = 262, $H_2/CO = 1$
e $P = 2.1 \times 10^3$ kpa, HGSV = 200, $H_2/CO = 0.67$
f Olefin yields are reported as wt% of total product

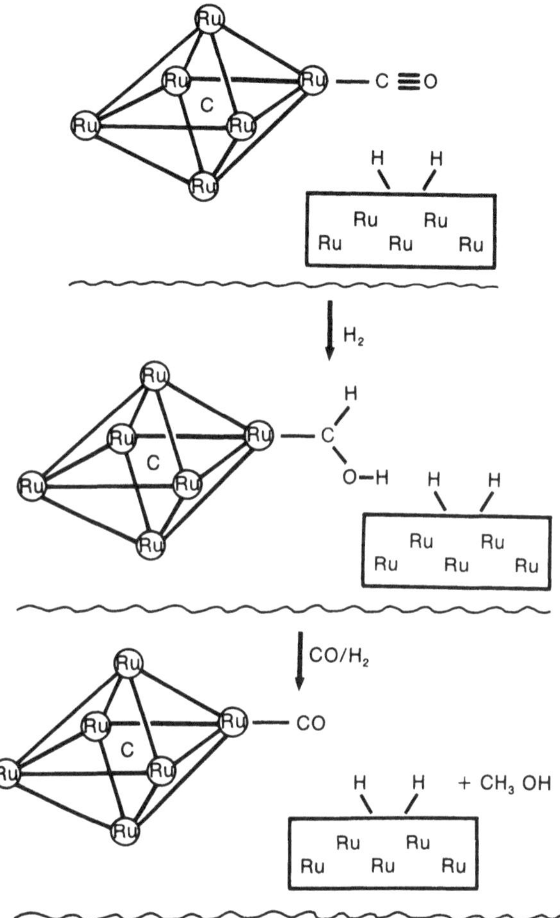

Fig. 1. Proposed Mechanism for CO Hydrogenation Over $[Ru_6C(CO)_{16}]^{2-}$/ Al_2O_3 Formed from $[H_4Ru_4(CO)_{12}]/Al_2O_3$

hydrocarbons were formed. Infrared spectral analysis after reaction has shown that the cluster was still intact on the surface. Upon air oxidation, $[Os_3(CO)_{12}]$ decomposed on the surface to Os(II) carbonyls which were also active for CO hydrogenation. During reaction with CO and H_2 the trinuclear cluster was reformed on the surface from the mononuclear Os(II) species.

Okuhara et al.[19] have found that K promoted Ru catalysts prepared from $[Ru_3(CO)_{12}]$ were more active and more selective than Ru/Al_2O_3

prepared by more conventional methods. They found that the cluster derived catalyst resulted in olefin selectivities as high as 80 wt% of the product. Conventionally prepared catalysts resulted in selectivities that were significantly lower as well as CH_4 yields that were higher than observed with the cluster catalysts.

We have recently studied the use of $[Ru_3(CO)_{12}]$ on reducible oxides to determine the differences between catalysts prepared using metal cluster precursors and those made with conventional precursors such as $RuCl_3$ and also to study the catalytic differences arising from using different methods of preparing the supported metal cluster. A series of catalysts were made by the incipient wetness technique to incorporate $[Ru_3(CO)_{12}]$, $[H_4Ru_4(CO)_{12}]$, or $RuCl_3$ onto CeO_2. A second series of catalysts were prepared by decomposing $[Ru_3(CO)_{12}]$ in a hydrocarbon oil directly onto the support and also by dry grinding the $[Ru_3(CO)_{12}]$ with the CeO_2. The results of H_2/CO reactions run over these catalysts are summarized in Table III. The catalysts prepared by incipient wetness using $[Ru_3(CO)_{12}]$ or $[H_4Ru_4(CO)_{12}]$ were selective for the formation of olefinic products. In a typical reaction, the selectivity toward α-olefins was 55-60 wt% of the total product formed. The methane yields obtained over this catalyst were only in the 9-12 wt% range. The catalyst prepared from $RuCl_3$, on the other hand, gave product distributions that were significantly different from those obtained with catalysts prepared from the $[Ru_3(CO)_{12}]$. These catalysts gave methane yields of 35 wt% and typically had low olefin yields, although yields as high as 40 wt% were observed. There was also a significant dependence of the product distribution on the ruthenium loading for the $RuCl_3$ catalysts. For example, a catalyst with a 0.6 wt% ruthenium loading resulted in little or no olefin selectivity but significant selectivity to oxygenated products was observed. On the other hand, the catalyst with a 4.0 wt% ruthenium loading resulted in significant olefin selectivity but not a high selectivity toward oxygenates. This type of dependence on the metal loading was not observed for the catalyst prepared from ruthenium carbonyls. As Table III shows, the catalyst containing 0.4 wt% ruthenium and the catalyst containing 2.3 wt% ruthenium gave essentially the same distribution when the cluster was the precursor. The dry grinding and

TABLE III

Product Distributions over CeO_2 Supported Catalysts[a]

Catalyst	Ru wt%	T (°C)	CO Conv	Product Distribution (wt%) C_1	C_2–C_4	C_5–C_{10}	C_{11}+	Olefins[b]	Oxygenates
$Ru_5(CO)_{12}/CeO_2$ (Impregnated)	2.32	343	28.1	12.5	43.3	37.6	6.1	55.5	0.6
$H_4Ru_4(CO)_{12}/CeO_2$ (Impregnated)	0.38	346	20.8	12.7	38.5	31.5	13.7	54.3	3.6
$Ru_3(CO)_{12}/CeO_2$ (Slurry Prep I)	2.44	350	42.5	40.3	34.3	13.7	1.7	21.5	9.9
$Ru_3(CO)_{12}/CeO_2$ (Slurry Prep II)	2.70	348	62.1	56.1	28.2	11.5	1.5	4.9	2.9
$Ru_3(CO)_{12}/CeO_2$ (dry grinding)	0.86	353	52.1	24.8	43.3	19.6	2.6	23.3	9.8
$RuCl_3/CeO_2$	0.6	351	84.6	35.7	11.9	1.7	0	1.6	39.6
$RuCl_3/CeO_2$	4.11	313	6.9	36.8	49.4	7.2	0	36.7	6.7

[a] Conditions P = 2.1×10^3 kPa, GHSV - 200–700, $H_2/CO = 1$

[b] Olefin yields are reported as wt% of total product

decomposition in a hydrocarbon oil techniques for preparation resulted in catalysts that were generally unselective, giving in fairly high methane formation and low olefin selectivities.

In an attempt to understand the differences between the catalysts prepared from $RuCl_3$ and $[Ru_3(CO)_{12}]$ several methods of characterization were employed. Hydrogen chemisorption results on the catalysts prepared from $[Ru_3(CO)_{12}]$ and $[H_4Ru_4(CO)_{12}]$ by the incipient wetness technique show that after reaction the catalyst exhibits the suppression of hydrogen chemisorption that is typical for catalysts in the SMSI state.[19] $RuCl_3$ catalysts, however, exhibit very little suppression of the hydrogen chemisorption with a particle size calculated to be about 22 angstroms for these materials. XPS results show that on all catalysts the Ru is reduced to Ru(O) and that the CeO_2 is also reduced in the presence of the synthesis gas. The reduction of the CeO_2 occurs only when the metal is present on the surface. Table IV summarizes the results of the XPS studies on these catalysts. The surface Ru/Ce ratios calculated from the XPS data for the catalysts also suggest that the catalyst prepared from $[Ru_3(CO)_{12}]$ is more highly dispersed than the catalyst prepared from $RuCl_3$. An alternate explanation for this could be that when preparing the $RuCl_3$ catalysts some dissolution of the support could occur and the CeO_2 then redeposits onto the surface of the Ru metal resulting in a lower surface Ru/Ce ratio. Similar results on the conversion of synthesis gas over Ru/CeO_2 catalysts, prepared from $RuCl_3$, were reported by Kikuchi _et al._[21] who found low metal dispersions and high CH_4 yields for these catalysts.

III. CONCLUSIONS

The use of supported metal clusters to prepare catalysts results in materials that are different from those described previously. Decomposition of the metal clusters into metal particles can result in catalysts that exhibit narrow particle size distributions and thus have a significant effect on the product selectivities observed. Likewise, metal clusters can be stabilized on some supports for short periods of time and produce selectivities that are also uncharacteristic of the particular metal. In the case of the Ru clusters cited above, the intact clusters resulted in oxygenate selectivities that are

TABLE IV

XPS Data for Ru/CeO_2 Catalysts

Catalyst	B.E. (ev)[e]			Ru/Ce	Ce^{4+}/Ce^{3+}
	Ru $3d_{5/2}$	Ru 3p	Ce $3d_{5/2}$		
4% $RuCl_3/CeO_2$ [a]	278.8	460.1	881.8/885.5	0.27	1.1
4% $RuCl_3/CeO_2$ [b]	278.7	460.1	881.3/885.2	0.16	1.1
4% $Ru_3(CO)_{12}/CeO_2$ [c]	279.0	460.3	881.5/885.4	3.35	1.4
CeO_2 [d]	--	--	881.5	--	--

[a] Reduced in H_2, 450°

[b] Pretreated in H_2 + CO, 320°C

[c] Pretreated in H_2 + CO, 350°C

[d] Treated in H_2, 450°C

[e] Referenced to Al_{2p} = 74.7 ev.

not observed for Ru metal under similar reaction conditions. It, therefore, appears that the intact metal clusters behave very much like their homogeneous analogs, except that the pressures required for the reaction on the support are significantly lower. Typically high pressures are required in solution to stabilize the metal complex.[22] Another example shows that supported metal clusters can be used to generate strong metal support interactions on interacting supports such as CeO_2. These strong metal support interactions permit one to prepare catalysts that are significantly more selective than catalysts prepared using conventional techniques without a high temperature reduction, which could lead to the sintering of the metal.

The chemistry and catalysis by supported metal clusters is a fairly new area, and the full understanding of the cluster surface organometallic chemistry and the interaction of clusters on supports and how they relate to catalysis is still not that well explored. The catalytic chemistry of reactions other than synthesis gas conversion is also unexplored and the use of supported clusters in other reactions might prove to be a fruitful area.

IV. EXPERIMENTAL

A. Catalyst Preparations - All catalysts were protected from air and moisture.

[$Ru_3(CO)_{12}$]/CeO_2 and [$H_4Ru_4(CO)_{12}$]/CeO_2 (Impregnated). The cluster (0.5g) was dissolved in 8 mL of hexane and added to 10g of CeO_2 (Cerac, predried at 200°C in vacuo). The solvent was then removed in vacuo. Analysis gave a Ru value of 2.32%.

[$Ru_3(CO)_{12}$]/CeO_2 (Slurry Preparation I). [$Ru_3(CO)_{12}$] (2.00g) and 40.0g of CeO_2 were suspended in 250 mL of Witco #40 oil and heated under a CO/H_2 purge for 2 hr. The solid was then filtered in a glove box and used without drying. A second catalyst (Slurry Prep II) was prepared by drying the oil in vacuo at 260°C. Anal.; Ru, 2.44% (Slurry Prep I): Ru, 2.70% (Slurry Prep II).

[$Ru_3(CO)_{12}$]/CeO_2 (Dry grinding). CeO_2 (30.0g) and 1.5g of [$Ru_3(CO)_{12}$] were ground together and then heated in synthesis gas for 3 hr. at 200°C. Anal.; Ru, 0.86%.

$RuCl_3/CeO_2$. These catalysts were prepared by adding an aqueous solution of $RuCl_3 \cdot 3H_2O$ to CeO_2 using the incipient wetness technique.

B. Catalyst Testing

All reactions were carried out in a downflow tubular reactor with on-line gas chromatography. The GC analyses were performed on a Porapak R (C_1-C_5 + oxygenates), SP2100 (C_6-C_{30}, Porapak N(CO_2,CH_4) and MS5A (H_2, CO) columns. Selectivities are reported as:

$$\text{Selectivity} = \frac{\text{Wt Product } i}{\sum_i \text{Wt Product}} \times 100\%$$

C. Catalyst Characterization

Chemisorption results were obtained on a Micrometrics digisorb. XPS results were obtained on a Physical Electronics PHI 560 spectrometer. Elemental analyses were performed by Schwartzkopf microanalytical laboratories.

ACKNOWLEDGMENTS

The author would like to thank Ellen Valagene, Scott Madison, Ilse Kingsley, Eugene Karwacki, Andrew Nordquist, and Paul Dyer for their experimental work and many helpful discussions. The author wishes to express his gratitude to Air Products and Chemicals and the DOE, Fossil Energy Division for funding under Contract DE-AC22-80PC30021.

REFERENCES

1. (a) R.B. Anderson, in Catalysis (P.H. Emmett ed) Reinhold, New York, 1956 Vol. 4; (b) R.B. Anderson and Y. Chan, Prep. Div. Petrol. Chem., ACS, **23**, 578 (1978).
2. (a) H.H. Nijs and P.A. Jacobs, J. Catal., **65**, 328 (1980); (b) P.A. Jacobs in Catalysis by Zeolites (B. Imelik et al. eds) Elsevier, Amsterdam, 1980, p. 293; (c) R.C. Revel and C.H. Bartholomew, J. Catal., **85**, 78 (1984).
3. DC. Bailey and S.H. Langer, Chem. Rev., **81**, 109 (1981); (b) B.C. Gates and J. Lieto, Chemtech., **10**, 195, 248 (1980); (c) J. Phillips and J.A. Dumesic, Appl. Catal., **9**, 1 (1984).

4. V. Kuznetsov, A.T. Bell and Y.I. Yermakov, J. Catal., **65**, 374 (1980).

5. J.G. Goodwin and C. Naccache, J. Mol. Catal., **14**, 259 (1982).

6. (a) R. Psaro, R. Ugo, G.M. Zanderighi, B. Besson, A.K. Smith and J.M. Basset, J. Organomet. Chem., **213**, 215 (1981); (b) M. Deeba and B.C. Gates, J. Catal., **67**, 303 (1981); (c) M. Deeba, J.P. Scott, R. Barth and B.C. Gates, J. Catal., **71**, 373 (1981); (d) H. Knozinger and Y. Zhao, J. Catal., **71**, 337 (1981); (e) H. Knozinger, Y. Zhao, B. Tesche, R. Barth, R. Epstein, B.C. Gates and J.P. Scott, Faraday Disc. Chem. Soc., **72**, 53 (1982); (f) A.K. Smith, B. Besson, J.M. Basset, R. Psaro, A. Fusi and R. Ugo, J. Organomet. Chem., **192**, C31 (1980); (g) R. Ugo, R. Psaro, G.M. Zanderighi, J.M. Basset, A. Theolier and A.K. Smith in Fundamental Research in Homogeneous Catalysis (M. Tsutsi, ed.) Vol. 3, 1979, p. 579.

7. A. Theolier, A. Choplin, L. D'Ornelas, J.M. Basset and G. Zanderighi Polyhedron, **2**, 119 (1983).

8. R. Pierantozzi, E.G. Valagene, A.F. Nordquist and P.N. Dyer, J. Mol. Catal., **21**, 189 (1983).

9. (a) D. Commereoc, Y. Chauvin, F. Hugues, J.M. Basset and D. Olivier, J. Chem. Soc., Chem. Commun., 154 (1980); (b) F. Hugues, B. Besson and J.M. Basset, J. Chem. Soc., Chem. Commun., 719 (1980); (c) F. Hugues, B. Besson, P. Bussiere, J.A. Dalmon, J.M. Basset and D. Oliver, Nouv. J. Chim., **5**, 207 (1981); (d) F. Hugues, J.A. Dalmon, P. Bussiere, A.K. Smith, J.M. Basset and D. Olivier, J. Phys. Chem., **86**, 5136 (1982); (e) F. Hugues, A.K. Smith, Y.B. Taarit, J.M. Basset, D. Commereuc and Y. Chauvin, J. Chem. Soc., Chem. Commun. 68, (1980).

10. J.R. Budge, J.P. Scott and B.C. Gates, J. Chem. Soc., Chem. Commun., 342 (1983).

11. J.E. Crawford, G.A. Melson, L.E. Makovsky and F.R. Brown, J. Catal., **83**, 454 (1983).

12. (a) D. Ballivet-Tkatchenko and G. Coudurier, Inorg, Chem., **18**, 558 (1979); (b) P. Gelin, G. Coudurier, Y. Bentaarit and C. Naccache, J. Catal., **70**, 32 (1981); (c) J.G. Goodwin, Jr., C. Naccache, J. Mol. Catal., **14**, 259 (1982); (d) S. Namba, T. Komatsu and T. Yashima,

Chem. Lett., 115, (1982); (e) D.G. Blackmond, J.G. Goodwin, Jr., J. Chem. Soc., Chem. Commun., 125, (1981); (f) T. Huang and J. Schwartz, J. Am. Chem. Soc., **104**, 5244 (1982); (g) D. Ballivet-Tkatchenko, G. Coudurier and H. Mozzanega in **Catalysis by Zeolites** (B. Imelik et al. eds.) Elsevier, Amsterdam, 1980, p. 309; (h) P. Gelin, Y. Benntaarit and C. Naccache in **New Horizons in Catalysis**, Elsevier, Amsterdam, 1981, p. 898; (i) K. Tanaka, K.L. Watters, R.F. Howe and S.L.T. Anderson, J. Catal., **79**, 251 (1983); (j) R.L. Schneider, R.F. Howe and K.L. Watters, J. Catal., **79**, 298 (1983); (k) S. Abdo and R.F. Howe, J. Phys. Chem., **87** 1713, 1722 (1983).

13. M. Iwamoto, H. Kusano and S. Kagawa, Inorg. Chem., **22**, 3365 (1983).
14. J.M. Basset and A. Choplin, J. Mol. Catal., **21**, 95 (1983).
15. K. Tanabe, in **Solid Acids and Bases**, Academic Press, N.Y., 1970, p. 45.
16. J.S. Kristoff and D.F. Shriver, Inorg. Chem., **13**, 499 (1974).
17. G. Collier, D.J. Hunt, S.D. Jackson, R.B. Moyes, I.A. Pickering and P.B. Wells, J. Catal., **80**, 154 (1983).
18. D.L. King, J. Catal., **51**, 386 (1978).
19. T. Okuhara, K. Kobayashi, T. Kimura, M. Misono and Y. Yoneda, J. Chem. Soc., Chem. Commun., 114 (1981).
20. S.J. Tauster, S.C. Fung and R.L. Garter, J. Am. Chem. Soc., **100**, 170 (1978).
21. E. Kikuchi, H. Nomura, M. Matsumoto and Y. Morita, Appl. Catal., **7**, 1 (1983).
22. See for example, D.R. Fahey, J. Am. Chem. Soc., **103**, 136 (1981).

8

CO Hydrogenation on Alkali-Promoted Nickel Catalysts

John L. Falconer, Keith M. Bailey, and Paul D. Gochis
Department of Chemical Engineering
University of Colorado, Box 424
Boulder, Colorado 80309

ABSTRACT

The modification of catalytic activity and selectivity by alkali promoters for CO hydrogenation on nickel was studied for four oxide supports. Steady-state kinetics were measured in a differential reactor (3:1 mixture of H_2:CO) with gas chromatographic detection. Temperature-programmed reaction (TPR) with mass spectrometric detection was carried out in excess hydrogen. In the differential reactor studies a potassium promoter decreased the activity but increased the olefin selectivity when nickel was supported on SiO_2, Al_2O_3, or TiO_2. The extent of activity decrease was a function of the support, however, and may be due to different distributions of promoter between the nickel and the supports. For Ni on $SiO_2 \cdot Al_2O_3$, overall activity increased and the selectivity to higher paraffins increased while olefin selectivity was unchanged. An interaction between the promoter and the $SiO_2 \cdot Al_2O_3$ support caused the increased rates of hydrogenation. The TPR results were in good agreement with the differential reactor studies; decreased activities for methane and ethane formation on Ni/SiO_2 and Ni/TiO_2 were due to decreased hydrogenation rates. On Ni/Al_2O_3, the small rate decrease was mainly due to decreased dispersion. TPD results showed that hydrogen was more weakly bound to the promoted Ni/SiO_2 catalyst. For promoted $Ni/SiO_2 \cdot Al_2O_3$, the hydrogenation rates either increased or decreased only slightly in TPR, a fact also in agreement with the differential reactor studies.

I. INTRODUCTION

Alkali promoters are used to modify catalytic activity and selectivity of supported metal catalysts, but their effect on CO hydrogenation has not been studied in detail. Most studies have reported that alkali promoters decrease methanation rates.[1-4] Huang and Richardson,[5] however, found a maximum in the methane turnover number for sodium-promoted $Ni/SiO_2 \cdot Al_2O_3$. The range of activities measured in these previous studies indicates that the oxide support may modify the role of the promoter. These studies used different preparation methods, however, so direct comparisons are difficult. Thus, to study the role of the support, promoted nickel catalysts were prepared by the same method on four supports: SiO_2, Al_2O_3, TiO_2, and $SiO_2 \cdot Al_2O_3$. Promoter pre-impregnation was used since Huang and Richardson[5] reported a rate increase with this preparation method. Carbon monoxide hydrogenation to methane and higher hydrocarbons was studied for similar promoter loadings on each catalyst.

A steady-state, differential reactor was used to measure activities and product distributions. Temperature programmed reaction (TPR) was used to measure specific activities and the role of site blocking, and to determine whether additional sites were created by the addition of promoters. Steady-state rection and TPR studies were carried out on some of the same catalysts to determine if TPR is a valid method for comparing catalytic activities. Good agreement was obtained between the two methods. On most catalysts, alkali promoters decreased specific activities and increased olefin selectivities. On Ni/Al_2O_3, the rate decreased a small amount, whereas on Ni/SiO_2 and Ni/TiO_2 large decreases were seen. On $Ni/SiO_2 \cdot Al_2O_3$, however, the activity increased slightly. The amount of promoter present on the nickel and the reaction of the promoter with the support may explain these differences.

TABLE I

Activities at 548 K and Activation Energies: Differential Reactor

Catalyst	CH_4 Activity (μmol/g·Ni·s)	E_{CH_4} (kJ/mol)	Total Activity (μmol/g·Ni·s)	E_{CO} (kJ/mol)
16% Ni/ SiO_2	83	103	110	82
11% Ni/0.8% K/ SiO_2	1.9	126	2.0	148
9.6% Ni/ Al_2O_3	120	128	220	112
10% Ni/0.8% K/ Al_2O_3	35	119	84	100
9.4% Ni/ $SiO_2 \cdot Al_2O_3$	110	110	130	77
9.7% Ni/0.8% K/ $SiO_2 \cdot Al_2O_3$	110	125	150	96
9.7% Ni/ TiO_2	220	133	560	105
10% Ni/0.8% K/ TiO_2	4.6	122	14	96

II. RESULTS AND DISCUSSION

A. Differential Reactor Measurements

Carbon monoxide hydrogenation at low conversion on supported nickel yielded C_1-C_4 hydrocarbons; methane was the predominant product with all catalysts. As expected from previous studies on unpromoted catalysts, the activity and selectivity depended on the support. The changes in activity and selectivity with the addition of potassium, however, also depended on the support. As shown in Table I, with nickel on SiO_2, Al_2O_3, or TiO_2, methane activities and total activities (per gram nickel) decreased when potassium was added. Olefin activities decreased to a lesser extent or increased, and higher paraffin activities decreased. Thus olefin selectivities increased significantly, but this increase depended on the support (Figs. 1 and 2). Comparisons of promoted and unpromoted catalysts at the same conversions showed that the increased olefin selectivities on the promoted catalysts were not due to lower conversions.

On $Ni/SiO_2 \cdot Al_2O_3$, potassium increased the CO activity. Methane activity did not change (Table I), but C_2-C_4 paraffin activities increased, so selectivities for higher paraffins increased (Fig. 2). Olefin activities

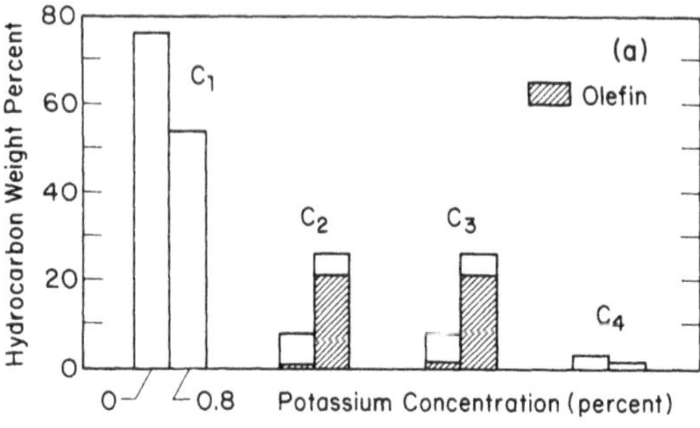

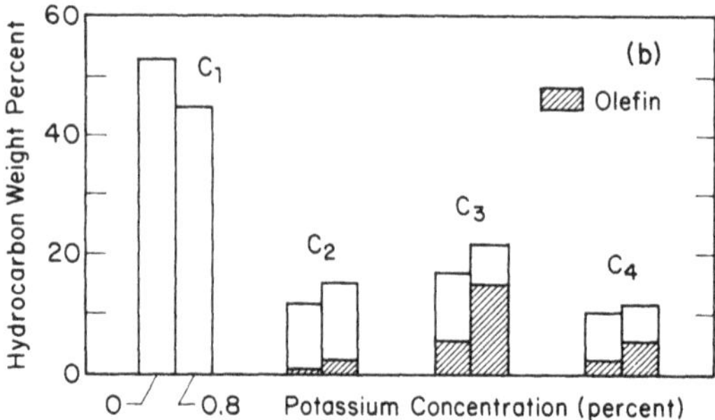

Fig. 1. Differential-reactor product distributions at 548 K for unpromoted and potassium-promoted (a) Ni/SiO_2 and (b) Ni/Al_2O_3.

increased only slightly. Measurements on $Ni/SiO_2 \cdot Al_2O_3$ catalysts which had a range of potassium concentrations showed that a maximum in methane activity occurred at potassium concentrations below 0.25.[6]

B. Temperature Programmed Reaction

Temperature programmed reaction (TPR) was carried out on the catalysts used in the differential reactor studies and also on a few sodium-promoted catalysts. For the TPR studies carbon monoxide was adsorbed to saturation coverage at room temperature, and the catalyst was then heated in hydrogen to 773 K. Methane, the predominant product, formed in a relatively narrow temperature range with high yields on all catalysts. On unpromoted Ni/Al_2O_3, a second, broader methane peak of similar area was seen at higher temperatures. On unpromoted Ni/TiO_2 a second, small methane peak was seen at higher temperatures.

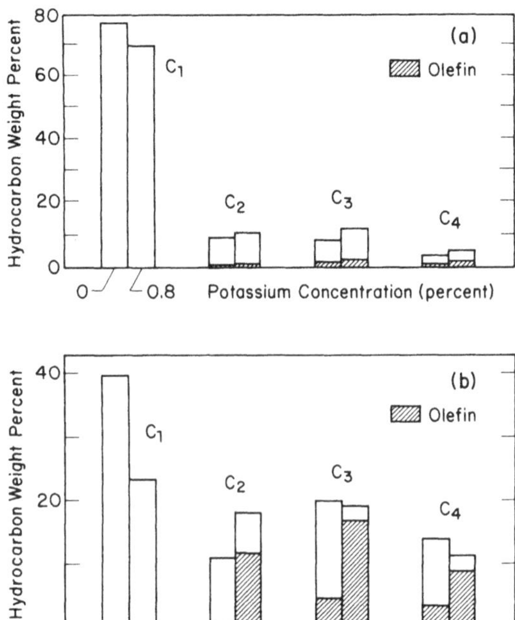

Fig. 2. Differential-reactor product distribution at 548 K for unpromoted and potassium-promoted (a) $Ni/SiO_2{\cdot}Al_2O_3$ and (b) Ni/TiO_2.

Ethane formed at lower temperatures than did methane and was approximately one percent of the methane on all catalysts except Ni/Al_2O_3. Ethane was not detected on Ni/Al_2O_3. Unreacted carbon monoxide desorbed in a single peak at low temperatures, and CO desorption was complete by 550 K. On the Ni/SiO_2 catalysts, water formed with a peak temperature and shape almost identical to that of methane.[7] For nickel on the other supports, readsorption on the support delayed water detection to higher temperature.[8] The detailed hydrocarbon spectra depended on the promoter as well as the support, as will be discussed below.

<u>Ni/SiO_2 and Ni/TiO_2 Catalysts</u>: Sodium and potassium promoters increased the peak temperatures of methane and ethane formation over Ni/SiO_2 catalysts (Fig. 3 and Table II). A higher potassium to nickel ratio caused a larger increase in peak temperature (Table II). Potassium increased the methane and ethane peak temperatures on Ni/TiO_2 also (Fig. 4). Since

TABLE II

Peak Temperatures from the TPR of Carbon Monoxide

Catalyst	Peak Temperature (K) CH_4	C_2H_6
Ni/SiO_2		
unpromoted	498	474
0.5% Na	530	515
0.6% K	538	503
0.9% K	569	513
Ni/TiO_2		
unpromoted	447	440
0.8% K	495	488
Ni/Al_2O_3		
unpromoted	466,510	--
0.8% K	466,540	--
$Ni/SiO_2{\cdot}Al_2O_3$		
unpromoted	480	470
0.2% Na	478	457
0.3% Na	473,536	443
0.8% K	490	481

readsorption is not significant for CH_4 on nickel,[9] a higher peak temperature corresponds to a lower areal rate (specific activity) for the temperature range in which the peaks occur.

The methane peak halfwidths for the Ni/SiO_2 catalysts were between 40 and 48 K, except for the 0.9% potassium catalyst, which had a halfwidth of 89 K (Fig. 3). The fraction of adsorbed CO that was converted to CH_4 was relatively constant between 70 and 80% on the Ni/SiO_2 catalysts. The amount of methane formed corresponded to the number of sites available

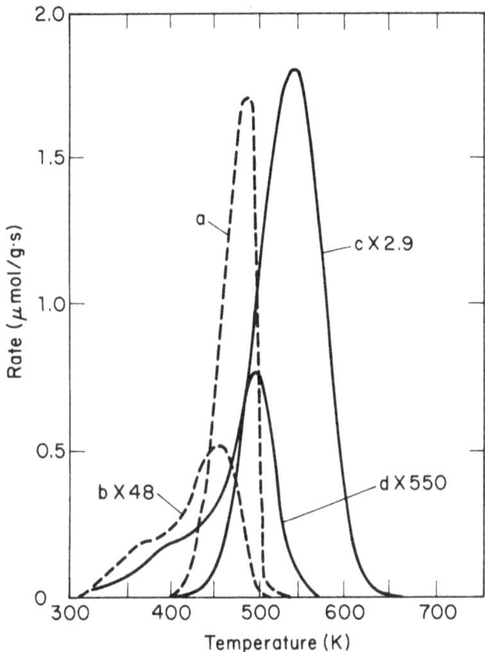

Fig. 3. Hydrocarbon products from TPR of CO on Ni/SiO_2 (a) CH_4 (b) C_2H_6 and on Ni/SiO_2 with 0.9 % potassium (c) CH_4 (d) C_2H_6.

for reaction and, thus, was an excellent measure of active surface area. The methane amount indicated that addition of 0.9% potasium to Ni/SiO_2 decreased the active surface area to 68% of that for the unpromoted Ni/SiO_2.

On Ni/TiO_2, potassium did not change the methane peak halfwidth, but the second, small methane peak was significantly reduced (Fig. 4). The fraction of unreacted CO also decreased, and CO desorption was shifted to higher temperatures. The amount of methane formed during TPR decreased by a factor of three (note scale factors in Fig. 4), apparently due mainly to decreased active surface area of the promoted Ni/TiO_2.

Ni/Al_2O_3 Catalysts: The Ni/Al_2O_3 catalyst behaved somewhat differently upon addition of potassium. On unpromoted Ni/Al_2O_3, two overlapping peaks were seen at 466 and 510 K; the peak at 510 K was smaller in amplitude but broader. Potassium decreased the magnitude of the 466 K

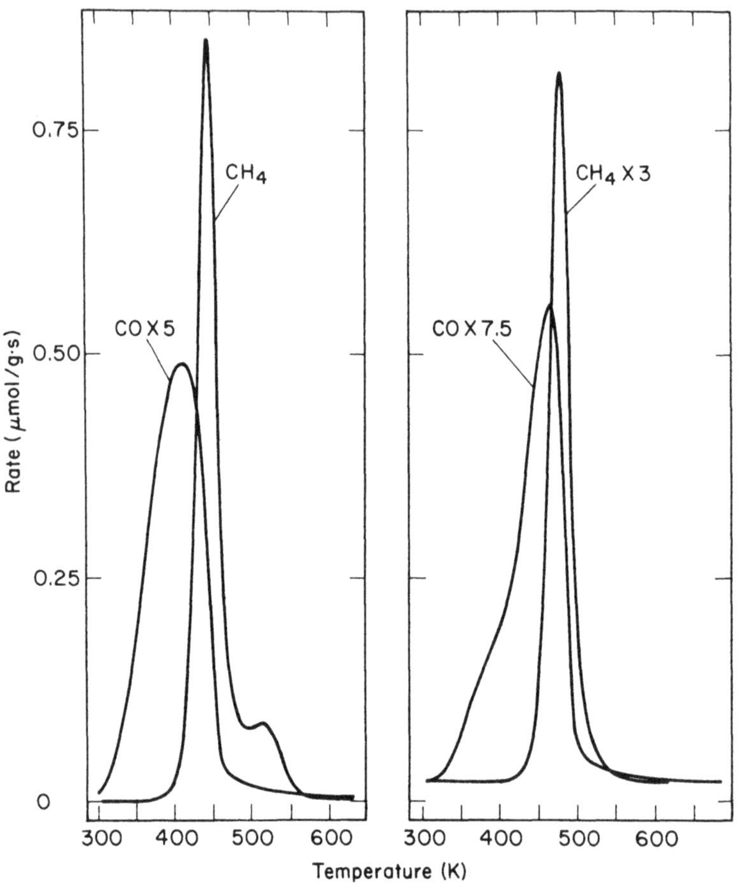

Fig. 4. Methane and unreacted CO from TPR of CO on (a) Ni/TiO_2 and (b) Ni/TiO_2 with 0.8% potassium.

peak by approximately 50%, but it did not change the peak temperature. That is, the specific activity for methanation did not change on Ni/Al_2O_3. The main effect of potassium was to shift the 510 K peak to 540 K, and to reduce its amplitude by 75%.

Ni/SiO_2·Al_2O_3 Catalysts: Addition of sodium to Ni/SiO_2·Al_2O_3 catalysts decreased the CH_4 peak temperature, indicating that specific activities increased. A higher sodium concentration caused a larger decrease in the methane peak temperature, and it also decreased the ethane peak

TABLE III

Kinetic Parameters for TPR Methanation

Catalyst	$E_{1/2}$ (kJ/mol)	$E_{3/4}$ (kJ/mol)	$\nu_{1/2}$*	k(s^{-1}) at 548 K
Ni/ SiO_2				
unpromoted	102	109	1.5×10^9	0.47
0.5% Na	95	83	9.4×10^7	0.083
0.6% K	107	111	1.1×10^9	0.069
0.9% K	64	61	1.8×10^4	0.014
Ni/ TiO_2				
unpromoted	140	147	1.9×10^{15}	87
0.8% K	156	161	2.2×10^{15}	3.0
Ni/ Al_2O_3				
unpromoted	104	121	3.1×10^{10}	3.8
0.8% K	138	151	2.7×10^{14}	19
Ni/ $SiO_2{\cdot}Al_2O_3$				
unpromoted	96	93	1.7×10^9	1.2
0.2% Na	88	71	2.3×10^6	0.94
0.3% Na	96	98	2.5×10^9	1.8
.08% K	99	94	2.1×10^8	0.78

*Preexponential factor calculated from the halfwidth activation energy ($E_{1/2}$)

temperature. Potassium caused a small increase in peak temperature (Table II), but essentially no change in surface area. These small changes in peak temperatures were reproduced on several samples. The fraction of adsorbed CO that reacted to give CH_4 decreased from 88% on the unpromoted catalyst, to 77% for 0.2% sodium to 53% for 0.3% sodium. A second methane peak, approximately one-third the size of the peak at 473 K, formed at 536 K when 0.3% sodium was added.

C. TPR Methanation Activities

During TPR, adsorbed CO is both desorbing and reacting, while the hydrogen surface coverage is changing. This complicated process prevents an accurate determination of kinetic parameters without a detailed model of the mechanism. Previous studies on Ni/SiO_2, however, found that methane formation during TPR is first order in CO coverage.[7] Peak temperatures and curve widths were thus used to estimate activation energies for methanation.[10] The agreement between activation energies from halfwidth and from three-quarter width measurements was excellent (see Table III). Preexponential factors in Table III were calculated from Redhead's[11] equation, using the more accurate halfwidth activation energies.

The kinetic parameters in Table III were used to estimate methanation rate constants at 548 K, the temperature where steady-state measurements are presented in Table I. These rate constants (Table III) were used to compare TPR and steady-state kinetics.

Accurate halfwidths could not be obtained for the unpromoted Ni/Al_2O_3 catalyst because of the presence of the overlapping, high-temperature peak. This peak broadened the main peak at 466 K, and resulted in a lower activation energy. Thus the TPR rate constant for the unpromoted Ni/Al_2O_3 is not meaningful.

D. Comparison of TPR and Steady-state Kinetics

The ratios of TPR rate constants for promoted to unpromoted catalysts on a given support are listed in Table IV. The ratios of steady-state rates, per gram of nickel, are also given in Table IV. Areal rates were calculated by dividing steady-state rates by the methane yields in TPR, and ratios of these areal rates are given in Table IV. Using the methane yields (proportional to the number of sites active for methanation) probably give a better measure of areal rates than using static chemisorption of hydrogen. For Ni/SiO_2 the ratio of rates from TPR is the same as the ratio of areal rates from steady state. The ratios from the two methods differ by less than a factor of two for Ni/TiO_2 and $Ni/SiO_2{\cdot}Al_2O_3$. The overlapping peaks for the unpromoted Ni/Al_2O_3 make its TPR rate constant incorrect. Direct comparison of the TPR spectra for promoted and unpromoted Ni/Al_2O_3, however, show that

TABLE IV

Methanation Ratios of Unpromoted to Promoted Catalysts

	TPR Ratios		Steady-State Ratios	
Catalysts	$k(s^{-1})$	Yield (μmol/g)	Rate (μmol/g)	Areal Rate*
Ni/SiO_2	0.030	0.68	0.023	0.034
Ni/TiO_2	0.034	0.29	0.021	0.072
Ni/Al_2O_3	5.0	0.46	0.29	0.63
$Ni/SiO_2{\cdot}Al_2O_3$	0.65	0.97	1.0	1.0

*The areal rate was estimated by dividing the rate per gram from steady-state by the methane yield in TPR.

the specific rate did not change with addition of promoter; the peak temperatures were the same. Since the ratio of areal rates from steady-state kinetics is 0.6, agreement is also good for Ni/Al_2O_3.

Table IV shows that both TPR and differential, steady-state kinetics yield similar results: the effect of alkali promoters is dependent on the support. For the same promoter concentration, the rate can decrease a factor of 30 (Ni/SiO_2, Ni/TiO_2), or less than a factor of 2 (Ni/Al_2O_3, $Ni/SiO_2{\cdot}Al_2O_3$). The good agreement between the techniques shows that the same reaction process is being studied by both techniques.

A direct comparison of apparent activation energies from the two techniques was not made, since inhibition of the steady-state reaction by gas-phase CO can significantly modify apparent activation energies. Carbon monoxide coverage also varied over a larger range in TPR, from saturation to zero coverage, than in steady-state reaction. The effective H_2:CO ratio in TPR is also much larger. The combination of these effects is responsible for the differences in activation energies.

E. Comparison to Literature

Few studies have compared the effects of promoters for different supports under the same experimental conditions. Schoubye[1] and Kruissink et al.[2] saw decreases in methanation rates on Ni/Al_2O_3. Gonzalez and Miura[4] observed that potassium decreased the turnover number for both methane and higher hydrocarbon formation and increased the selectivity to olefins on Ru/SiO_2. They also observed agreement between TPR and differential reactor studies.[4]

Similarly, the results on $Ni/SiO_2{\cdot}Al_2O_3$ are in agreement with Huang and Richardson's results.[5] They reported that sodium increased the rate of methanation over $Ni/SiO_2{\cdot}Al_2O_3$ at low concentrations, and a maximum in rate was seen. Recent studies for a range of potassium concentrations on $Ni/SiO_2{\cdot}Al_2O_3$ catalysts show a similar maximum in methanation rate.[6]

F. Promoter-support Interaction

Nickel on silica, a relatively inert support, displays behavior similar to unsupported Ni(100): decreased activity and increased olefin selectivity.[12] Thus, the results for Ni/SiO_2 indicate a direct alkali metal-nickel interaction. Changes in the Ni/TiO_2 catalyst with promotion were similar to those observed with Ni/SiO_2, indicating a similar alkali-nickel interaction on titania.

The Ni/Al_2O_3 and $Ni/SiO_2{\cdot}Al_2O_3$ catalysts may be different because the alkali reacts with the support. This has two effects. Less of the potassium is on the nickel and more is on the support than for Ni/SiO_2. Thus, a smaller decrease in activity occurs due to a direct alkali-nickel interaction than is seen on Ni/SiO_2. The support properties are changed by potassium, and a metal-support effect then modifies the nickel properties and increases the activity. This increase compensates for the decrease due to the direct alkali-nickel interaction, so little net change in rate is seen.

Huang and Richardson[5] indicated their increased activity was caused either by solution of sodium into nickel or induced effects from the silica-alumina support. They said the loss of support acidity on $SiO_2{\cdot}Al_2O_3$ due to sodium may have modified the nickel by electron transfer across the metal-support interface. Ino _et al._[13] also interpreted their results as being due to electron transfer between alkali and supported iron, with the electron transfer occurring through the support. They found that supported iron had increased rates for CO hydrogenation at ten atmospheres when potassium was added, but the increase depended on the support.

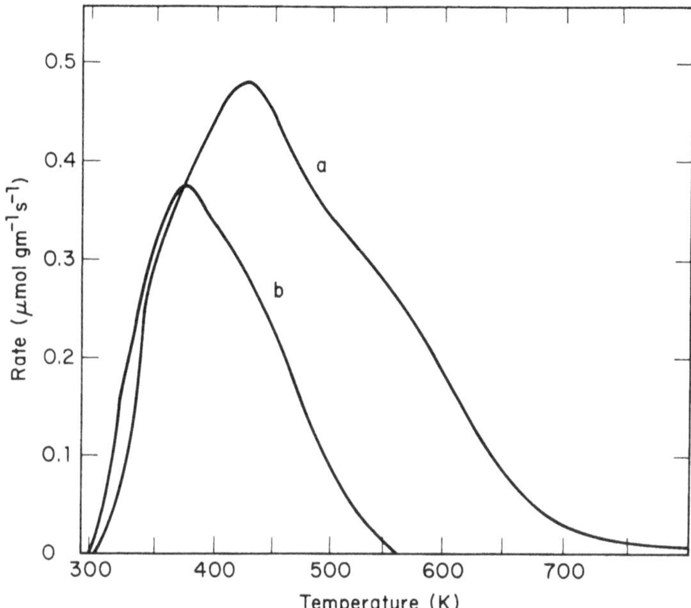

Fig. 5. Hydrogen desorption (TPD) for hydrogen adsorbed at room temperature on (a) Ni/SiO_2 and (b) Ni/SiO_2 with 0.9 % potassium.

G. Reaction Mechanism

It is unclear how alkali metals change the reaction rate, but one of the most notable effects of alkali is on hydrogen desorption in TPD. A 0.9% potassium promoter concentration on Ni/SiO_2 significantly weakened hydrogen bonding so that the hydrogen desorption peak temperature decreased 60 K and desorption was complete by 550 K. Desorption was not complete until 700 K on unpromoted Ni/SiO_2 (Fig. 5). In contrast, on sodium- and potassium-promoted $Ni/SiO_2{\cdot}Al_2O_3$, desorption was not complete even at 773 K. The weaker hydrogen bonding on promoted Ni/SiO_2 results in decreased hydrogenation rates and increased olefin selectivities. Hydrogen bonding strength was not weakened on $Ni/SiO_2{\cdot}Al_2O_3$, so the hydrogenation rate is not decreased.

III. CONCLUSIONS

Changes in areal rates of CO hydrogenation upon addition of alkali promoters to supported nickel were measured by temperature programmed reaction (TPR) and steady-state kinetics. Good agreement was obtained between the two techniques, indicating that the same reaction process was measured. On Ni/SiO_2, addition of 1% potassium decreases methanation activity by approximately a factor of 30 and significantly increases selectivity to olefins. Potassium decreases hydrogenation rates on Ni/SiO_2 by weakening hydrogen bonding. On Ni/Al_2O_3 and $Ni/SiO_2{\cdot}Al_2O_3$ the activities are essentially unchanged by 1% potassium. On $Ni/SiO_2{\cdot}Al_2O_3$, low sodium concentrations increase methanation activity slightly. The dependence of activity changes on the support may be because the alkali distribution between the nickel and the support depends on the support. On Ni/Al_2O_3 and $Ni/SiO_2{\cdot}Al_2O_3$ a strong promoter-support interaction keeps more of the promoter on the support. The reaction of the promoter with the support then modifies the nickel properties through a metal-support interaction. Alkali promoters change activity and selectivity by altering the nickel sites, and in general, promotion also decreases the number of sites available for methanation.

EXPERIMENTAL

Steady-state kinetics were measured in a differential reactor (below 5% conversion) with a 3:1 mixture of H_2:CO. The reactor contained 100 mg of catalyst and was heated by a fluidized sand-bath heater. Products were analyzed with an HP 5793A gas chromatograph that separated C_1-C_4 paraffins and olelfins. Kinetic data were obtained at 4-8 temperatures for each catalyst; between exposure to the H_2:CO mixture, a pure H_2 stream was used to clean the catalyst.

Temperature-programmed reaction (TPR) was carried out in an apparatus similar to one described in the literature.[14] Repeat CO pulses from a pulse valve were used to saturate the catalyst at room temperature in a hydrogen carrier. The catalyst (100 mg) was then heated in hydrogen flow to 773 K at 1 K/s, and the effluent immediately downstream from the reactor was continuously analyzed with a computer-controlled, UTI quadrupole mass spectrometer. The spectrometer was located in a turbomolelcular-pumped, ultrahigh vacuum system. A 0.25-mm thermocouple, whose tip was located in the catalyst sample, provided the temperature measurement and control. Calibrations of adsorption and desorption amounts were carried out by injections with gas-tight syringes.

A. Catalysts

Nickel catalysts were prepared by impregnation with nickel nitrate using procedures described in the literature.[7] Davison silica (Grade 57) and silica-alumina (Grade 970) were used for supports. The alumina was Kaiser A-201, and the titania was Degussa P-25. The supports were soaked in aqueous solutions of NaCl or KCl to add the promoters.[5] The supports were then dried and calcined in air for 2 hr before addition of nickel nitrate. After drying and reduction in H_2 at 773 K, the catalysts were passivated by a low concentration of oxygen in a helium stream. Before differential reactor or TPR measurements, the catalysts were pretreated in H_2 for 2 hr at 773 K.

Metal and alkali loadings were measured by atomic absorption.

ACKNOWLEDGMENTS

We gratefully acknowledge support by the Department of Energy, Office of Basic Energy Sciences. We also thank the Mobil Foundation for support in the initial stages of this work.

REFERENCES

1. P. Schoubye, J. Catal., **14**, 238 (1969).
2. E.C. Kruissink, H.L. Pelt, J.R.H. Ross, and L.L. Van Reyen, Appl. Catal., **1**, 23 (1981).
3. G.B. McVicker and M.A. Vannice, J. Catal., **63**, 25 (1980).
4. R.D. Gonzalez and H. Miura, J. Catal. **77**, 338 (1982).
5. C.P. Huang and J.T Richardson, J. Catal., **51**, 1 (1978).
6. G.Y. Chai and J.L. Falconer, submitted to J. Catal.
7. J.L. Falconer and A.E. Zagli, J. Catal., **62**, 280 (1980).

8. S.Z. Ozdogan, P.D. Gochis, and J.L. Falconer, J. Catal., **83**, 257 (1983).
9. A.E. Zagli and J.L. Falconer, Appl. Catal., **4**, 135 (1982).
10. C.M. Chan, R. Aris, and W.H. Weinberg, Appl. Surf. Sci., **1**, 360 (1978).
11. P.A. Redhead, Vacuum, **12**, 203 (1962).
12. C.T. Campbell and D.W. Goodman, Surf. Sci., **123**, 413 (1982).
13. T. Ino, H. Watanabe, E. Kikuchi, and Y. Morita, Waseda Daigaku, Tokyo, Rikogaku Kenkyusha, Waseda Daigabu Rikogau, Kenkyusho Hokoku, **93**, 49 (1980).
14. J.L. Falconer and J.A. Schwarz, Catal. Rev., **25**, 141 (1983).

9

1,4-Butanediol via Rhodium Carbonyl-Catalyzed Hydroformylation of Allyl Alcohol and Allyl t-Butyl Ether

W.E. Smith,[a] G.R. Chambers,[a] R.C. Lindberg,[b] J.N. Cawse[b]
A.J. Dennis,[c] G.E. Harrison,[d] and D.R. Bryant[e]

[a]General Electric Company
Corporate Research and Development
P.O. Box 8
Schenectady, NY 12301

[b]General Electric Company
Plastics Business Group
Pittsfield, MA 01201

[c]Davy McKee Research and Development
Stockton-on-Tees
Cleveland, U.K.

[d]Davy McKee (London) Ltd.
London, U.K. NW1 2 PG

[e]Union Carbide Corporation
Technical Center
South Charleston, WV 25303

ABSTRACT

The rhodium carbonyl-catalyzed hydroformylation of allyl alcohol affords primarily the relatively unstable intermediates, 2-hydroxytetrahydrofuran and 2-(hydroxymethyl)propionaldehyde, which on hydrogenation are converted to 1,4-butanediol and 2-methyl-1, 3-propanediol. In alcohol solvents the 2-hydroxytetrahydrofuran oxo intermediate can be diverted to the corresponding 2-alkoxytetrahydrofurans, which on hydrolysis-hydrogenation are also converted to the 1,4-diol. In this paper these processes are compared to a novel alternative approach based on the conversion of allyl alcohol to allyl _t_-butyl ether by acid resin-catalyzed reaction with isobutylene, a hydroformylation-hydrogenation sequence to produce the 1,4- and 1,3-ether alcohols, and a catalystic de-etherification step to produce the diols and isobutylene for recycle. The results from bench and pilot plant-scale evaluation of these processes are presented, with principal focus on the reaction chemistry and catalysis aspects.

I. INTRODUCTION

1,4- Butanediol is today a chemical of substantial commercial importance, with a total production capacity worldwide on the order of 500 million lbs. It is the immediate precursor to tetrahydrofuran, and, in a rapidly growing use, is a key intermediate in the production of poly(butylene terephthalate) resin. Presently, virtually all butanediol is derived from acetylene and formaldehyde in the Reppe process.[1] A number of other processes based on alternative feedstocks such as butadiene,[2] maleic anhydride,[3] and propylene[4] have been developed but have not in any significant way displaced the Reppe technology. The propylene-based process, developed at General Electric in the 1970's, involves the cobalt carbonyl-catalyzed hydroformylation of allyl acetate. It thus bears some formal relationship to the principal subject of the present paper, which is a discussion of new routes to butanediol based on the rhodium carbonyl-catalyzed hydroformylation of allyl alcohol and allyl _t_-butyl ether as outlined in Fig. 1.

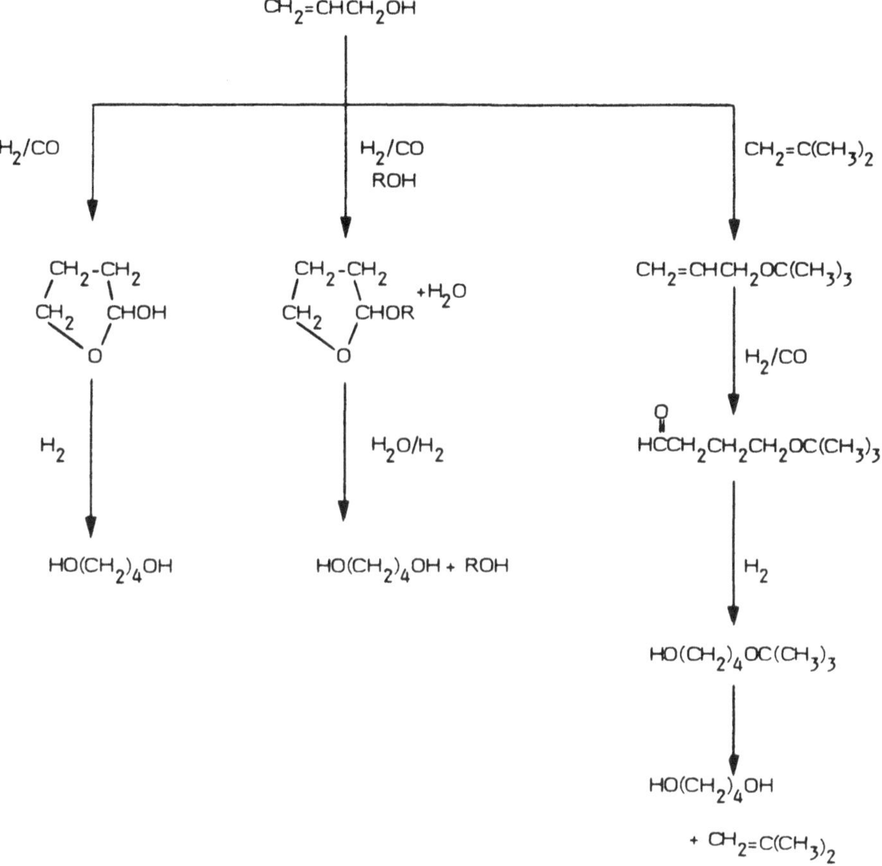

Fig. 1. Butanediol via hydroformylation of allyl alcohol and allyl _t_-butyl ether

II. RHODIUM CARBONYL-CATALYZED HYDROFORMYLATION OF ALLYL ALCOHOL

In the presence of a rhodium carbonyl/triphenylphosphine catalyst, allyl alcohol undergoes hydroformylation under mild conditions to produce the "linear" and branched isomers 2-hydroxytetrahydrofuran and 2-(hydroxymethyl)propionaldehyde, as well as varying proportions of rearrangement, decomposition, condensation, and hydrogenation products (Fig. 2). The system is amenable to some n/iso ratio optimization with temperature, H_2/CO pressure and composition, and catalyst composition

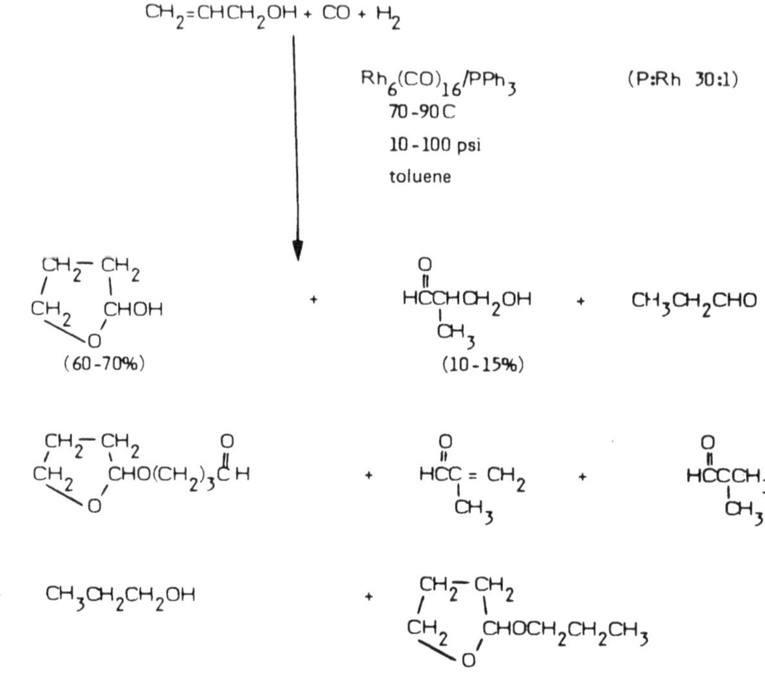

Fig. 2. Hydroformylation of allyl alcohol

(triphenylphosphine to rhodium ratio), factors which also influence catalyst activity and stability.[5]

Methacrolein is a poison for the rhodium catalyst; its formation by cleavage of the branched hydroxyaldehyde results in significant inhibition of the hydroformylation reaction, an effect that can be minimized but not eliminated with allyl alcohol as the substrate. Shimizu and Kurashiki[6] have claimed a rhodium oxo process for converting allyl alcohol to butanediol precursors which involves aqueous extraction for their separation from the catalyst, a step which may remove methacrolein as well.

A. Hydrogenation to Diols

After separation from the catalyst by volatilization under reduced pressure, the hydroformylation products, 2-hydroxytetrahydrofuran and 2-hydroxymethyl)propionaldehyde, are readily hydrogenated to 1,4-butanediol and 2-methyl-1,3-propanediol (Fig. 3).

$$\left[\text{2-hydroxytetrahydrofuran} \rightleftharpoons HO(CH_2)_3CHO \right] + HCOCH(CH_3)CH_2OH$$

$$\downarrow H_2,\ Ni/SiO_2$$

$$HO(CH_2)_4OH + HOCH_2CH(CH_3)CH_2OH + \text{2-}[HO(CH_2)_4O]\text{-tetrahydrofuran}\ (< 1\%)$$

Fig. 3. Hydrogenation of hydroxyaldehydes

III. 1,4-DIOLS VIA 2-ALKOXYTETRAHYDROFURANS

A. Hydroformylation of Allylic Alcohols in Alcohol Solvents[7]

The detection of 2-propoxytetrahydrofuran and other 2-alkoxytetrahydrofuran derivatives as minor by-products in the hydroformylation of allyl alcohol and the subsequent hydrogenation step suggested the possibility that the 2-hydroxy intermediate could be diverted and stabilized as a 2-alkoxy compound by employing an alcohol as the oxo reaction medium. This is portrayed mechanistically in Fig. 4 and is, in fact, possible, although the required process conditions differ substantially from those of the more conventional hydroformylation, beyond the use of an alcohol solvent. Some representative examples are illustrated in Fig. 5.

As indicated in Fig. 5, higher temperatures and pressures are required to effect this transformation. The hydroformylation reaction itself is inhibited by the alcohol medium, and the alkoxylation step is even slower under the milder conditions in absence of its own catalyst. The overall

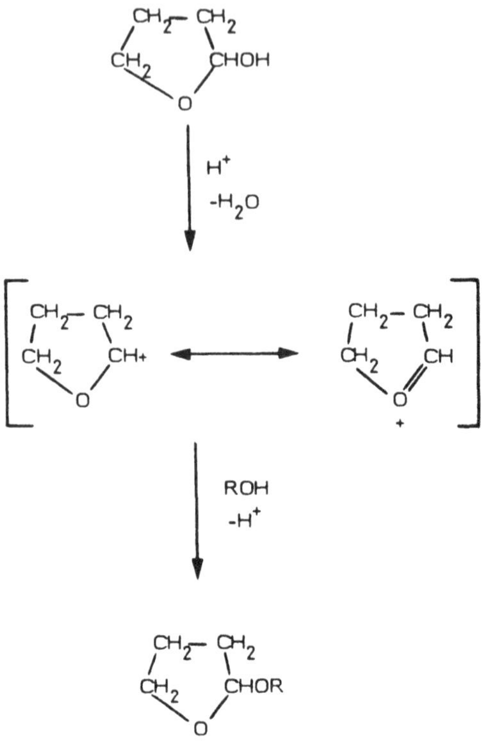

Fig. 4. In situ alcoholysis of 2-hydroxytetrahydrofuran

process can be brought to completion in one hour or less with higher temperature, with substitution of a triphenylarsine-modified hydroformylation catalyst (which apparently imparts some acidic character to the system), or with the presence of a phthalic acid co-catalyst (1 mole % relative to allyl alcohol). The phthalic acid (pK_a 2.89) has its own inhibiting effect on the hydroformylation catalyst, which is overcome at the higher pressures and temperatures. At best, however, the yields of 1,4-butanediol precursors achieved in the oxo-alkoxylation sequence only marginally approach those achieved conventionally when allyl alcohol is the substrate.

B. Hydroformylation of Substituted Allylic Alcohols in Alcohol Solvents[7]

Substituted allylic alcohols present a less demanding n/iso selectivity situation in the hydroformylation reaction. This is reflected in the excellent

Solvent	Catalyst	H_2/CO, psi	T, °C	Product (% Yield)	
CH_3OH	$Rh_6(CO)_{16}/Ph_3P$	300-500	125-150	2-methoxytetrahydrofuran (CH_2–CH_2, CH_2, $CHOCH_3$, O ring)	(45)
"	$Rh_6(CO)_{16}/Ph_3As$	900-1200	130-135	"	(53)
"	$Rh_6(CO)_{16}/Ph_3P$/ o-$C_6H_4(COOH)_2$	800-1200	115	"	(60)
$CH_3CH_2CH_2OH$	$Rh_6(CO)_{16}/Ph_3As$	900-1200	125	CH_2–CH_2, CH_2, $CHOCH_2CH_2CH_3$, O ring	(48)
$(CH_3)_2CHCH_2OH$	$Rh_6(CO)_{16}/Ph_3P$/ o-$C_6H_4(COOH)_2$	800-1200	125	CH_2–CH_2, CH_2, $CHOCH_2CH(CH_3)_2$, O ring	(49)

Fig. 5. Hydroformylation of allyl alcohol in alcohol solvents

yields achieved in the hydroformylation-alkoxylation sequence with this class of substrates (Fig. 6). The conversions of 2-methyl-3-buten-2-ol are particularly interesting. The oxo intermediate, 2-hydroxy-5,5-dimethyltetrahydrofuran, is quite stable and can be readily isolated in high purity. In the presence of the "acidic" triphenylarsine-modified rhodium carbonyl hydroformylation catalyst and absence of an alcohol, it undergoes the unusual self-condensation reaction shown in Fig. 6.

C. Hydrolysis-Hydrogenation of 2-Alkoxytetrahydrofurans[8]

The 2-alkoxytetrahydrofurans readily undergo hydrolysis-hydrogenation to the corresponding 1,4-diols and alkanols. As indicated in Fig. 7, the process is applicable to the formation of 1,4-butanediol, 2-methyl-1,4-butanediol, and

$$CH_2{=}C(CH_3)CH_2OH + CO + H_2 + CH_3OH \xrightarrow[\substack{150C \\ 900\text{-}1200\ psi}]{Rh_6(CO)_{16}/Ph_3P} \underset{(80\%)}{\text{CH}_3\text{CH-CH}_2 / \text{CH}_2\ \text{CHOCH}_3 / \text{O}} + H_2O$$

$$CH_2{=}CHC(CH_3)_2OH + CO + H_2 + CH_3OH \xrightarrow[\substack{150C \\ 800\text{-}1200\ psi}]{Rh_6(CO)_{16}/Ph_3P} \underset{(94\%)}{\text{CH}_2\text{-CH}_2 / (\text{CH}_3)_2\text{C}\ \text{CHOCH}_3 / \text{O}} + H_2O$$

$$CH_2{=}CHC(CH_3)_2OH + CO + H_2 + CH_3CH_2CH_2OH \xrightarrow[\substack{125C \\ 800\text{-}1200\ psi}]{Rh_6(CO)_{16}/Ph_3As} \underset{(90\%)}{\text{CH}_2\text{-CH}_2 / (\text{CH}_3)_2\text{C}\ \text{CHOCH}_2\text{CH}_2\text{CH}_3 / \text{O}} + H_2O$$

$$CH_2{=}CHC(CH_3)_2OH + CO + H_2 \xrightarrow[\substack{125C \\ 800\text{-}1200\ psi}]{Rh_6(CO)_{16}/Ph_3As} \underset{(23\%)}{\text{CH}_2\text{-CH}_2 / (\text{CH}_3)_2\text{C}\ \text{CHOH} / \text{O}}$$

$$+ \underset{(67\%)}{\text{CH}_2\text{-CH}_2\ \ \text{CH}_2\text{-CH}_2 / (\text{CH}_3)_2\text{C}\ \text{CHOCH}\ \text{C(CH}_3)_2 / \text{O}\ \ \text{O}} + H_2O$$

Fig. 6. Hydroformylation of substituted allylic alcohols in alcohol solvents

4-methyl-1,4-pentanediol from the various 2-alkoxytetrahydrofurans derived from allyl alcohol, methallyl alcohol, and 2-methyl-3-buten-2-ol. The optimal catalysts for this conversion are authentically bifunctional, i.e., the hydrolysis step is promoted by an active support such as kieselguhr or alumina.

$$\underset{\text{(ring: } CH_2\text{-}CH_2\text{-}CH_2\text{-}CH(OCH_3)\text{-}O)}{CH_2\text{-}CH_2,\ CH_2\ CHOCH_3,\ O} + H_2O + H_2 \xrightarrow[\text{125C, 800-1000 psi}]{Ni/SiO_2} HO(CH_2)_4OH + CH_3OH \quad (97\%)$$

$$\text{(ring: } CH_2\text{-}CH_2\text{-}CH_2\text{-}CHOCH_2CH_2CH_3\text{-}O) + H_2O + H_2 \xrightarrow[\text{150C, 800-1200 psi}]{Ni/SiO_2} HO(CH_2)_4OH + CH_3CH_2CH_2OH \quad (90\%)$$

$$\text{(ring: } CH_2\text{-}CH_2\text{-}CH_2\text{-}CHOCH_2CH(CH_3)_2\text{-}O) + H_2O + H_2 \xrightarrow[\text{125C, 800-1000 psi}]{Ni/SiO_2} HO(CH_2)_4OH + (CH_3)_2CHCH_2OH \quad (100\%)$$

$$\text{(ring: } CH_3CH\text{-}CH_2\text{-}CHOCH_3\text{-}O\text{-}CH_2) + H_2O + H_2 \xrightarrow[\text{125C, 800-1000 psi}]{Co/Al_2O_3} HOCH_2CH(CH_3)CH_2CH_2OH \ (94\%) + CH_3OH$$

$$\text{(ring: } (CH_3)_2C\text{-}CH_2\text{-}CH_2\text{-}CHOCH_3\text{-}O) + H_2O + H_2 \xrightarrow[\text{125C, 800-1000 psi}]{Co/Al_2O_3} HOC(CH_3)_2CH_2CH_2CH_2OH \ (92\%) + CH_3OH$$

Fig. 7. Hydrolysis-hydrogenation of 2-alkoxytetrahydrofurans

IV. 1,4-BUTANEDIOL VIA ALLYL-t-BUTYL ETHER[9]

In an effort to circumvent the product and catalyst instability problems presented with allyl alcohol as the hydroformylation substrate, the isobutylene cap/uncap strategy shown in Fig. 8 was adapted. It was anticipated that the linear and branched t-butyl ether aldehydes would be considerably more stable in the reaction medium and during isolation from the catalyst. This was, in fact, the case. Successful realization of the

$$CH_2=CHCH_2OH \quad + \quad CH_2=C(CH_3)_2$$

$$\downarrow H^+$$

$$CH_2=CHCH_2OC(CH_3)_3$$

$$\downarrow H_2/CO$$

$$HC(=O)CH_2CH_2CH_2OC(CH_3)_3 \quad + \quad HC(=O)CH(CH_3)CH_2OC(CH_3)_3$$

$$\downarrow H_2$$

$$HO(CH_2)_4OC(CH_3)_3 \quad + \quad HOCH_2CH(CH_3)CH_2OC(CH_3)_3$$

$$\downarrow E^+$$

$$HO(CH_2)_4OH \quad + \quad HOCH_2CH(CH_3)CH_2OH$$

$$+ \; CH_2=C(CH_3)_2$$

Fig. 8. Butanediol via allyl t-butyl ether

strategy, however, requires high efficiency in the isobutylene capping and ether cleavage steps, as well as in the hydroformylation and hydrogenation reactions.

A. Allyl t-Butyl Ether from Allyl Alcohol and Isobutylene

The etherification of allyl alcohol with isobutylene is a simple acid-catalyzed reaction which can be brought about at high rate and conversion with very low levels of by-products. Several acidic catalysts were evaluated, with a sulfonated polystyrene resin found to be most effective. This kind of catalyst is widely employed in the conversion of methanol and isobutylene to methyl t-butyl ether, a gasoline additive.

The etherification reaction is reversible and exothermic ($\Delta H = -17$ Kcal/mole), so operation at lower temperatures results in higher equilibrium conversion but diminished reaction rates (Fig. 9). High reaction rate and

$$CH_2=CHCH_2OH + CH_2=C(CH_3)_2 \xrightleftharpoons{H^+} CH_2=CHCH_2OC(CH_3)_3$$

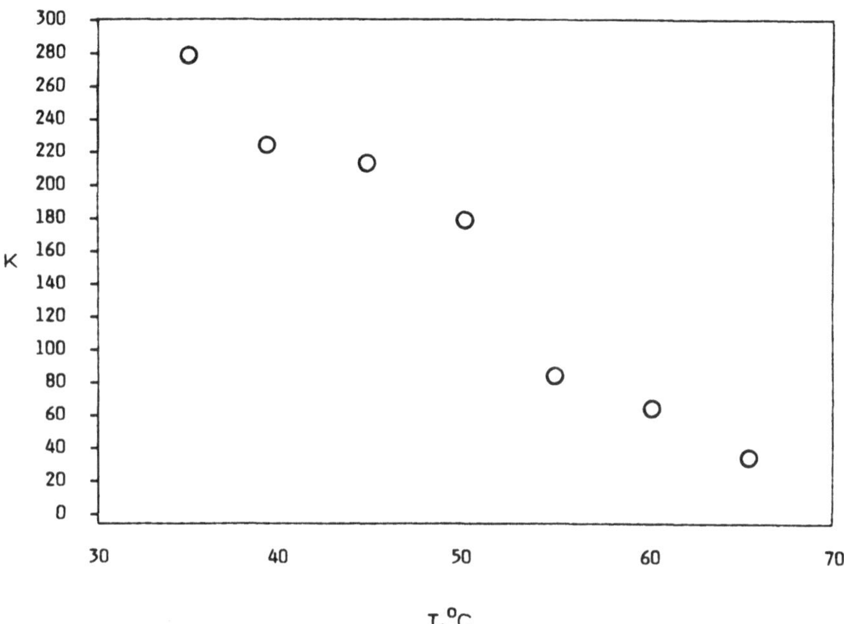

Fig. 9. Etherification of allyl alcohol. Equilibrium constant vs temperature

conversion can be simultaneously achieved by operating sequential column reactors.

The by-products in the etherification process, though small in proportion, must be controlled and preferably removed from the allyl t-butyl ether feedstock for the hydroformylation. Dimers of isobutylene are formed in a reversible process (< 1%) and diallyl ether is formed to a very small extent (< 0.1%). Isomerization of allyl alcohol to propionaldehyde and of allyl t-butyl ether to the propenyl-t-butyl ether are very minor processes (< 0.1%). These compounds as well as allyl alcohol, t-butyl alcohol, and water form low-boiling azeotropes with allyl t-butyl ether and are easily removed.

$$CH_2=CHCH_2OC(CH_3)_3 + H_2 + CO \xrightarrow[\substack{10-20\ \text{psi} \\ 60-70\ \text{C}}]{HRh(CO)_x(PPh_3)_y} \underset{(75\text{-}80\%)}{HC(=O)CH_2CH_2CH_2OC(CH_3)_3} + \underset{(15\text{-}20\%)}{HC(=O)CH(CH_3)CH_2OC(CH_3)_3}$$

$$+ CH_3CH_2CH_2OC(CH_3)_3 + CH_3CH=CHOC(CH_3)_3$$

Fig. 10. Hydroformylation of allyl t-butyl ether

B. Hydroformylation of Allyl t-Butyl Ether

The rhodium carbonyl-catalyzed hydroformylation of allyl t-butyl ether proceeds at a high rate with up to 80% selectivity to the 1,4-butanediol precursor, 4-t-butoxy-n-butyraldehyde (Fig. 10). Other products include the branched oxo isomer 3-t-butoxy-2-methylpropionaldehyde, propyl t-butyl ether by substrate hydrogenation, and propenyl-t-butyl ether (which undergoes hydroformylation only very slowly) by isomerization.

As with allyl alcohol (and olefins in general),[5] the relative yields of the linear and branched aldehydes are affected by the synthesis gas composition and partial pressures, temperature, and catalyst composition (excess triphenylphosphine). Lower partial pressure of carbon monoxide disfavors the branched product but leads to more isomerization. This is consistent with the idea of a common intermediate in the two latter cases.

While traces of methacrolein (presumably formed by cleavage of the branched ether aldehyde) can be detected in the reaction medium and presumably have some catalyst-inhibiting effect, the hydroformylation of allyl t-butyl ether can be operated continuously over hundreds of hours with barely discernible decrease in reactor productivity. This performance is

$$\mathrm{HC(=O)CH_2CH_2CH_2OC(CH_3)_3} + \mathrm{HC(=O)CH(CH_3)CH_2OC(CH_3)_3}$$

$$\xrightarrow[\text{125C, 500 psi}]{H_2,\ Ni/SiO_2 \text{ or } Ni/Al_2O_3}$$

$$\underset{(97\text{-}99\%)}{\mathrm{HO(CH_2)_4OC(CH_3)_3} + \mathrm{HOCH_2CH(CH_3)CH_2OC(CH_3)_3}}$$

$$+ \underset{(<1\%)}{\text{2-(t-butoxy)tetrahydrofuran: } \mathrm{CH_2\text{-}CH_2,\ CH_2,\ CHOROC(CH_3)_3,\ O}} + \underset{(<1\%)}{\text{Aldol condensation products}}$$

Fig. 11. Hydrogenation of t-butyl ether aldehydes

much like that attainable with conventional olefins,[5] but very different from the case in which allyl alcohol is the substrate.

C. Hydrogenation of t-Butyl Ether Aldehydes

The mixed oxo ether aldehydes, 4-t-butoxy-n-butyraldehyde and 3-t-butoxy-2-methylpropionaldehyde, readily undergo hydrogenation over a variety of catalysts to produce the hydroxy ethers, 4-t-butoxy-n-butanol and 3-t-butoxy-2-methyl-n-propanol, in nearly quantitative yield (Fig. 11). Process temperature and the nature of the catalyst support are important considerations in preventing premature cleavage of the t-butyl ethers, which can result in the formation of 2-alkoxytetrahydrofuran compounds (precursors to troublesome impurities at the diol isolation stage) as well as isobutylene loss. With a nickel catalyst supported on either silica or alumina and an operating temperature kept below 150°C, high efficiency in this respect was attained. In principle, the addition of water to the aldehyde ether stream should serve to convert any 2-alkoxytetrahydrofuran

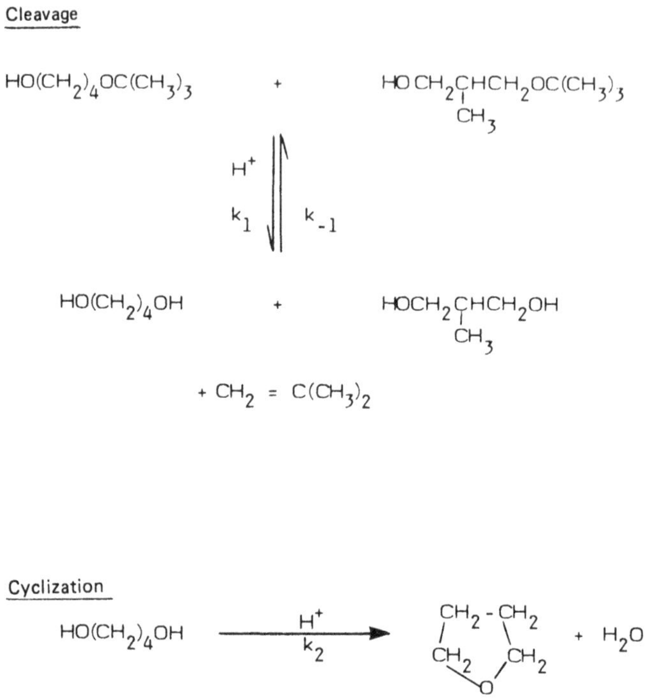

Cleavage

$$HO(CH_2)_4OC(CH_3)_3 + HOCH_2CH(CH_3)CH_2OC(CH_3)_3 \underset{k_{-1}}{\overset{H^+,\ k_1}{\rightleftharpoons}} HO(CH_2)_4OH + HOCH_2CH(CH_3)CH_2OH + CH_2=C(CH_3)_2$$

Cyclization

$$HO(CH_2)_4OH \xrightarrow[k_2]{H^+} \text{tetrahydrofuran } (CH_2-CH_2-CH_2-CH_2-O \text{ ring}) + H_2O$$

Fig. 12. Acid-catalyzed ether cleavage and diol cyclization processes

compounds formed to 1,4-butanediol and the hydroxy ethers.[8] Introduction of water would have deleterious effects, however, on the hydrogenation rate and, unless it is removed beforehand, in the subsequent ether cleavage process.

D. Selective Cleavage of Ether Alcohols

The conversion of the hydrogenation products, 4-t-butoxy-n-butanol and 3-t-butoxy-2-methyl-n-propanol, to the "uncapped" products, 1,4-butanediol and 2-methyl-1,3-propanediol, presents a significant catalyst selectivity challenge in that "conventional" acid catalysts also serve to promote the cyclization of 1,4-butanediol to tetrahydrofuran. The two processes, depicted in Fig. 12, are in fact competitive when the acid resin catalyst employed in the formation of allyl t-butyl ether is used for the ether cleavage reaction. (A direct cyclization of the 4-t-butoxy alcohol with

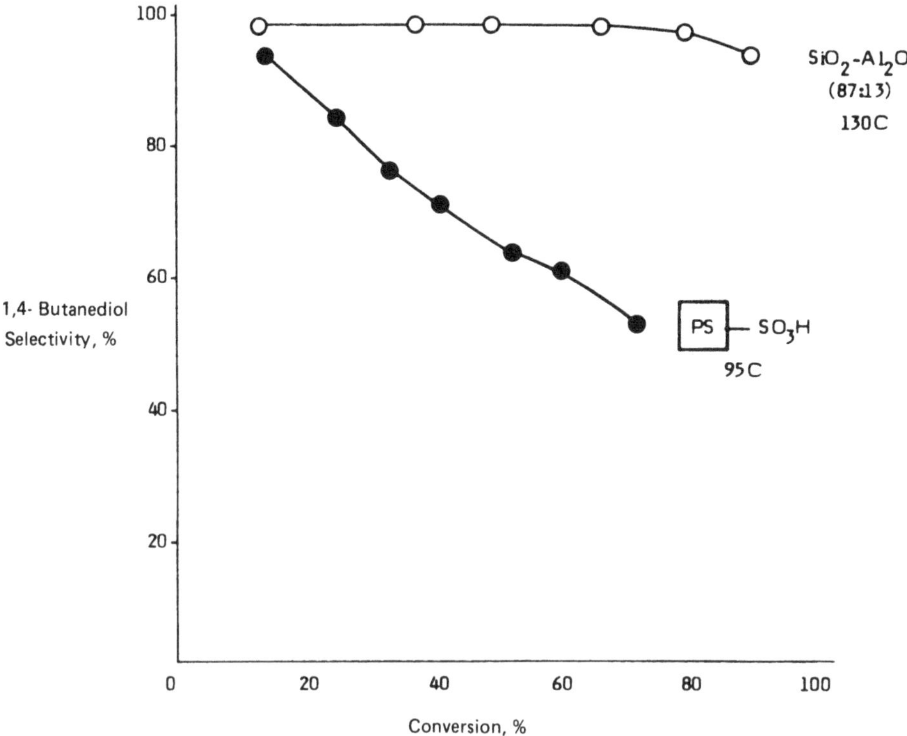

Fig. 13. *t*-Butyl ether cleavage promoted by acid resin and silica-alumina catalysts

elimination of *t*-butanol may also be operative.) The poor selectivity achieved with the acid resin catalyst is illustrated in Fig. 13. Similar results were obtained with a variety of other homogeneous and heterogeneous protonic catalysts, including H_2SO_4, H_3PO_4, and acidified montmorillinite clay (10% H_2SO_4 loading).

In marked contrast, a class of silica-alumina catalysts promotes the ether cleavage with a very high level of selectivity to 1,4-butanediol, despite the higher temperature required to attain activity comparable to that of the sulfonated polystyrene resin catalyst (130° vs 95°C). As illustrated in Fig. 13, an 87:13 SiO_2-Al_2O_3 catalyst brings about isobutylene elimination with very little tetrahydrofuran formation (2-3%) over a broad range of conversions. Above 90% conversion the selectivity declines, as does the reactor productivity because of the reaction equilibrium effects. Similar performance was achieved with a 25:75 SiO_2-Al_2O_3 catalyst composition. Interestingly, a number of single-constituent silica and alumina catalysts were found to have low activity for the cleavage process at temperatures as high as 190°C. A number of zeolites did have activity in the 150-175°C range, with selectivities to 1,4-butanediol on the order of 90%. Overall, an extensive survey of potential catalysts for the process served to point up the remarkable performance of the silica-alumina compositions.

In practice, the silica-alumina catalyzed cleavage process was operated at 130°C and 4 atmospheres pressure (to facilitate liquid-phase recovery of the isobutylene), in a tubular reactor, with a conversion to 1,4-butanediol of about 80%. Under these conditions rates on the order of 2-3 gram moles per liter catalyst per hour were maintained over hundreds of hours, with selectivity to 1,4-butanediol at about 97% and very little evidence of decreasing catalyst activity.

V. CONCLUSIONS

The practical advantages of the cap/uncap approach to converting allyl alcohol to 1,4-butanediol via hydroformylation have been fully substantiated. The hydroformylation process itself can be brought about with significantly greater efficiency when allyl t-butyl ether is the substrate, compared to the allyl alcohol case where the catalyst stability and kind of selectivity associated with the rhodium carbonyl-catalyzed hydroformylation of terminal olefins is not readily attainable. The t-butyl ether formation and cleavage steps can be carried out with minimal losses. Overall 1,4-butanediol yields of about 75% have been achieved in an integrated, sustained and fully workable pilot plant operation.

REFERENCES

1. W. Reppe et al., Ann., **596**, 1 (1955).
2. T. Onoda et al., U.S. Patent 3,922,300 (1975).
3. U.S. Patent 2,772,291 (1956)(DuPont).
4. W.E. Smith, J.L. Webb, J.E. Corn, N. Kirk, D.F. Aycock, and R.J. Gerhart, 174th American Chemical Society Meeting, Industrial and Engineering Chemistry Division, August, 1977.

5. R.L. Pruett, Advances in Organometallic Chemistry, **17**, 1-6 (1979); U.S. Patent 3,527,809.
6. T. Shimizu and O. Kurashiki, German Patent 2,538,364.
7. W.E. Smith, U.S. Patent 4,123,444.
8. W.E. Smith, U.S. Patent 4,091,041.
9. N. Harris, A.J. Dennis, and G.E. Harrison, U.S. Patents 4,287,127, 4,383,125 and 4,414,420.
10. G.R. Chambers and W.E. Smith, U.S. Patent 4,254,290.

PART III

HETEROGENEOUS CATALYSIS

10

Single Crystals as Model Catalysts

D.W. Goodman
Surface Science Division
Sandia National Laboratories
Albuquerque, NM 87185

ABSTRACT

An important new application of surface analytical methods is the study of catalytic reactions using well defined materials under conditions approaching those typically found in process environments. Recent studies combining kinetic and surface analytical studies have provided new information regarding the critical steps of certain reactions as well as the mechanisms of catalyst poisoning and promotion. These studies have demonstrated the appropriateness of metal single crystals as realistic models for a variety of catalytic reactions.

I. INTRODUCTION

Catalytic processing by metals is of considerable importance in the chemical and petroleum industries yet few surface reactions are well understood at the molecular level. The past twenty years have produced an array of surface techniques which can be brought to bear routinely on this most intriguing and technologically important area. The surface science of catalysis is relatively new, yet it promises to be an extremely valuable complement to the more traditional approaches to fundamental catalytic research. During the two decades which produced the currently available surface techniques, a large inventory of information has accumulated regarding surface atomic and electronic structure as well as molecular adsorption and reaction. A logical extension of these kinds of studies has been into the area of kinetics at elevated pressures (~1 atm.) combined intimately with surface analytical techniques. Although the integration of kinetics under these conditions with vacuum surface techniques has been limited to a few groups,[1-4] the results, thus far, have been extremely useful in detailing certain reaction mechanisms. These studies have established this hybrid approach as an important interface between the traditional surface physics and catalytic communities. This discussion will summarize the highlights of some of the studies, primarily those of the author and coworkers.

II. RESULTS AND DISCUSSION

An apparatus, described in Section IV, combining a microcatalytic reactor and surface analysis chamber, has been used to study reactions (Eq. 1-8) over single crystals of nickel, ruthenium, and rhodium.

Methanation

$$3H_2 + CO \longrightarrow CH_4 + H_2O \quad (1)$$

$$4H_2 + CO_2 \longrightarrow CH_4 + 2H_2O \quad (2)$$

Hydrogenation

$$\text{ethylene} \longrightarrow CH_2CH_3 \quad (3)$$

$$\text{cyclopropane} \longrightarrow CH_3CH_2CH_3 \quad (4)$$

Oxidation

$$CO + 1/2\ O_2 \longrightarrow CO_2 \quad (5)$$

Hydrogenolysis

$$\text{ethane} \longrightarrow 2CH_4 \quad (6)$$

$$\text{n-butane} \longrightarrow CH_4 + CH_3CH_3 + CH_3CH_2CH_3 \quad (7)$$

$$\text{cyclopropane} \longrightarrow CH_4 + CH_3CH_3 \quad (8)$$

These are examples of the two major categories of heterogeneously catalyzed reactions: structure insensitive (facile) (Eq. 1-5) and structure sensitive (demanding) reactions (Eq. 6-8). The results of these studies are summarized in Sections A, B, and C.

A. Kinetics of Structure Insensitive Reactions Over Clean Metal Surfaces

A critical question in model studies concerns the appropriateness of the model to reflect the parameters of the more realistic system. In this case the point of interest is the ability of metal single crystals to reproduce the catalytic chemistry generally found for supported catalysts. Sufficient data have been collected for a number of reactions to show that single crystal catalysts reproduce in many cases the chemistry of the corresponding supported catalyst.[4-7] For those reactions listed above which are considered to be structure insensitive (CO and CO_2 methanation, hydrogenation, and CO oxidation) the rates and activation parameters measured for the single crystal catalysts are in excellent agreement with values for supported systems. These results support the use of these idealized catalysts as relevant models and suggest that surface characterization following reaction will, in many respects, correspond to the metal particles of supported catalysts. Surface analysis of active single crystal catalysts have been particularly useful in detailing the important surface species in the methanation reaction. Auger spectroscopy has been used to characterize two distinct forms of surface carbon on nickel[8] and ruthenium,[9] a carbidic or active form and a graphitic or inactive type. The kinetics associated with the reactivity of these carbon types with hydrogen have been studied[8,9] in detail and the relationship of the surface concentration of these carbon types with catalytic activity has been documented.[8-10]

B. Kinetics of Structure Sensitive Reactions Over Clean Single Crystal Surfaces

Reactions known to be responsive to changes in catalyst morphology are also amenable to study using model single crystals. For example, alkane hydrogenolysis, a well known example of this category of reactions, has been studied over nickel single crystals.[6,11,12] Two important observations have been made regarding the clean surface measurements. First, ethane, butane, and cyclopropane hydrogenolysis have been found to be structure sensitive in that these reactions proceed at a significantly faster rate over the Ni(100) surface than over the Ni(111) surface. The results for butane

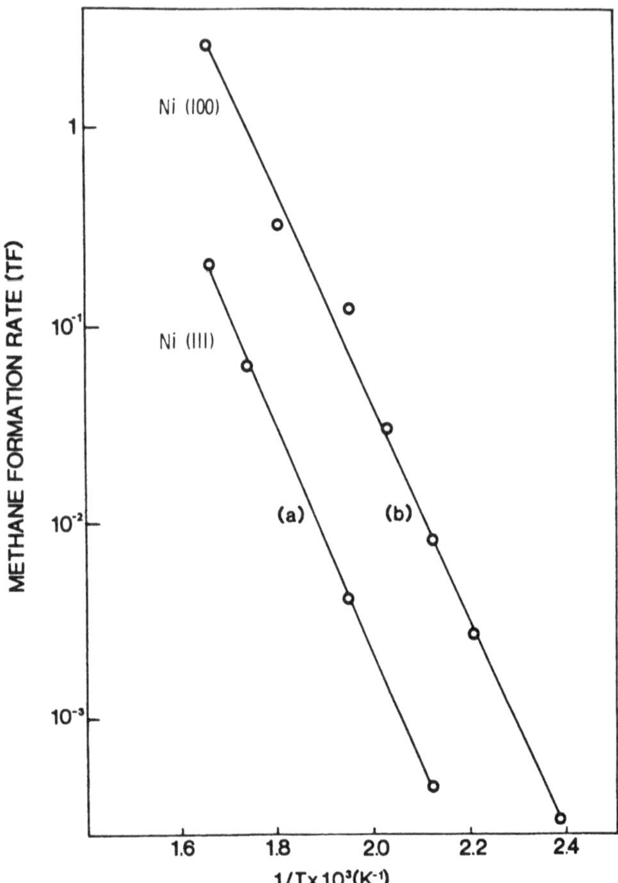

Fig. 1. A comparison of n-butane hydrogenolysis over single crystals of nickel. Pressure = 100 Torr. H_2/n-butane ratio = 100.

hydrogenolysis are shown in Fig. 1. These data suggest that the structure sensitivity in alkane hydrogenolysis is due mainly to an intrinsic activity difference between facets. Particle size effects result then from facet redistribution during sintering and not from intrinsic electronic differences among metal particles.

A second observation is that methane production occurs via the same mechanism over a Ni(100) catalyst regardless of the carbon source (CO, ethane, butane, cyclopropane).[6] This mechanism is one involving a reactive surface carbon or a carbidic type carbon.[4,8,9]

C. Kinetics Over Chemically Modified Single Crystal Surfaces

The modification of catalytic behavior by surface impurities is of crucial importance to most catalytic processing. At present, the mechanisms responsible for surface chemical changes induced by surface additives are poorly understood. However, the current interest and activity in this area of research promises an emerging understanding of the fundamentals by which impurities alter surface chemistry. A pivotal question concerns the relative importance of steric (local) versus electronic (extended or "long-range") effects. A general answer to this question will critically influence the degree to which we will ultimately be able to tailor-make efficient catalysts by fine-tuning material electronic structure. If, indeed, low surface impurity concentrations can profoundly alter the surface electronic structure and thus catalytic activity, then the possibilities for the systematic manipulaltion of these properties via additive selection would appear limitless. On the other hand, if steric effects dominate the mode by which surface additives alter the catalytic chemistry then a different set of considerations for catalyst alteration come into play, a set which will most certainly be more constraining than the former. In the final analysis, an understanding will include a component of an "electronic" and a "steric" effect, the relative importance to be assessed for a given reaction and conditions. A major emphasis of our research has been in the area of addressing and partitioning the importance of these two effects in the influence of surface additives in catalysis.

Catalyst deactivation and promotion are extremely difficult questions to address experimentally.[13] For example, the interpretation of related data on dispersed catalysts is severely limited by the uncertainity concerning the structural characterization of the active surface. Specific surface areas cannot always be determined with adequate precision. In addition, a knowledge of the crystal orientation, the concentration and the distribution of impurity atoms, as well as their electronic states is generally poor. The degree of contamination may vary considerably along the catalytic bed and the impurity may alter the support as well as the metal. Moreover, the active surface may be altered in an uncontrolled manner as a result of sintering or faceting during the reaction itself.

The use of metal single crystals in catalytic reaction studies essentially eliminates the difficulties mentioned above and allows, to a large extent, the utilization of a homogeneous surface amenable to study using modern surface characterization regarding surface structure and composition. Carefully controlled, single-crystal catalytic surfaces are particularly suited to the study of impurity effects on catalytic behavior because of the ease with which impurity atoms can be uniformly introduced to the surface. In many cases, the impurity atom position is known. To date relatively few studies have incorporated both surface science techniques with kinetics at elevated pressures ($\sim$1 atm). Kinetics, however, are an essential link between these kind of model catalytic studies and the more relevant catalytic systems, establishing the crucial connection between the reaction rate parameters. Although the studies to date are few, the results appear promising in addressing the fundamental aspects of catalytic poisoning and promotion.

Impurities whose electronegativities are greater than those for transition metals generally poison a variety of catalytic reactions, particularly those involving H_2 and CO. Of these poisons sulfur is the best known and technologically the most important. The first step in the systematic definition of the poisoning mechanism of this category of impurities is the study of the influence of these impurities on the adsorption and desorption of the reactants.

The effect of preadsorbed electronegative atoms Cl, S, and P on the adsorption-desorption of CO and H_2 on Ni(100) has been extensively studied[14] using temperature programmed desorption (TPD), low energy electron diffraction (LEED), and Auger electron spectroscopy (AES). It has been found that the presence of electronegative atoms causes a reduction of the adsorption rate, the bond strength, and the adsorption capacity of the Ni(100) surface for CO and H_2 adsorption. The poisoning effect becomes more prominent with increasing electronegativity of the preadsorbed atoms.

Adsorption of Cl, S, and P on nickel causes a reduction in both hydrogen and CO adsorption and a shift of the TPD peak maxima to a lower temperature.[14] The effect of preadsorbed Cl, S, and P on the CO TPD behavior is to decrease markedly the amount of CO adsorbed. The effect of

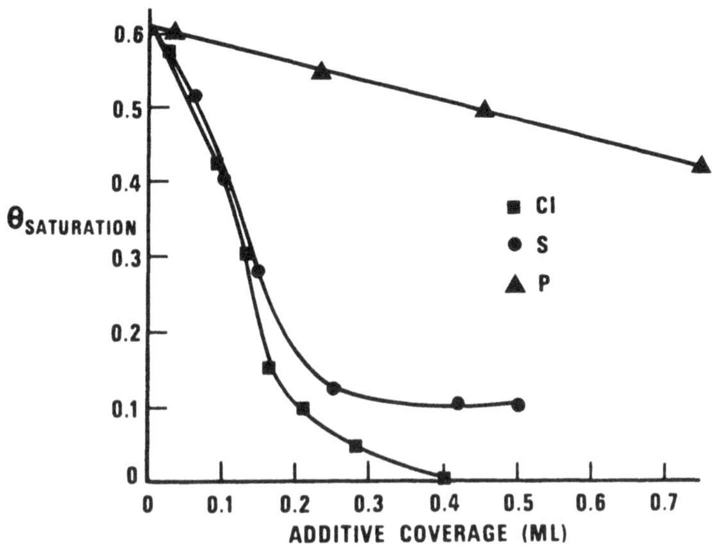

Fig. 2. Dependence of total CO adsorption on additive precoverage.

P, however, is much less pronounced than that observed for Cl or S. Fig. 2 shows the observed dependence of the total CO adsorption on impurity coverage.

Similarly, Cl, S, and P cause a reduction of hydrogen adsorption and a shift of the TPD peak maxima to a lower temperature.[14] At higher impurity levels a lower temperature state emerges. The extent of this effect increases in the sequence P, S, Cl. As seen in Fig. 3, the reduction of H_2 coverage is most apparent in the presence of Cl atoms. The similarity in the atomic radii of Cl, S, and P (0.99, 1.04, and 1.10A, respectively[15]), suggests a relationship between electronegativity and the poisoning of chemisorptive properties by these surface impurities.

Related studies have been carried out in the presence of C and N.[16] These impurities have the same electronegativities as S and Cl, 2.5 and 3.0, respectively. The comparison between the results for C and N and those for S and Cl are consistent with the interpretation that electronegativity effects dominate poisoning of chemisorption by surface impurities with close atomic size, occupying the same adsorption sites. In the case of adsorbed impurities with the same electronegativity but with different

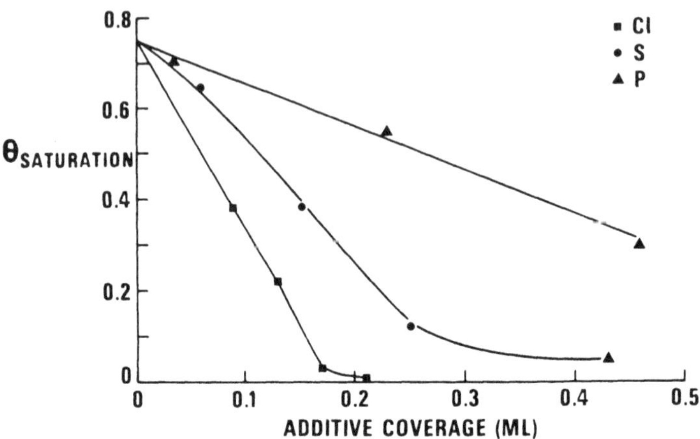

Fig. 3. Dependence of H_2 adsorption on additive precoverage.

atomic radii (S and C, Cl and N), the effect becomes less pronounced with decreasing atomic radius.

Particularly notewothty in the above studies is the general observation that those impurities strongly electronegative with respect to nickel, Cl, N, and S, modify the chemisorptive behavior far more strongly than would result from a simple site blocking model. The initial effects of these impurities shown in Figs. 2 and 3 suggest than a single impurity atom successfully poisons as many as ten nickel atoms, a result supporting an interaction that is electronic in nature.

Kinetic studies have been carried out for several reactions as a function of sulfur coverage over single crystals of nickel,[7,17-19] rhodium,[20,21] and ruthenium.[22] A result common to these studies is that sulfur effectively poisons catalytic activity at coverages less than 10% of the surface metal concentration (or 0.1 monolayers, ML). Fig. 4a shows the relationship between the sulfur coverage and the methanation rate at 600 K. A sharp drop in catalytic activity is observed for low sulfur coverages. The poisoning effect quickly maximizes with little reduction in reaction rate at sulfur levels exceeding 0.2 monolayers. The activity attenuation at the higher sulfur coverages is in excellent agreement with that found for supported Ni/Al_2O_3 by Rostrup-Nielsen and Pedersen.[23] The initial change in the poisoning depicted in Fig. 4a suggests that more than ten nickel atom

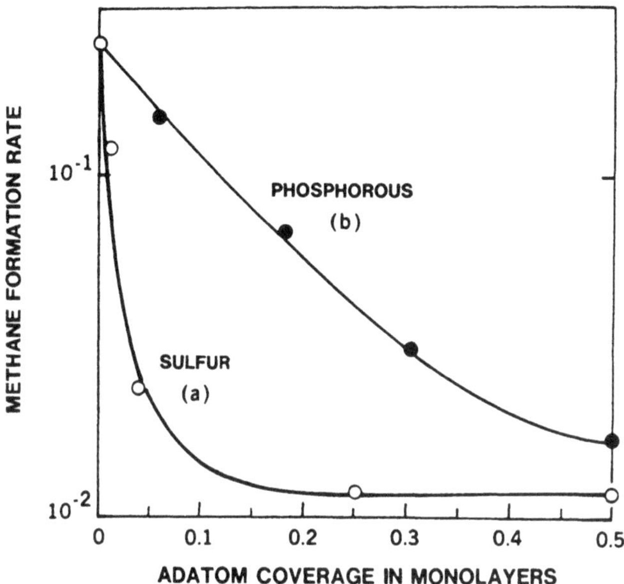

Fig. 4. Methanation rate as a function of phosphorous and sulfur coverage on a Ni(100) catalyst. Pressure = 120 Torr, H_2/CO = 4, reaction temperature = 600 K.

sites are deactivated by one sulfur atom. There are two possible explanations for this result: (1) a "long-range" electronic effect or ligand effect as discussed in the Introduction or (2) an ensemble effect, the requirement that a certain collection of surface atoms are necessary for the reaction to occur. Experimentally these two possibilities can be distinguished. If an ensemble of more than ten nickel atoms is required for methanation then altering the electronic character of the impurity should produce little change in the degree to which the impurity poisons the catalytic activity. That is, the impurity serves merely to block a single site in the reaction ensemble, nothing more. On the other hand, if electronic effects are playing a significant role in the poisoning mechanism, then the reaction rate should respond to a change in the electronic character of the impurity. Substituting phosphorus for sulfur in a similar set of experiments results in a marked change in the magnitude of poisoning at low coverages

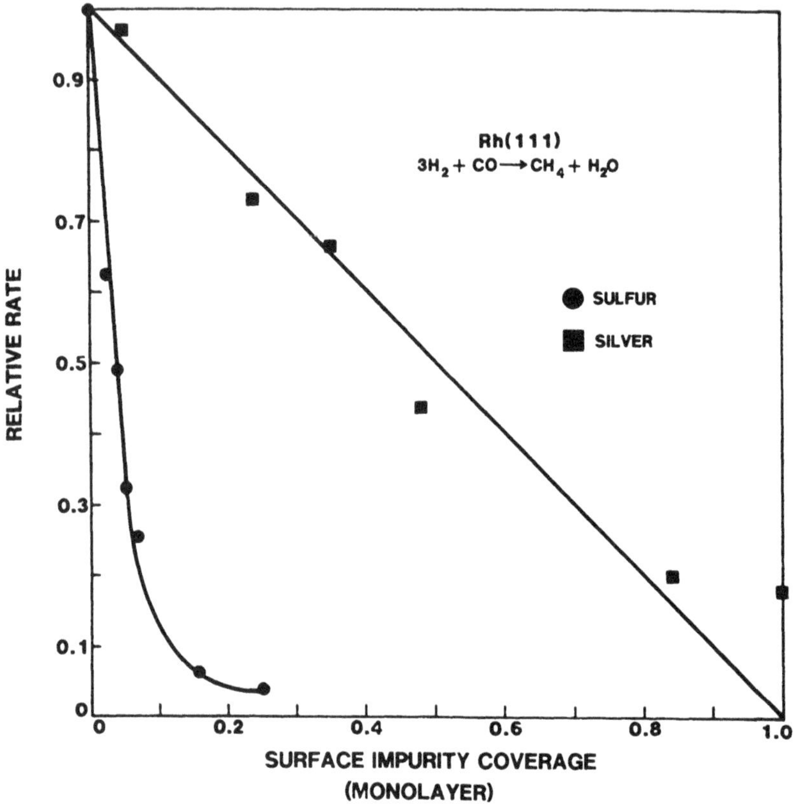

Fig. 5. Methanation rate as a function of silver and sulfur coverage on a Rh(111) catalyst. Pressure = 120 Torr, $H_2/CO = 4$, reaction temperature = 600 K.

as shown in Fig. 4b. Phosphorous, because of its less electronegative character, effectively poisons only the four nearest neighbor metal atom sites.

Recent results for methanation over rhodium are completely consistent with the conclusions drawn from the nickel data.[21] The results for rhodium are summarized in Fig. 5. Poisoning by sulfur for the methanation reaction is remarkably similar to the results for nickel, one sulfur deactivates more than ten rhodium surface atoms. In contrast, addition of silver to rhodium poisons only one surface site per silver atom.[21]

Silver, with an electronegativity similar to rhodium, is believed to be an inactive diluent of rhodium. The results for silver on rhodium suggest an ensemble size for the critical step in methanation to be one metal surface atom site. An ensemble requirement of this size is consistent with methanation being a structure-insensitive or facile reaction.

Effective poisoning of catalytic activity at sulfur coverages less than 0.1 ML has been observed for other reactions including ethane and cyclopropane hydrogenolysis,[12] ethylene hydrogenation,[21] and CO_2 methanation.[5] These studies indicate that the sensitivity of the above reactions to sulfur poisoning are generally less than CO methanation. The rate attenuation is, nevertheless, strongly nonlinear at the lower sulfur levels. A direct consequence of the relative molecular sizes of the reactants involved in the reactions investigated is that electronic effects rather than ensemble requirements dominate the catalytic poisoning mechanism for these experimental conditions.

Little theoretical work has been undertaken to address directly the predicted magnitude of the near surface electronic perturbations by impurity atoms. Work by Grimley and others[24] has suggested that atom-atom interactions through several lattice spacings do occur, although these studies did not specifically address the effects of sulfur. Calculations expressly addressing the surface electronic perturbation by sulfur have recently been reported.[25] These results indicate a correlation between the electronegativity of the impurity and its relative perturbation of the density of states near the Fermi level.

Recent studies using high resolution electron energy loss and photoelectron spectroscopy to investigate the effect of sulfur on the CO/Ni(100) system are consistent with an extended effect by the impurity on the adsorption and bonding of CO.[26] Sulfur levels of a few percent of the surface nickel atom concentration were found sufficient to significantly alter the surface electronic structure as well as the CO bond strength.

A direct consequence of interpreting the poisoning effects of electronegative impurities in terms of electronic surface modification is that additives with electronegativities less than that of the metal should promote a different chemistry reflecting the donor nature of the additive.

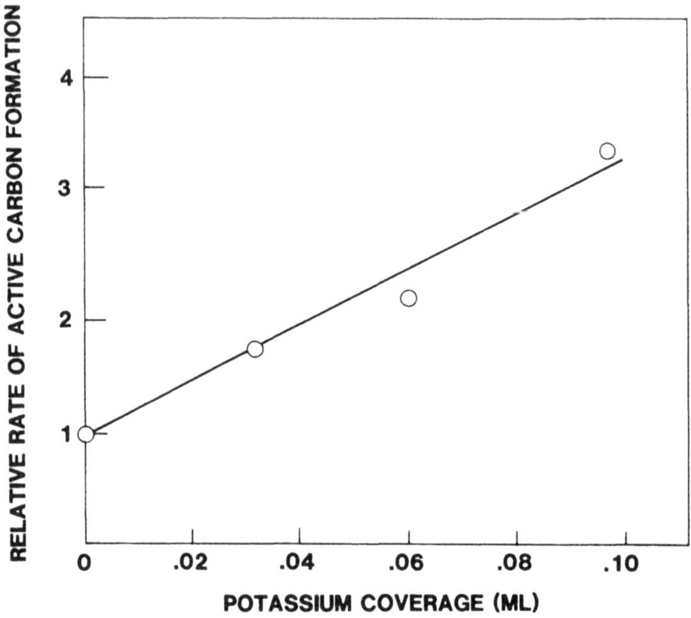

Fig. 6. The relative initial rate of reactive carbon formation from CO disproportionation as a function of potassium coverage. P_{CO} = 24 Torr, Temperature = 500 K.

For example, alkali atoms on a transition metal surface are known to exist in a partially ionic state, donating a large fraction of their valence electron to the metal, resulting in a work function decrease. This additional electron density on the transition metal surface atoms is thought to be a major factor in alkali atoms altering the chemisorptive bonding of molecules such as N_2[17] or CO,[28] and in promoting catalytic activity in ammonia synthesis.[29] These results are consistent with the general picture that electron receptors tend to inhibit CO hydrogenation reactions whereas electron donors typically produce desirable catalytic effects, including increased activity and selectivity. Recent surface and kinetic studies[30] support the relationship between the electron donor-acceptor properties of the impurity and its effect on the catalytic behavior.[30]

Adsorbed potassium causes a marked increase in the rate of CO dissociation on a Ni(100) catalyst.[30] The increase of the initial formation rate of "active" carbon or carbidic carbon from CO is illustrated in Fig. 6. The relative rates of CO dissociation were determined for the clean and potassium-covered surfaces by observing the growth in the carbon Auger signal with time in a CO reaction mixture, starting from a carbon-free surface. Of particular significance is the reduction of the activation energy of reactive carbon formation from $\sim$23 kcal mole^{-1} for the clean Ni(100) surface to $\sim$10 kcal mole^{-1} for a 10% potassium covered surface.

Kinetic measurements of methanation over a Ni(100) catalyst containing well-controlled submonolayer quantities of potassium show a decrease in the steady-state rate under a variety of reaction conditions.[30] The presence of potassium did not alter the apparent activation energy associated with the kinetics; however, the potassium did change the steady-state coverage of active carbon on the catalyst. This carbon level changed from 10% of a monolayer on the clean catalyst to 30% on the potassium covered catalyst.

Adsorbed potassium causes a marked increase in the steady-state rate and selectivity of nickel for higher hydrocarbon synthesis.[30] At all temperatures studied, the overall rate of higher hydrocarbon production was faster on the potassium-dosed surface showing that potassium is a promoter with respect to Fischer-Tropsch synthesis. This increase in higher hydrocarbon production is attributed to the increase in the steady-state active carbon level during the reaction, a factor leading to increased carbon polymerization.

Potassium impurities on a nickel catalyst, then, cause a significant increase in the CO dissociation rate and a decrease in the activation energy at low carbon coverages. These effects can be explained in terms of an electronic effect, whereby the electro-positive potassium donates extra electron density to the nickel surface atoms, which in turn donate electron density to the adsorbed CO molecule. This increases the extent of π-backbonding in the metal CO complex, resulting in an increased metal-CO bond strength and a decrease in C-O bond strength. This model satisfactorily explains the decrease in the activation energy for carbide build-up brought about by potassium.

Since a local effect by the potassium is sufficient to account for the kinetics, the spatial extent of the effect of potassium in these experiments cannot be assessed. However, recent high resolution electron energy loss spectroscopy (HREELS) of CO on potassium doped platinum[31] and ruthenium[32] indicate an alkali influence significantly larger than a simple potassium radius. These results taken together suggest that extended electronic perturbations are effectively altering the surface chemistry via a similar mechanism as that invoked for the poisoning results discussed above.

III. CONCLUSIONS

The examples of the model studies presented show how the meshing of modern surface techniques with reaction kinetics can provide detailed insight into the mechanisms of surface reactions and serve as a valuable complement to the more traditional techniques. Close correlations between these two areas holds great promise for a better understanding of many subtleties of heterogeneous catalysis.

Model studies using metal single crystals are also profitable in developing an understanding of the mechanisms by which poisons and promoters alter catalytic performance. Kinetic measurements in conjunction with these studies are particularly useful in linking the surface analytical measurements to practical catalysts. Because of the importance of surface chemical modification in catalysis and many related technological areas, much more work should be invested in defining in detail the physics and chemistry associated with the changes induced by surface impurities. Of particular interest are the specific bonding sites on and the electronic interaction of the impurity with the substrate. Also the influence of the impurity on the chemisorptive behavior of the reactants as well as the bond strengths of the reactants are key pieces of information needed. These kinds of data are currently assessable using as array of modern surface techniques. These studies in parallel with studies on supported catalysts promise to be most revealing regarding the basic mechanisms of surface chemical modification.

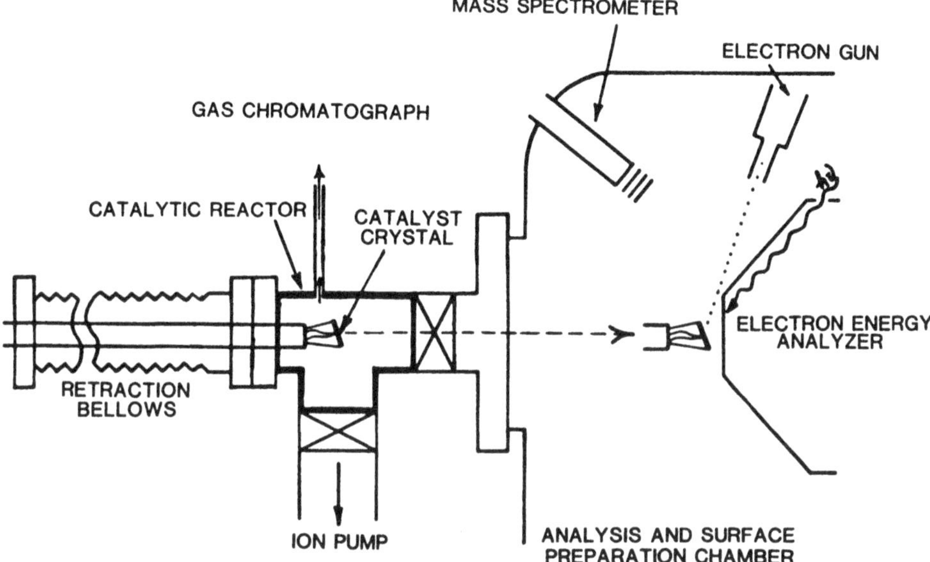

Fig. 7. An ultrahigh vacuum apparatus for studying single crystal catalysts before and after operation at atmospheric pressure in a catalytic reactor.

IV. EXPERIMENTAL

The studies discussed were carried out utilizing the specialized apparatus shown in Fig. 7 and discussed in detail in Ref. 4. This device consists of two distinct regions, a surface analysis chamber and a microcatalytic reactor. The custom-built reactor, contiguous to the surface analysis chamber, employs a retraction bellows that supports the metal crystal and allows translation of the catalyst in vacuo from the reactor to the surface analysis region. Both regions are of ultrahigh vacuum construction, bakeable, and capable of ultimate pressures of less than 10^{-10} Torr. The catalyst samples are mounted on tungsten leads and heated resistively. The reactor is operated in a batch mode with sampling subsequent to reaction into an on-line gas chromatograph. Analysis is via a flame ionization detector. The single crystals of nickel ($\sim$1.0 cm^2 surface area) were cleaned by oxidation at 1400 K in 10^{-6} Torr O_2 followed by reduction at 800 K in 5 Torr H_2. The details of this procedure together with supporting surface analysis are given in Ref. 4. The rhodium samples were cleaned by repeated ion sputtering with the crystal held at 600 K followed by an *in vacuo* flash to 1200 K to eliminate surface carbon residuals. All reactants were initially of high purity. Further purification procedures are generally used to improve the gas quality. These typically include multiple distillations for condensables and/or cryogenic scrubbing using a low conductance glass wool-packed trap at 80 K.

ACKNOWLEDGEMENT

We acknowledge with pleasure the partial support of this work by the Department of Energy through the Division of Basic Energy Sciences under contract #DE-AC04-76DP00789.

REFERENCES

1. G.A. Somorjai, **Chemistry in Two Dimensions**, Cornell University Press, 1981.
2. H.J. Krebs, H.P. Bonzel, and G. Gafner, *Surf. Sci.*, **88**, 269 (1979).
3. C.T. Campbell and M. Paffett, *Surf. Sci.*, **139**, 396 (1984).
4. D.W. Goodman, R.D. Kelley, T.E. Madey, and J.T. Yates, Jr., *J. Catal.*, **63**, 226 (1980).
5. D.E. Peebles, D.W. Goodman, and J.M. White, *J. Phys. Chem.*, **87**, 4378 (1983).
6. D.W. Goodman, Proceedings of the Eighth International Conference on Catalysis, Berlin, 1984.
7. D.W. Goodman, *Acc. Chem. Res.*, **17**, 194 (1984).
8. D.W. Goodman, R.D. Kelley, T.E. Madey, J.T. Yates, Jr., and J.M. White, *J. Catal.*, **64**, 479 (1980).
9. D.W. Goodman and J.M. White, *Surf. Sci.*, **90**, 201 (1979).
10. R.D. Kelley and D.W. Goodman, *Surf. Sci.*, **123**, L743 (1982).
11. D.W. Goodman, *Surf. Sci.*, **123**, L679 (1982).
12. D.W. Goodman, *J. Vac. Sci. Technol.*, **2A**, 873 (1984).
13. **Metal-Support and Metal-Additive Effects in Catalysts**, (B. Imelik, C. Naccache, G. Coudurier, H. Praliaud, P. Meriaudeau, P. Gallezot, G.A. Martin, J. Vedrine, eds.) Elsevier, Amsterdam, 1982, p. 315.

14. M. Kiskinova and D.W. Goodman, Surf. Sci., **108**, 64 (1981).
15. C. Kittel, **Introduction to Solid State Physics**, John Wiley & Sons, Inc., New York, 1971
16. M. Kiskinova and D.W. Goodman, Surf. Sci., **109**, L555 (1981).
17. D.W. Goodman and M. Kiskinova, Surf. Sci., **105**, L265 (1981).
18. D.W. Goodman, Proceedings AICHE Meeting, San Francisco, 1984.
19. R.D. Kelley and D.W. Goodman, in **The Chemical Physics of Solid Surfaces and Heterogeneous Catalysts**, Elsevier, 1982, Vol. 4.
20. J.E. Houston, D.E. Peebles, and D.W. Goodman, J. Vac. Sci. Technol. A, **1**, 995 (1983).
21. D.W. Goodman, Proceedings IUCCP Conference, Texas A & M Univ., 1984.
22. C.H.F. Peden and D.W. Goodman, Proceedings of Symposium in Surface Science in Catalysis, ACS, 1984.
23. J.R. Rostrup-Nielsen and K. Pedersen, J. Catal., **59**, 395 (1979).
24. (a) T.B. Grimley and M. Torrini, J. Phys. Chem., **87**, 4378 (1973); (b) E. Einstein and J.R. Schrieffer, Phys. Rev. B, **7** 3629 (1973).
25. J. Feibelman and D.R. Hamman, Surf. Sci., in press.
26. J.E. Houston, J.W. Rogers and D.W. Goodman, J. Vac. Sci. Technol., **2A**, 882 (1984).
27. S. Anderson and U. Jostell, Surf. Sci., **46**, 625 (1974).
28. M. Kiskinova, Surf. Sci., **111**, 584 (1981).
29. G. Ertl, Catal. Rev., **21**, 201 (1980).
30. C.T. Campbell and D.W. Goodman, Surf. Sci., **123**, 413 (1982).
31. E.L. Garfunkel, J.E. Crowell, and G.A. Somorjai, J. Phys. Chem., **86**, 310 (1982).
32. F.M. Hoffman, Surf. Sci. Reports, **3**, 107 (1983).

11

Asymmetric Sites on Heterogeneous Catalysts

Wolfgang M.H. Sachtler
Department of Chemistry
Ipatieff Laboratory
Northwestern University
Evanston, Illinois 60201

ABSTRACT

Methyl acetoacetate (MAA) can be smoothly hydrogenated to methyl 3-hydroxybutyrate acid (MHB) in the gas or liquid phase, using unsupported or silica supported nickel catalysts. The catalyst is rendered enantioselective by chemisorption of "modifiers", i.e. asymmetric molecules. With (R,R)-tartaric acid this chemisorption is corrosive, i.e. the metal surface becomes partially covered with a Ni tartrate complex. Isolated Ni (R,R)-tartrate in the absence of metallic Ni is not a hydrogenation catalyst under the conditions used, but addition of this compound to Ni/SiO_2 results in a catalyst of high activity and some selectivity of the correct sign, in accordance with the dual-site model of enantioselective hydrogenation.

The enantiomeric excess obtained with tartaric acid modified Ni depends on the conditions of catalyst modification. Significant changes in selectivity are observed if alkali halides are present. Harada found an optical purity of 89% for MHB with catalysts modified in the presence of NaBr. This effect has been the subject of recent work. Our results do not support an earlier model assuming preferential poisoning of non-selective Ni sites by alkali halide. They suggest, rather, a stereochemical interaction of the alkali ion with the tartrate complex. For electroneutrality reasons the number of adsorbed halide ions on the catalyst surface must be equal to the number of alkali ions; therefore, the effect on enantioselectivity also depends on the heat of adsorption of the halide anion.

I. INTRODUCTION

Two compounds, whose molecules are each other's exact mirror images, have identical free energies. Whenever a molecule with a center of asymmetry is formed from reactants lacking any element of asymmetry, both enantiomers should be formed in exactly equal quantities, as their transition states have identical free energies.

An excess of one enantiomer can arise, however, if molecules with a prochiral center first form an adduct with an auxiliary component, C_{as}, containing an element of asymmetry with defined chirality. A reaction of this adduct which transforms the prochiral center into an asymmetric center, converts the adduct into a complex for which two diastereomeric forms are possible. These are not each other's mirror image and will, in general, be formed in unequal quantities, as the transition states leading to either diastereomer have different free energies. If, subsequently, the new complex dissociates such that the auxiliary component, C_{as}, is regenerated without change of its chirality, C_{as} can be used again and the process is catalytic.

For a macroscopic asymmetric catalyst the presence of such asymmetric groups, C_{as}, is essential. A further condition is that the two possible chiral forms of C_{as} are present in significantly different numbers. The latter condition is not fulfilled in the Ziegler-Natta catalyst, α-$TiCl_3$, although this compound is known to have asymmetric centers, C_{as}, at its surface. A polypropylene chain, R-$(CH_2$-$CHCH_3)_n$-R, formed on such a center has all $\underline{n}$ asymmetric C atoms in the same configuration. However, the α-$TiCl_3$ crystal has an equal number of centers, C_{as}, in the R and the S-configuration, therefore the catalyst is not asymmetric.

If the C_{as} group is present on the surface of a solid and the reactants are liquid or gaseous, the catalyst is called heterogeneous. Heterogeneous asymmetric catalysis has the practical advantage that the enantiomeric product is formed in one macroscopic reaction (although at least three elementary steps can be discerned), and that the separation of products from the catalyst is trivial.

The first attempts to study asymmetric heterogeneous catalysis were carried out by Schwab _et al._[1] who used reactions catalyzed by nickel, deposited on either dextro- or laevorotary surfaces of asymmetric quartz. The yields were extremely low.

In 1956, Akabori _et al._[2] used palladium deposited on natural silk. With this system, the hydrogenation of oximes and oxazolones was found to be enantioselective; optical yields of 36% were achieved. In 1958, Isoda _et al._[3] used Raney nickel which had been made enantioselective by "modification" with an asymmetric molecule, such as an amino acid.

This has turned out to be a very fruitful approach. Higher yields were obtained using the hydrogenation of methyl acetoacetate (MAA)

$$CH_3\text{-}CO\text{-}CH_2COOCH_3 + H_2 \longrightarrow CH_3\text{-}C(OH)H\text{-}CH_2COOCH_3$$

to methyl 3-hydroxybutyrate as a test reaction with (R,R)-tartaric acid as the modifier. Izumi[4], Tai[5], Harada[6-8] and their coworkers have published a large number of papers describing this and related reactions, varying the conditions of both the liquid phase hydrognation and of the catalyst modification. In 1978, Harada _et al._[7] reported that further improvements of the optical yield were obtained by adding sodium bromide to the aqueous solution of tartaric acid used for modifying the Raney nickel catalyst. With the above test reaction a yield of 89% and with related test reactions virtually pure enantiomers were obtained.

While this research group in Osaka concentrated virtually all research on modified Raney nickel catalysts, the group of Klabunovskii and Vedenyapin[9,10] in Moscow reported enantioselectivity for other transition metals. Yasumori[11,12] in Tokyo and the present writer's group in Leyden[13-15] showed that enantioselective hydrogenation of MAA can also be carried out in the gas phase. Upon replacing Raney nickel by silica supported nickel we were able to use catalysts which also exposed a high surface area after extensive reduction at 400°C, at which temperature Raney nickel would lose a large part of its surface area. Very low enantioselectivity was found for

Ni[13] or Ru[16] supported on some aluminas, and Harada et al.[17] confirmed that the aluminum content of most Raney nickel had an adverse effect on the enantioface differentiating ability of the modified catalyst. These authors, therefore, recommend removal of aluminum by leaching, and they ascribe part of the high efficiency of tartrate modifiers to the removal of aluminum by tartaric acid.

II. CATALYTIC SURFACE POPULATION IN STEADY STATE

In the following we shall use the symbol H_2TA for tartaric acid and NiTA for nickel tartrate. Amino acids and H_2TA are strongly adsorbed by nickel. The part of the surface which is occupied by these modifiers is not available for the dissociative adsorption of dihydrogen. It had, therefore, been expected that the rate of hydrogenation would be much lower for modified than for unmodified catalysts. The experimental observations did not, however, confirm this expectation. The rate of hydrogenation was of the same order, sometimes even slightly higher for the modified catalyst. The cause for this apparent lack of poisoning by a potential strong adsorptive is the equally strong adsorption of MAA. Our IR spectra have revealed that MAA adsorbed on Ni/SiO_2 forms stable chelates **1** and **2**. Upon evacuating

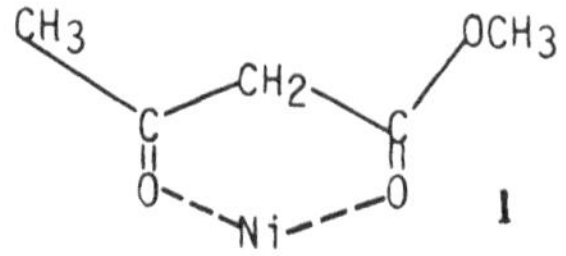

CH_3, CH_2, OCH_3, C, C, O, O, Ni — **1**

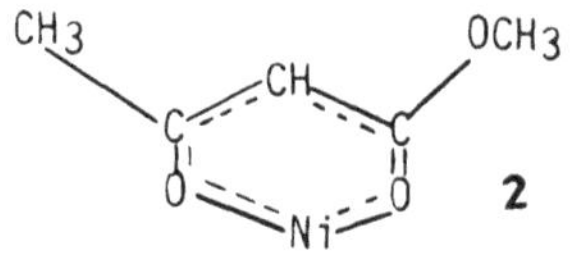

CH_3, CH, OCH_3, C, C, O, O, Ni — **2**

the catalyst at 75°C, the IR bands assigned to complex **2** become more intense, indicating that formation of this enolate is not reversible under these conditions. Recently we have studied used catalysts by Auger Electron Spectroscopy and found that the samples were covered with significant quantities of carbon containing material, the C/Ni signal ratios being of similar values for unmodified and modified catalysts.

In the enolate complex, **2**, the nickel is formally a positive ion and the same holds, of course, for NiTA. The oxidation of the Ni atoms can occur by oxygen adsorption. Indeed, we found for catalysts reduced under rigorous conditions and modified under exclusion of air an extremely low enantioselectivity, while traces of oxygen had a very positive effect on the enantioselectivity. In agreement with our spectroscopic data we believe, therefore, that the NiTA complex and possibly also the enolate complex of MAA are abundant on the catalyst surface under steady state conditions. For Pd catalysts Poels _et al._[18] report chelation of Pd ions with acetylacetone. They make use of this phenomenon to leach Pd ions out of incompletely reduced Pd catalysts. The existence of rather stable MAA chelates of Ni does not necessarily imply that they are the most reactive species of adsorbed MAA or reaction intermediates in hydrogenation. From the reaction order in MAA (+0.4 for gas phase hydrogenation, see below) we would rather assume that under steady state conditions the reactive adsorbate covers only part of the sites available to it; and these sites may even include Ni ions in Ni complexes created by the preceding corrosive chemisorption.

Corrosive chemisorption implies, by definition,[13,19] that metal atoms have left their original positions in the metal surface. In the presence of sufficient H_2TA this process continues and NiTA is dissolved. This dissolution has been confirmed in various ways. If sufficient oxygen is present and the pH is low, the modification solution turns greenish and the loss of nickel from a supported catalyst can be measured by chemical analysis. X-Ray scanning shows that for modified catalyst pellets the Ni has been preferentially leached from the edges of the pellet. While not explicitly using our concept of corrosive adsorption, Harada[20] confirms that after modification at low pH a nickel salt of the modifying acid is formed.

Before steady state is reached, the activity of the catalysts changes with time. For hydrogenation in the gas phase with unmodified catalysts we have usually observed a decrease of the reaction rate during the first few hours on stream; this observation is in accordance with corrosive chemisorption of MAA and formation of an overlayer. For modified samples the rate rather increases initially, as expected for a superficially oxidized metal catalyst which has to be reduced by the reactants to attain high activity.

Adsorbed MAA forms bonds with the NiTA complexes covering the catalyst surface. This is concluded by Yasumori[11,21] from two observations:

(1) the reaction order in MAA is lower for modified than for unmodified catalysts; (2) TDS shows that the desorption peak of MAA is shifted to higher temperatures on tartrate modified surfaces. An interaction between the tartrate and substrates containing C=O groups has also been deduced from elegant work by Tai et al.[5] These authors used substrates which contain one asymmetric carbon atom and a prochiral center. A second asymmetric center is created by hydrogenation over modified nickel. It was found that in the diastereomers thus formed the two conformations of the second asymmetric carbon atom are present in unequal quantities and their relative concentration depends on the nature of the catalyst modifiers. Tai concluded that substrate and modifier must have formed a complex prior to the hydrogenation proper.

In conclusion, we can state that the well-modified portion of Ni catalysts in the steady state of MAA hydrogenation is covered to a large extent with Ni complexes of both the modifier and the enolate of MAA. As formation of these complexes is of the corrosive type, a large fraction of the Ni atoms which were present at the surface of the freshly reduced catalyst have left their positions in the metal lattice. Hydrogen adsorption can take place on the bare parts of the nickel surface where dissociative adsorption is possible. Hydrogen atoms might also be bonded to vacant ligand positions of the Ni ions in the tartrate or MAA enolate complexes. It is conceivable that dissociative adsorption of hydrogen on bare Ni, followed by spillover to the complexes, forms a kinetically favorable path to populate the vacant positions of chelate nickel ions with hydrogen ligands.

As catalyst modification and MAA hydrogenation in the liquid phase are processes which involve diffusion of fairly large molecules through narrow pores, it is possible that only part of the nickel particles in a Ni/SiO_2 or a Raney nickel catalyst are well modified and covered with MAA enolate complexes. To estimate the contribution of this part to the total catalysis, we shall have to inspect the effect of transport phenomena first.

III. EFFECTS OF TRANSPORT PHENOMENA ON KINETIC PARAMETERS AND ON CATALYST MODIFICATION

In every catalytic reaction molecules must be transported to the active site and products must leave the place of their formation. If these transport phenomena are not infinitely rapid in comparison to the rate of the chemical process at the active site, they will have an effect on the overall reaction kinetics. The diffusion of molecules is slower in liquids than in the gas phase because of the large difference of the mean free paths. Diffusion is particularly slow in narrow pores as present in Ni/SiO_2 and Raney nickel catalysts. If the reactants are present in a non-stoichiometric ratio such that there is a large excess of one reactant, the concentration profile of the minority component will determine the degree of catalyst utilization. For a component present at low concentration in the homogeneous liquid, any catalytic conversion will lead to a further depletion resulting in a negative concentration gradient from the pore mouth towards the deeper regions. If the rate of chemical conversion at the pore mouth is large compared to the rate of diffusion, this reactant will be consumed in areas near the pore mouth and catalyst surface segments at large distances from the mouth will not be utilized.

This situation prevails when MAA is hydrogenated in the liquid phase with porous catalysts. The minority component is then the dissolved hydrogen which is consumed near the pore mouths. In the limit of a complete pore diffusion controlled reaction the apparent activation energy becomes equal to one-half of the true activation energy of the reaction at the active site. In a previous paper[15] we have shown that the liquid phase hydrogenation, as carried out by most authors, is approximated by this model. The consequences are that the apparent activation energies published are too low and the apparent reaction orders (1 for hydrogen, 0.2 for MAA) reflect the influence of transport phenomea. A conclusion that the addition of hydrogen is the rate limiting step may not be based on these reaction orders. A simple phenomenological test for transport controlled reactions is to study the effect of pellet crushing on the rate. For the liquid phase hydrogenation of MAA, we found the rate increased by a factor of nine, when the pellets were crushed from an initial diameter of 2.7 mm to a final diameter of 0.1 to 0.01 mm.

If hydrogenation is carried out in the gas phase, the effect of pore diffusion is smaller, but not necessarily negligible at high conversions.

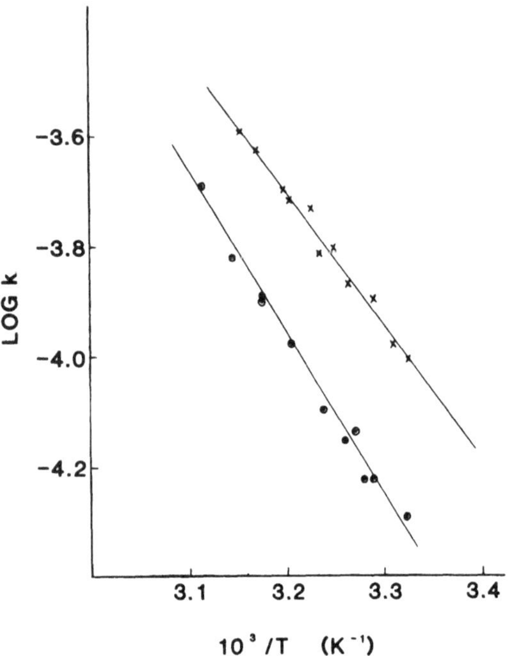

Fig. 1. Arrhenius lines for MAA hydrogenation on a Ni/SiO_2 catalyst. (-0-) non-modified; (-x-) modified with (R,R)-tartaric acid.

Under differential conditions we obtained[14] an activation energy of 61 ± 5 kJ/mole^{-1}. At higher conversion, pore diffusion can no longer be ignored, even in the gas phase. As under these conditions the concentration ratio H_2/MAA is in the order of 1.7 x 10^3, the partial pressure of hydrogen remains constant inside the pores, but the partial pressure of MAA decreases with the distance from the pore mouth. Typical reaction orders are 0.40 for MAA and 0.20 for H_2.

In order to compare the activation energies for modified and unmodified catalysts, the true rates should be available in a sufficiently wide range of temperatures. In the past we tried this by working under differential conditions where conversions can easily be converted to reaction rates even if the reaction orders are unknown. However, the necessity of maintaining low conversion even at elevated temperatures forced us to use catalysts with low nickel loading and these also display a low enantioselectivity after modification. Therefore, the Arrhenius lines for modified and unmodified catalysts obtained in that work were, not surprisingly, parallel.[14] More recently we have used samples which contained, initially 14w% of Ni on SiO_2; after modification with H_2TA the nickel content had dropped to 11w% due to leaching. The modified samples displayed an enantioselectivity of 37 ± 7% in the gas phase hydrogenation of MAA. The rates were calculated from the conversions, using the integrated rate equation and the measured reaction orders. As shown in Fig. 1 the Arrhenius lines for modified and unmodified catalysts are not parallel. The modified catalyst with 11w% Ni displays a higher rate and an activation energy of 46 ± 3 kJ/mole, the unmodified catalyst with 14w% of Ni has a higher activation energy of 57 ± 4 kJ/mole in agreement with our previous result.[14] The difference in activation energy is not large, but outside experimental scatter.

Diffusion phenomena are of considerable influence during catalyst modification. This operation is usually carried out in an aqueous solution, containing typically 2w% of H_2TA, i.e. the molar fraction of H_2TA is 2.4×10^{-3}. As the acid is consumed when reacting with the surface of nickel particles, insufficient diffusion rate results in a decreasing concentration profile for H_2TA from pore mouth towards the interior of a pellet. As mentioned, X-ray analysis has shown that under somewhat severe modification conditions, the nickel particles at the periphery of the pellet are dissolved. The NiTA formed in the process of modification has to diffuse out of the pore. Near the pore mouth the NiTA concentration will be small, as this part communicates efficiently with the surrounding solution. At the far end of a pore the concentration of dissolved NiTA is, again, small, because its formation is limited by the low local concentration of H_2TA. In the intermediate region the NiTA concentration will be large, perhaps near the saturation value of NiTA in the solution. Here we have the desired conditions that the catalyst surface remains covered with adsorbed NiTA. The extent of the region inside each pore where these conditions prevail will depend on the H_2TA concentration, the temperature and the loading of the catalyst with nickel. If the two former factors are kept constant, the portion of each pore, where metallic nickel is in contact with a nearly saturated solution of NiTA, will be larger for high nickel loadings. The general observation that higher nickel loadings result in higher enentioselectivities may have its cause in this circumstance. An alternative interpretation was suggested by Nitta,[22] who assumed that the larger particles at higher metal loading expose more "smooth" crystal faces. To verify this hypothesis, one should change the size of the Ni particles while keeping the Ni loading of the catalyst constant. We plan to do this by using Ostwald ripening of nickel in Ni/SiO_2 samples.

Another result which becomes understandable when considering the transport effects is the difference in enantioselectivity for modified catalysts examined in the liquid or the gas phase respectively. If modification is carried out under circumstances where the mentioned concentration profiles of H_2TA and NiTA are established, the outer parts of a catalyst pellet display a higher enantioface differentiating behavior than the inner parts. If hydrogenation is carried out in the liquid phase, i.e. in the regime of pore diffusion controlled reaction rates, the outer parts of the catalyst pellet are preferentially utilized and the measured selectivity is high. In the gas phase also the inner segments of a catalyst pellet are also utilized and selectivity is lower. Qualitatively, the same result can be achieved in the liquid phase by reducing the effective length of the pores. Indeed, we found[15] that crushing a catalyst pellet resulted in a decrease of enantiomeric excess from 58 to 44%.

IV. STEREOCHEMICAL INTERACTIONS

Much research has been done to understand the stereochemical interaction between the modifier on the catalyst surface and the adsorbed substrate. Ideally, a model of this interaction should provide a rationalization of the sign of the enantiomeric excess obtained (why is it that L-hydroxy acids and L-amino acids as modifiers induce an enantiomeric excess of opposite sign?) and it should rationalize the effects of the various parameters of reaction (temperature, concentration, presence of third substances) and the modification (pH, temperature, concentration, duration, presence of oxygen, presence of third substances). In spite of numerous models proposed, none has been able to answer all these questions. The huge number of conceivable possibilities has, however, been reduced by some experimental facts. In this section we mention three sets of data of possible relevance.

A. The role of mixed complexes

As chelating molecules such as β-diketones and α-hydroxy acids form complexes with Ni, it is possible that mixed complexes result. The asymmetric (R,R)-tartate ligand can either act as a template to an MAA ligand attached to the same Ni^{+2} ion, or it can interfere stereochemically with MAA molecules adsorbed on different but adjacent centers. The present data do not permit the dismissal of one of these possibilities. We have tried to decide whether NiTA alone, in the absence of Ni metal is sufficient to catalyze the hydrogenation of MAA. The results of these experiments show that NiTA alone is insufficient, even when it has been spread on a high surface area support. Conversely, impregnating a nickel catalyst with NiTA resulted in a high hydrogenation rate and a selectivity of the correct sign.[13] We conclude that a NiTA complex and metallic nickel are both necessary for enantioface differentiating hydrogenation of MAA. A possible rationalization of this finding is provided by the "dual site model", which assumes that dihydrogen is dissociated on the surface of nickel metal, the H atoms migrate to the adsorbed NiTA complex where an MAA molecule is adsorbed and the formation of the O-H and C-H bonds takes place while the tartrate ligand(s) act as template(s).[23]

B. The structure of NiTA

The stereochemical description of the interaction of two surface complexes with each other or of two ligands in the case of a mixed complex, requires a knowledge of the structure of the complex(es). Since no data are available for the structure of a two-dimensional layer of NiTA, we had, in the past, used the assumptions (i) a surface complex of NiTA will have a structure similar to NiTA in the interior of its own crystal; (ii) the unknown crystal structure of NiTA will be similar to that of copper tartrate.[24]

The second of these assumptions prompted us to prepare crystalline nickel(II) (R,R)-tartrate monohydrate and to identify its structure in collaboration with Prof. Reedijk and his coworkers of the National

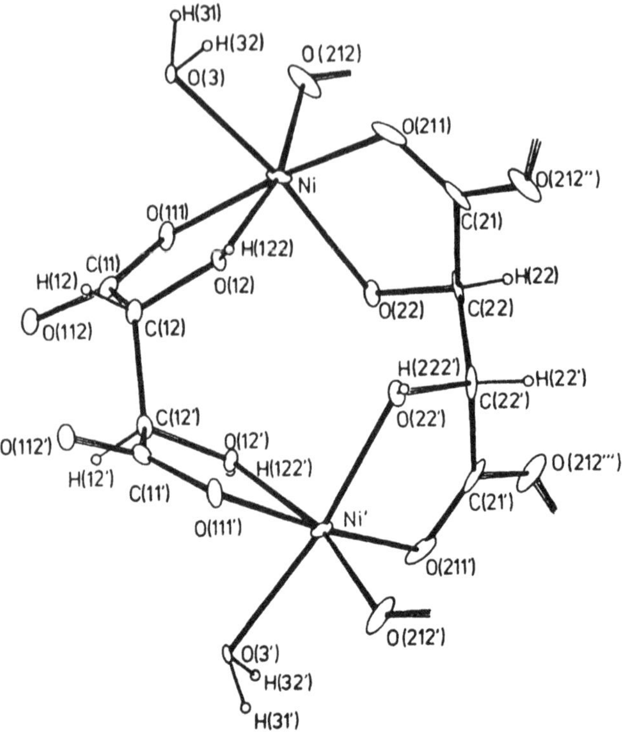

Fig. 2. Dimeric unit of Ni_2(R,R tartrato)$_2$(H_2O)$_2$ in crystalline compound (Non-coordinating H_2O molecules omitted).

University of Leyden, The Netherlands, by means of X-ray diffraction. The structure analysis which is described in a separate paper[25] revealed that the basic building block of this compound is a dimeric entity [Ni_2(R,R)-tartrate)$_2$] (H_2O)$_2$. This unit is shown in Fig. 2. Each Ni^{2+} ion is coordinaated by two halves of two different tartrate dianions, via chelation through the alcohol and carboxylate groups. The other half of the two tartrate anions is coordinated in the same way to a second Ni^{2+} ion. The somewhat distorted octahedral geometry around each Ni^{2+} ion is completed by a water molecule (0(3) in the Fig. 2) and by a non-chelating carboxylate-oxygen atom of another dimer. The crystal can be described as consisting of polymeric corrugated sheets of the dimeric units and water molecules. The three-dimensional network is held together via hydrogen bonds and the mentioned coordination bond between each nickel ion and the "third" tartrate ion. A similar dimeric unit as now identified for NiTA had been reported before for copper tartrate, justifying our assumption (ii). The fact that the compounds form corrugated layers also suggests that on the surface of a nickel crystal a high coverage with NiTA will result in a similar layer structure. Nickel ions exhibiting a sufficient number of available ligand positions required for adsorbing MAA and at least one hydrogen atom will then be located at the edges and at imperfections of this overlayer.

C. The role of alkali halides

As mentioned in the Introduction, Harada et al.[17] had observed that addition of NaBr during the modification of the Ni catalyst with H_2TA induces an enhancement of the enantioselectivity of the catalyst. They assumed that NaBr poisons both the selective and the non-selective sites of the surface, but preferentially non-selective sites were blocked. In 1979 we published data[26] which were evidently at variance with this view; under appropriate conditions the enantioselectivity was significantly increased, but the reaction rate remained virtually unchanged. This observation seemed to exclude a site poisoning model as the only explanation of the important effect of alkali halide. We therefore concluded that NaBr also interferes with stereochemistry of the templates on the surface and we decided to study the interaction of alkali halides with the surface of modified nickel catalysts by physical methods. The results of this work will be published in a separate paper[27] and are briefly summarized in the next section.

V. INTERACTION OF ALKALI HALIDES WITH THE CATALYST SURFACE

Addition of alkali halides to the aqueous H_2TA solution used in modifying the catalysts also results in significant changes of the enantioselectivity in our experiments. However, we found that this effect also depends on the washing procedure which is usually applied after modification of the catalyst and before bringing it into the reactor. After extensive washing with methanol, the selectivity of catalysts modified with H_2TA and NaBr dropped to the same value as obtained when modification was carried out in

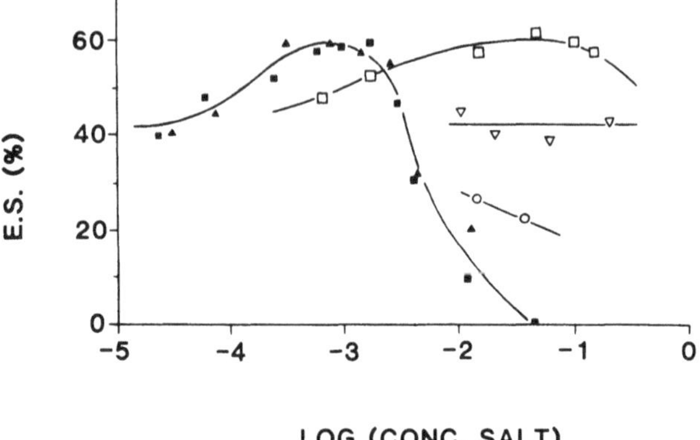

Fig. 3. Enantioselectivity of liquid phase hydrogenation of MAA with tartrate modified Ni/SiO_2 catalyst at varying concentrations of alkali halides in solution.

-▽- LiBr; -□- NaBr; -○- RbBr; -■- NaI; -▲- KI

the absence of NaBr. We conclude that this halide is reversibly adsorbed on the catalyst surface, unlike the H_2TA which remains strongly attached and survives extended reaction times without appreciable loss in selectivity.

We then decided to study liquid phase hydrogenation of MAA in methanol to which controlled concentrations of alkali halides had been added. The bromides of Li^+, Na^+ and Rb^+ and the iodides of Na^+ and K^+ were tested. In typical experiments the reaction mixture contained 50 vol% of MAA and 50 vol% of methanol. Hydrogenation was carried out at 1 bar of H_2 in three parallel reaction vessels containing portions of the same catalyst batch. In two of the vessels alkali halide was present, the third served as a blank. After measuring the catalytic conversions and the enantioselectivities, the catalyst was separated from the reaction mixture by centrifugation, volatile components were removed by distillation and the halide content of the catalyst was analyzed by means of X-ray fluorescence.

It was found that the effects of different halides on the selectivity were markedly different (see Fig. 3). At very high concentration, all halides had a negative effect on the catalyst activity, and, with the possible exception of LiBr, also on the enantioselectivity. At low concentrations,

NaBr, NaI, and KI had a positive effect on the enantioselectivity; this parameter therefore passes through a maximum when the concentration of any of these halides is increased. A selectivity maximum for NaBr had been reported by Murakami et al.[28] with respect to the NaBr concentration in the modifying liquid. We found the concentration required to achieve maximum selectivity enhancement to depend in a rather dramatic way on the nature of the halide anion. For the iodides, the maximum is observed at a concentration of 10^{-3} molar, whereas for NaBr 10^{-1} molar is required for the rather shallow maximum.[27]

The suggestion from these data that the iodide ion is more strongly adsorbed than the bromide ion is confirmed by the analysis of the used catalysts. In the case of bromides, samples which were well washed after use did not contain detectable quantities of Br^-; and for the samples which had not been washed, the amount of Br^- detected did not differ significantly from the quantity estimated for the dissolved bromide in the liquid filling the pore volume. For the iodides, however, the amount of I^- detected significantly exceeded the quantity dissolved in the pores.

While the chemical nature of the anion thus controls the quantity of halide adsorbed, it is not probable that an interaction of the anion with the tartrate enhances the selectivity. For electroneutrality reasons we have to assume that the concentration of cations brought to the surface by halide adsorption must be equal to the concentration of the anions; and additional alkali cations are present from the modification with H_2TA if the pH of the latter has been adjusted by alkali hydroxide. We have carried out modifications in methanol in the absence of alkali and obtained moderate enantioselectivity showing that, while the alkali ions are not necessary prerequisites to enantioselectivity, they can, however, be beneficial. Tanabe et al.[29] had reported that the nature of the alkali hydroxide used to adjust the pH during modification has a significant influence on the enantioface differentiating ability of the catalyst, and our present results with added halides show that Na^+ and K^+ are beneficial, while Li^+ and Rb^+ do not seem to have a favorable effect on the enantioselectivity.

As to the precise stereochemical interaction of the alkali cation with the adsorbed tartrate we are left to speculations. It appears conceivable that these ions become coordinated to oxygen donor atoms of adsorbed tartrate and MAA. For Na^+ and K^+ which tend to coordinate octahedrally, the interaction improves the template effect. Rb^+ is known to prefer ninefold coordination with oxygen, and appears to be unfavorable for the enantioselectivity. Li^+ which ordinarily favors tetrahedral or tetragonal pyramidal coordination, seems to have no effect on the enantioselectivity. The anions apparently interact with the templates in a way that is unfavorable for the enantioselectivity and this effect predominates at high concentration; but a low concentration the main function of adsorbed anions may be to fix an equivalent number of cations on the surface.

VI. CONCLUSION

Enantioselective hydrogenation with modified heterogeneous catalysts has been found to be useful in preparing enantiomers with high yield. The field still awaits further research, as many questions still cannot be answered. While corrosive chemisorption of H_2TA appears to be well established, the site where the hydrogenation proper takes place has still to be identified. In the "dual site model" it is assumed that the metal ion in the tartrate complex is decisive;[23] the substrate adsorbed on this ion will stereochemically interact with the tartrate ligand(s), and the transition states for adding a hydrogen atom to the prochiral C atom will have different free energies for the two possible conformations. Other models assume that the metal surface is the site of hydrogenation and that sites adjacent to a tartrate complex will lead to the preferred formation of one enantiomer. The recent finding that tartrate modified nickel catalysts, although containing less nickel, display a higher activity and a lower activation energy than unmodified nickel, appears to favor the dual site model, but more research will be required before the question can be settled definitely. In one approach under study we are trying to separate the two catalytic functions of the dual site model. In this work we are fixing Pt particles in the cavities of a zeolite, while NiTA complexes are present at its exterior surface. The geometry of the zeolite does not permit MAA molecules to enter the cavities, but hydrogen dissociating on the Pt particles may spill over to the NiTA complexes. We may have to wait for the results of this research, before the dual site model can be regarded as proven or disproven.

ACKNOWLEDGEMENT

The writer is greatly indebted to L.J. Bostelaar and A. Hoek from whose Ph.D. theses numerous data and ideas quoted in this paper have been taken, and to M. Hopman who determined the Arrhenius parameters for modified and unmodified catalysts in the course of his Master's thesis. The author is indebted to the Leidsch Universiteits Fonds for enabling him to do this work in Leyden and to the National Science Foundation for support in continuing this research at Northwestern University.

REFERENCES

1. G.M. Schwab, F. Rost, and L. Rudolph, Kolloid-Zeitschrift, **68**, 174 (1934).
2. S. Akabori, Y. Izumi, Y. Fujii, and S. Sakurai, Nature, **178**, 323 (1956).
3. T. Isoda, A. Ichikawa, and T. Shimamoto, J. Sci. Res. Inst. (Riklen hokoku), **34**, 134 (1958).
4. see: Y. Izumi, Adv. Catal., **32**, 215 (1983).
5. A. Tai, H. Watanabe, and T. Harada, Bull. Chem. Soc. Jpn., **52**, 1468 (1979).
6. T. Harada, Y. Hiraki, Y. Izumi, J. Muraoka, H. Ozaki, and A. Tai, Proc. Int. Congr. Catal., 6th (London, 1976) 1024 (1977).
7. T. Harada and Y. Ixumi, Chem. Lett., 1195 (1978).
8. T. Harada, A. Tai, H. Yamamoto, H. Ozaki, and Y. Izumi, Proc. Int. Congr. Catal., 7th (Tokyo, 1980) 364, (1981).
9. E.I. Klabunlovskii and A.A. Vedenyapin, **Asymmetricheskii Kataliz, Gidrogenizatsiya na Metallakh,** (Nauka, Moscow, 1980).
10. E.I. Klabunovskii, A. A. Vedenyapin, E.I. Karpeiskaya, V.A. Pavlov, and N.D. Zelinskii, Proc. Int. Congr. Catal., 7th, (Tokyo, 1980), 390 (1981).
11. I. Yasumori, Y. Inoue, and K. Okabe, in **Relations Between Heterogeneous and Homogeneous Catalytic Phenomena,** (B. Delmon and G. Jannes eds.), Elsevier, Amsterdam, 1975, p. 141.

12. I. Yasumori, M. Yokozeki, and Y. Inoue, Disc. Faraday Div. Roy. Chem. Soc., Nottingham 1981, Paper 22.
13. A. Hoek and W.M.H. Sachtler, J. Catal., 276 (1979).
14. A. Hoek, H.M. Woerde and W.M.H. Sachtler, Proc. Int. Congr. Catal., 7th (Tokyo, 1981), 376 (1981).
15. H.M. Woerde, L.J. Bostelaar, A. Hoek, and W.M.H. Sachtler, J. Catal., **76**, 316 (1982).
16. A.A. Vedenyapin, E.I. Klabunovskii, E.I. Talanov, and N.P. Sokolova, Kin., Katal., **16**, 436 (1975).
17. T. Harada, M. Yamamoto, S. Onaka, M. Imaida, H. Ozaki, A. Tai, and Y. Izumi, Bull. Chem. Soc. Jpn., **54** 2323 (1981).
18. E.K. Poels and V. Ponec in **Catalysis** (G.C. Bond and G. Webb eds.) (Specialist Periodical Report), The Chemical Society, London, 1983, Vol. 6, p. 196.
19. A.A. Holscher and W.M.H. Sachtler, Disc. of the Faraday Soc., **41**, (1966).
20. T. Harada, Bull. Chem. Soc. Jpn., **53**, 1019 (1980).
21. I. Yasumori, Pure Appl. Chem., **50**, 971 (1978).
22. Y. Nitta, F. Sekine, T. Imanaka, and Sh. Teranishi, Bull. Chem. Soc. Jpn., **54**, 980 (1981).
23. W.M.H. Sachtler and L.J. Bostelaar, in **Spillover of Adsorbed Species**, (G.M. Pajonk, S.J. Teichner and J.E. Germain, eds.) Elsevier, Amsterdam, 1983 p. 207.
24. A. Hoek, **Tartrate Modified Nickel, an Enantioselective Heterogeneous Catalyst,** Thesis, Nat. University Leyden, Netherlands (1982).
25. L.J. Bostelaar, R.A.G. de Graaff, F.B. Hulsbergen, J. Reedijk, W.M.H. Sachtler, Inorg. Chem., **23**, 2294 (1984).
26. W.M.H. Sachtler, Farraday Disc., Chem. Soc., **72** (1981).
27. L.J. Bostelaar and W.M.H. Sachtler, J. Mol. Catal., **27**, 387 (1984).
28. S. Murakami, T. Harada, and A. Tai, Bull. Chem. Soc. Jpn., **53**, 1356 (1980).
29. T. Tanabe, K. Okuda, and Y. Izumi, Bull. Chem. Soc. Jpn., **46**, 514 (1973).

12

Organoactinide Complexes on Alumina

Robert L. Burwell, Jr. and Tobin J. Marks
Ipatieff Laboratory, Department of Chemistry
Northwestern University, Evanston, IL 60201

ABSTRACT

This paper surveys the surface chemical reactions which follow deposition of several organometallic complexes of uranium and thorium on alumina and the catalytic activities of the resulting materials for the hydrogenation of propylene at -63°C and the polymerization of ethylene at 25°C. The following complexes deposited on dehydroxylated alumina (but not on partially hydroxylated alumina) are active catalysts for both reactions (here Cp' is pentamethylcyclopentadienyl): $Cp'_2U(CH_3)_2$, $Cp'_2Th(CH_3)_2$, $Cp'_2Th(neopentyl)_2$, $[Cp'_2ThH_2]_2$, and, most active of all, $Cp'Th(benzyl)_3$. Adsorption of carbon monoxide to CO/M $\sim$0.03 poisons hydrogenation. Per molecule of complex, these materials are highly active hydrogenation catalysts, but per active site they are among the most active hydrogenation catalysts so far reported.

Lecture by the first author under the auspices of the Catalysis Society through its Robert Burwell Award sponsored by the Amoco Oil Company of which R.L.B., Jr., was the first recipient.

I. INTRODUCTION

This work was started with the hope that it might lead to some new chemistry, a hope which was supported by the following considerations. Although organometallic complexes supported on alumina and other oxides is an area in which there has been considerable recent activity,[1] to our knowledge no work on supported organoactinide complexes had been reported. The large size of actinide ions permits much larger coordination numbers than those usually found for complexes of transition elements. This provides a potential for new kinds of chemistry. Further, work by the second author had shown that organoactinide complexes are unusually labile in many reactions.

Actinide ions like Th(IV) and U(IV) are much more oxophilic than the species such as Pt(II) or Ni(II) from which one is apt to derive notions of the chemistry of organometallic complexes. For example, carbon monoxide reacts with $Cp'_2ThH(OCH_3)$ where Cp' is pentamethylcyclopentadienyl according to Eq. 1.[2] This oxophilic character is apparently a major factor in

$$2Cp'_2Th{<}^{H}_{OCH_3} + 2CO \longrightarrow Cp'_2Th{<}^{O-CH=CH-O}_{OCH_3\quad CH_3O}{>}ThCp'_2 \qquad (1)$$

leading the chemistry of organoactinide complexes to have the substantial resemblances to those of complexes of early transition elements like titanium and zirconium.

We started our investigations with $Cp'_2U(CH_3)_2$ and its thorium analog deposited on alumina. There was considerable background in the chemistry of such complexes derived from work done in this department.[3] The use of cyclopentadienyl leads to compounds having three cyclopentadienyl ligands, for example Cp_3UCH_3. However, only two of the more bulky Cp' ligands can enter the coordination sphere of U(IV) or Th(IV). Thus $Cp'_2Th(CH_3)_2$ and $Cp'_2U(CH_3)_2$ are, to some degree, coordinatively unsaturated molecules. The oxophilic character of the actinide ion leads alcohols to react readily at 25°C with these complexes in solution. In this paper, M will designate

$$Cp'_2M(CH_3)_2 + ROH \longrightarrow Cp'_2M(OCH_3)(CH_3) + CH_4 \qquad (2)$$

either U or Th. We can expect, then, that surface OH groups on alumina, σ-OH, will lead to protonolysis of M-CH_3 bonds with consequent attachment of the complex to the surface as in Eq. 3. However, as will appear, this

$$\text{M-CH}_3 + \sigma\text{-OH} \longrightarrow \text{M-O-}\sigma + \text{CH}_4 \quad (3)$$

reaction is disadvantageous. Further, $Cp'_2Th(CH_3)_2$ undergoes rapid hydrogenolysis in solution to form a dimer as shown in Eq. 4. It was thought

$$2\ \text{Cp}'_2\text{Th(CH}_3)_2 + 4\text{H}_2 \longrightarrow \text{Cp}'_2\overset{\text{H}}{\overset{|}{\text{Th}}}\langle\begin{smallmatrix}\text{H}\\ \text{H}\end{smallmatrix}\rangle\underset{|}{\underset{\text{H}}{\text{Th(CH}_3)_2}} \quad (4)$$

that the steric bulk of the surface might inhibit this dimerization and that hydrogenolysis of $Cp'_2M(CH_3)_2$ might then lead to a monomer.

Our investigation of supported $Cp'_2M(CH_3)_2$ complexes led to the immediate discovery that some of these materials were extremely active catalysts for the hydrogenation of propylene and the polymerization of ethylene.[4] The high activity for hydrogenation required us to work at -63°C and to use very small amounts of catalyst. Since we have been unable to achieve low conversions, we have not been able to obtain good kinetic data. Since the complexes we have studied are very sensitive to oxygen and water, we have had to work under extremely asceptic conditions. We have used a 99.99% pure γ-alumina, PHF of the American Cyanamid Company, surface area about 160 m^2g^{-1}, average pore diameter 12 nm and a pore volume of 0.6 $cm^3\ g^{-1}$.[5] An alumina prepared by wet methods is covered with surface OH groups (σ-OH), about 15 nm^{-2}. Upon heating in flowing helium, the reaction shown in Eq. 5 commences at about 200°C, and removal of surface oxygen

$$2\ \sigma\text{-OH} \longrightarrow \sigma\text{-O} + \text{H}_2\text{O(g)} \quad (5)$$

species results in the exposure to some degree of Al^{3+} to be designated Al^{3+}(cus). Surface O^{2-} formed by dehydroxylation is also (cus), i.e., of a degree of coordination lower than that in bulk. Here, (cus) means <u>coordinatively unsaturated surface</u>. Alumina dehydroxylated in flowing helium at 450°C will be designated PDA and that at 950°C, DA. We usually used 10-20 μmol of actinide complex in catalytic runs and 25-50 μmol in

TABLE I

Surface Groups Per nm^2 of Alumina

Group	Number	μmol/0.25 g
σ-OH in PDA	4[a]	270
σ-OH in DA	0.12[a]	8
20 μmol complex[b]	0.3	

[a]See Ref. 5

[b]Resulting from adding 20 μmol of actinide complex, from injecting 20 μmol of acetone, etc. to 0.25 g of Al_2O_3.

studies of surface chemistry. Table I shows the contents in σ-OH and in added complex per nm^2. As will appear, the degree of dehydroxylation of the alumina is of critical importance in the present work as it was in our studies of the results of heating $Mo(CO)_6/Al_2O_3$.[5]

We present below a review of some of our results dealing with the surface chemical processes which occur during the preparation and activation of actinide complex/Al_2O_3 catalysts and with the catalytic activity of these materials.

II. EXPERIMENTAL TECHNIQUES

General vacuum line techniques were employed.[5] Helium was initially freed from other gases by passage through a trap of Davison grade 62 silica gel at -196°C and hydrogen was supplied by an Elhygen electrolytic generator (Milton Roy) in which hydrogen diffuses through a Pd,Ag alloy. A MnO/SiO_2 trap just before the reactor reduced the content in oxygen in the helium and hydrogen streams and in C_3H_6 + H_2 and C_2H_4 + He feed streams to below 0.03 ppm.[5] Catalysts were prepared <u>in situ</u> by the addition of a solution of a complex in pentane to 0.25 g of alumina. The alumina in the fused silica reactor had been previously converted to PDA or DA. The pentane was evaporated in a flow of helium at 0°C for one hour. The pentane was removed from the helium stream in an empty trap at -196°C and other gases except hydrogen were trapped in silica gel at -196°C and then released as a pulse for gas chromatographic analysis by lowering the liquid nitrogen flask. To determine hydrogen, an additional trap of molecular sieve 5A at -196°C followed the silica gel trap. Removal of the liquid nitrogen liberated a pulse of hydrogen which was measured catharometrically after conversion to water by CuO at 500°C. After initial preparation, the catalyst was usually further treated in flowing helium or hydrogen at higher temperatures, where, again, any evolved gases were trapped and measured. A jacketed Carle injection valve permitted the injection of various gases or vapors into flowing helium or hydrogen. In flow hydrogenation of propylene, the H_2 + C_3H_6 mixture (H_2/C_3H_6 mole ratio = 4.68) was prepared by passage of hydrogen through a saturator maintained at -78°C. The saturator consisted of a tube packed with Filtros FS-140 the pores of which were filled with liquid propylene.

III. SURFACE CHEMISTRY

A. Activation of $Cp'_2M(CH_3)_2/Al_2O_3$

During the evaporation of pentane at 0°C in flowing helium which occupied one hour and which is designated He,0°,1 and during pretreatments in hydrogen or helium at up to 150°C, the only detectable gas evolved was methane. The amount of methane liberated during these steps, expressed as CH_4/M, was somewhat variable probably because of variation in the exact amount of σ-OH on the alumina, the amount of complex (this varied in the range 10-50 μmol), and possibly from impurity content. During activation of $Cp'_2U(CH_3)_2$ (to be designated as $U(CH_3)_2$) on DA, CH_4/M was usually in the range 0.11-0.20 during He,0°,1 and protraction of the time beyond one hour led to little further evolution of methane. A subsequent He,25°,1 led to a value of CH_4/M about 15% that of He,0°,1. He,100°,1 following He,0°,1 led to values of CH_4/M in the range 0.14-0.22, whereas H_2,100°,1 led to CH_4/M equal to about 0.3. $Th(CH_3)_2$/DA gave somewhat lower values of CH_4/M; He,0°,1 gave CH_4/M of about 0.1. Subsequent He,100°,1 gave ∿0.08, whereas H_2,100°,1 gave ∿0.3. Table II shows the effect of prolonging He,100°. Thus, He,0°,1 plus He,100°,7 led to the removal of about one-eighth of the methyl groups attached to Th.

TABLE II

Effect of Prolonging He,100° with 33.9 μmol of $Cp'_2Th(CH_3)_2$/DA

Pretreatment	CH_4/Th
He,0°,l	0.11
He,100°,l	0.086
He,100°,l	0.022
He,100°,l	0.010
He,100°,l	0.008
He,100°,l	0.006
He,100°,l	0.006
He,100°,l	0.004
Total	0.252

He,0°,l with $U(CH_3)_2$/PDA gave much larger yields of methane. In a typical experiment, CH_4/M was 1.20 during He,0°,l, 0.08 during He,25°,l, and 0.24 during He,100°,l. $Th(CH_3)_2$/PDA gave similar results. Silica gel which had been dehydroxylated at 400°C gave results similar to those obtained with PDA.

To determine the origin of the evolved CH_4, isotopic tracer experiments were run employing (a) $M(CH_3)_2/Al_2O_3$-d, (b) $M(CD_3)_2/Al_2O_3$-h, and (c) $M(CD_3)_2/Al_2O_3$-d.[6] Al_2O_3-d was made by treating the alumina D_2,475°,l which converts all σ-OH to σ-OD. Catalysts were activated He,0°,l, He,25°,l and He,100°,l in sequence and the evolved methane was collected and analyzed mass spectrometrically, The systems $U(CH_3)_2$/DA, $U(CH_3)_2$/PDA and $Th(CH_3)_2$/DA were examined.

Protonolysis by σ-OH, path S, Eq. 3 would give CH_3D in (a), CD_3H in (b), and CD_4 in (c). Reaction with a hydrogen atom of a ring methyl (Path R) would give CH_4 in (a) and CD_3H in (b) and (c). For an example of (c) where only one Cp' and only one its methyl groups is shown see Eq. 6.

$$\text{(ring)}-CH_3 \cdots M(CD_3)_2 \quad \sigma\text{-OD} \longrightarrow \text{(ring)}-CH_2-M(CD_3) \quad \sigma\text{-OD} + CD_3H \qquad \text{path R} \tag{6}$$

TABLE III

Isotopic Tracer Experiments Run to Characterize the Origin of Evolved Methane

Path	He,0°,l[a]	He,100°,l[b]
	$U(CH_3)_2PDA$	
S	0.81	0.70
M	0.12	0.29
R	0.016	0.012
(S+R+M)[b]	0.95	1.00
	$U(CH_3)_2/DA$	
S	0.45	0.19
M	0.45	0.73
R	0.07	0.11
(S+R+M)[b]	0.97	1.03
	$Th(CH_3)_2/DA$	
S	0.69	0.23
M	0.31	0.52
R	0.03	0.29
(S+R+M)[b]	1.03	1.04

[a]Opposite S, M and R are given the calculated fractions of methane liberated by each path during the pretreatment.

[b]The sum of the caluclated fractions which should ideally equal 1.00.

$$\text{(CH}_3\text{)M(CD}_3\text{)}_2 \quad \sigma\text{-OH} \longrightarrow \text{(CH}_3\text{)M=CD}_2 \quad \sigma\text{-OH} + CD_4 \qquad \text{path M} \tag{7}$$

Disproportionation (Path M) would give CH_4 in (a) and CD_4 in (b) and (c). For an example of (b) see Eq. 7. Ignoring kinetic isotopic effects, one can calculate the fractions of methane which result from paths S, R and M as given in Table III. Because of low yields of CH_4 in He,25°,l, accuracy in the calculated fractions is low and we omit these results. CH_4 does not exchange with DA-d or PDA-d under these conditions.

The "carbene" formed in Eq. 7 is presumably not a free one, but rather it is stabilized by interaction with Al^{3+}(cus)-O^{2-}(cus) pairs on the alumina by analogy with Group IVb carbenes stabilized by Al-X groups.[7,8] In any case, activated $M(CH_3)_2$/DA has only very weak activity for the metathesis of propylene. For example, a small pulse of propylene in helium passed at 100°C over $Th(CH_3)_2$/DA; He,0°,1; He,100°,1 (meaning a sample of $Th(CH_3)_2$ on DA pretreated by the succeeding successive two pretreatments) gave only trace yields of ethylene and butenes.

However, acetone pased over activated $U(CH_3)_2$/DA or $Th(CH_3)_2$/DA in helium carrier at 25°C led to the formation of isobutylene as one would expect from reactions between carbenes and acetone in solution.[9] For example, three pulses of 100 μmol of acetone were passed over 28.4 μmol $Th(CH_3)_2$/DA: He,0°,1; He,100°,1 at 25°C. Isobutylene/Th = 0.010 was formed. The ratio, carbene/M, is the product of the average fraction via path M and the ratio CH_4/M during the pretreatments. It was about three times larger than isobutylene/M.

The yields of methane from H_2,100°,1 indicate that another path for the formation of methane appears in the presence of hydrogen as shown in Eq. 8.

$$M\text{-}CH_3 + H_2 \longrightarrow M\text{-}H + CH_4 \qquad \text{path H} \qquad (8)$$

This was confirmed by treating $U(CD_3)_2$/DA: He,0°,1 by D_2,25°,1 to give methane (CH_4/U - 0.17) which was 86% CD_4 made mainly by path H and 14% CH_3D which was presumably made by path R. A subsequent D_2,100°,1 gave methane (methane/U=0.30) which was 76% CD_4.

B. Activation of $[Cp'_2ThH_2]_2$

During the preparation of ThH_2/DA, hydrogen was evolved during the evaporation of pentane at 0°C in helium flow, but little additional hydrogen was liberated by a following He,25°,1. With 10.7 μmol (as monomer) of ThH_2, the hydrogen liberated during He,0°,1 was 2.4 μmol which makes H_2/Th = 0.22. Presumably, this hydrogen results from protonolysis, Eq. 9. As shown

$$ThH + \sigma\text{-}OH \longrightarrow \sigma\text{-}O\text{-}Th + H_2 \qquad (9)$$

in Table I, the amount of σ-OH on 0.25 g of DA was about 9 mol or several times the amount of H_2 liberated.

C. Attempted Measurement of the Amount of M-H

Hydrogenolysis of M-CH_3 should form M-H Eq. 8. We attempted to measure the amount of M-H by exposing the catalyst to a pulse of methyl chloride. In solution, the reaction, Eq. 10. is essentially quantitative.[10] $M(CH_3)_2$

$$\text{actinide-H} + CH_3Cl \longrightarrow \text{actinide-Cl} + CH_4 \qquad (10)$$

catalysts were pretreated He,0°,1; H_2,100°,1; He,100°,1; He,25° and a 140-200 μmol pulse of methyl chloride was injected into flowing helium. The evolved products were collected on silica gel at -196°C and analyzed by GC. Results are shown in Table IV.

A sample of 12.1 μmol of ThH_2 (as monomer) on DA was pretreated He,0°,1 ; H_2,100°,1, used in a propylene hydrogenation run at -63°C, then treated He,25°,1, and 159 μmol of methyl chloride was passed over the catalyst. The CH_4/Th was 0.090.

TABLE IV

Reaction of CH_3Cl with $U(CH_3)_2$/DA and $Th(CH_3)_2$/DA at 25°C

	CH_4/Th from 49.4 μmol $Th(CH_3)_2$	CH_4/U from 23.1 μmol $U(CH_3)_2$
He,0°,1	0.07	0.12
D_2,100°,1	0.19	0.34
He,100°,1	0.07	0.06
He,25°,1	0.005	0.01
He,25°,1	0.002	0.002
CH_3Cl pulse[a]	0.020[b]	0.032[c]

[a]CH_4/M from CH_3Cl pulse in He carrier at 25°C.

[b]From 140 μmol pulse of CH_3Cl.

[c]From 207 μmol pulse of CH_3Cl.

TABLE V

Hydrogenation of Propylene on $Cp'_2U(CH_3)_2$ at -47°C

Pretreatment[a]	H_2,0°,l	H_2,50°,l	H_2,100°,l	H_2,150°,l
Flow rate of C_3H_6, $cm^3\ min^{-1}$	7	10.3	14.1	19.1
Conversion	0.965	0.997	0.999	1.000
$\underline{N}_t$[b], sec^{-1}	0.35	0.53	0.73	1.03

[a]Following evaporation of pentane in He,0°l. The catalyst was then treated H_2,0°,l, a hydrogenation was performed at -47°C, the catalyst was treated H_2,50°,l, etc.

[b]$\underline{N}_t$ per molecule of $U(CH_3)_2$.

IV. HYDROGENATION OF PROPYLENE

A. Hydrogenation on Various Supported Complexes

In toluene solution at 24°C and 1 atmosphere of H_2, $[Cp'_2Th(\mu\text{-}H)H]_2$ is moderately active for the hydrogenation of 1-hexene. The turnover frequency in sec^{-1} per atom of Th was 1.5×10^{-4}.

In the first runs,[4] it was necessary to lower the temperature of hydrogenation experiments to -47°C to get conversions of around 50%. A new apparatus containing a catalyst preparation-reactor unit especially designed for organoactinide complexes was then constructed. Results like those shown in Table V were obtained.

Accordingly, the temperature employed in the hydrogenation runs was lowered to -63°C, which is about the lowest temperature which can be employed without filling the catalyst pores with liquid propylene. Results of some sample runs on $U(CH_3)_2$/support are shown in Table VI. Conversions were usually greater than 50% on the more active catalysts. Furthermore, there may have been mass and heat transfer problems. Despite this, we give values of $\underline{N}_t$ to permit rough intercomparisons of activities. Because of the situation just described, we have been unable to study the kinetic orders of the hydrogenations.

TABLE VI

Hydrogenation of Propylene on $Cp'_2U(CH_3)X$/Support at -63°C

mol[a]	CH_4U[b]	Activation[c]	N_t sec^{-1}	CO/complex[d]
		$U(CH_3)_2DA$ at -63°C		
11.9	0.10	H_2,0°,1	0.58	
		H_2,100°,1[e]	0.89	
		H_2,150°,1[e]	0.92	0.019
13.8	0.17	H_2,100°,1	> 0.96[f]	
10.2	0.29	H_2,150°,1	0.45	
19.5	0.11	He,100°,1	0.89	
		$U(CH_3)_2$/PDA at -47°C		
13.8	1.12	H_2,100°,1	0.002	
		$U(CH_3)Cl/SiO_2$ at -47°C		
19.0	1.05	H_2,100°,1	0.0008	
		$U(CH_3)Cl$/DA at -63°C		
27.2	0.14[h]	H_2,100°,1	0.03	0.009

[a] μmol of complex on 0.25 g Al_2O_3.

[b] Methane liberated per U during He,0°,1.

[c] After He,0°,1.

[d] CO adsorbed per atom of U at 25°C from helium carrier.

[e] After preceding activation.

[f] Essentially 100% conversion.

[g] 10.8 μmol on 0.25 g of Davison grade 62 silica gel gave 0.03 sec^{-1} at 25°C.

[h] CH_4/U was 0.38 during the following H_2,100°,1.

Activated $U(CH_3)_2$/DA is a very active hydrogenation catalyst with N_t's approaching 1 sec^{-1} at -63°C. At -63°C on Pt/SiO_2, N_t per Pt_s would be 0.05 sec^{-1}.[11] Rhodium is usually considered to be the most active catalyst for hydrogenation of the group VIII metals. We measured the rate of hydrogenation on a 0.78 wt percent Rh/SiO_2. The catalyst had been made by Dr. V. Eskinazi by ion exchange of $Rh(NH_3)_5(H_2O)^{3+}$ with Davison Grade 62 silica gel. The percentage exposed of Rh was 60%. It was pretreated before hydrogenation runs by $\underline{O_2,300^o,0.5}$; $\underline{H_2,300^o,1}$; $\underline{He,300^o,1}$; cool to -63°C in He. $\underline{N}_t$ was 28 sec^{-1}, but such a value may have been distorted by effects of mass and heat transfer.

As shown by the data in Table VI, catalysts of $U(CH_3)_2$ on PDA and silica gel were of low activity and CH_4/U was large during $\underline{He,0^o,1}$.

$U(CH_3)Cl/DA$ had an activity of but a few percent of $U(CH_3)_2/DA$. However, in the absence of results with the latter system, one would have considered $U(CH_3)Cl/DA$ to be an unusually active catalyst.

Table VII shows results with some thorium complexes. Th(neopentyl)$_2$/DA was as active as $U(CH_3)_2/DA$; $Th(CH_3)_2/DA$ and ThH_2/DA were nearly as active. We also carried out a number of experiments with Cp'Th(benzyl)$_3$/DA (Cp'Th(bz)$_3$/DA), all activated $\underline{He,0^o,1}$; $\underline{He,130^o,1}$. A temperature of 130°C rather than 100°C was used in the second step to facilitate desorption of any toluene formed by protonolysis, Eq. 3. Toluene was indeed liberated as was shown by ultraviolet spectroscopic examination of the material condensed in the empty trap at -196°C (mostly pentane) during $\underline{He,0^o,1}$, $\underline{He,100^o,1}$. We have not as yet attempted to obtain quantitative measurements of toluene/Th. In an early experiment at -63°C, 8.0 μmol Cp'Th(bz)$_3$/DA at 25 cc C_2H_6 min^{-1} gave 100% conversion ($\underline{N}_t$ > 3 sec^{-1}). In our equipment, use of ∿ 1 μmol of complex gave poorly reproducible results presumably owing to poisoning of the very small amounts of complex which were employed. However, results with the more active samples indicate that $\underline{N}_t$ is greater than 10 sec^{-1} at -63°C.

Pulses of propylene in deuterium carrier were passed at 0°C over $Th(CH_3)_2/DA$; $\underline{He,0^o,1}$; $\underline{D_2,100^o,1}$. Hydrogenation was 100% and the product was $C_3H_4D_2$ except for ∿1% $C_3H_3D_3$ and a little C_3H_5D.

B. Poisoning of Hydrogenation by Carbon Monoxide

Actinide elements do not form simple carbonyl complexes as is true of those of late transition elements.[3] Nevertheless, carbon monoxide strongly poisons the hydrogenation of propylene.

At 25°C, small amounts of carbon monoxide adsorb on activated complexes supported on DA and the adsorbed carbon monoxide is not removed by passage of helium or hydrogen for many minutes at 25°C. DA

TABLE VII

Hydrogenation of Propylene on Supported Thorium Complexes at -63°C

μmol	CH_4/Th^a	Activation	$\underline{N}_t$	CO/Th
		$Th(CH_3)_2/DA$		
6.6	0.12	H_2,100°,l	0.54	0.027
32.6	0.10	H_2,100°,l	0.51	0.021
		$Th(CH_3)_2/PDA$		
27.6	0.59	H_2,100°,l	0.005	0.002[b]
		$[ThH_2]_2/DA$		
26.8[c]	—	H_2,100°,l	>0.66	0.085
12.1[c]	—	H_2,100°,l	0.44	
		$Th(neopentyl)_2/DA$		
10.0	—	He,100°,l	0.9	0.015[d]
		$Cp_3Th(n\text{-}butyl)/DA$		
13.8	—	He,100°,l	0.000[e]	—

[a]During He,0°,l.

[b]Zero to within the experimental error.

[c]As monomer.

[d]On another sample with 28.5 μmol of complex.

[e]$\underline{N}_t$ was 0.000 at 0°C and 0.06 sec^{-1} at 100°C. Cp_3UCH_3 had similar activity.

itself adsorbed about 0.3 μmol of CO per 0.25 g at 25°C, but DA exposed to pentane as in catalyst preparation and then treated He,0°,l; He,100°,l adsorbed no more than 0.1 as much CO. Some pentane remains on DA after He,0°,l and some of that is released when acetone is passed at 25°C over $U(CH_3)_2/DA$ and $Th(CH_3)_2/DA$.

The amounts of carbon monoxide adsorbed on $U(CH_3)_2/DA$ and $Th(CH_3)_2/DA$ directly after activation or after a subsequent hydrogenation run at -63°C were nearly equal and correspond to CO/M equal 0.02 to 0.03. Further, CO/M was nearly the same for adsorption of CO from helium as from hydrogen carrier. Some examples of values of CO/M are given in

Tables VI and VII. CO/M was larger for ThH_2/DA than for the other cases. Because of the high activity of Cp'Th(bz)$_3$, we could not measure CO/M on catalysts for which hydrogenations were incomplete. Catalysts with 8.0 and 6.0 μmol gave 100% hydrogenation following which CO/Th was measured to be 0.03. However, in ∿1 μmol catalysts, pulses of CO in which CO was several times Th led to at least 90% loss of activity for hydrogenation.

V. POLYMERIZATION OF ETHYLENE

In our studies of the polymerization of ethylene, a series of pulses of ethylene (C_2H_4/M = 1) were passed over a catalyst at 25°C in helium carrier and the amount of ethylene which passed through the catalyst was measured. The catalyst was usually activated He,0°,1; He,100°,1. In a run with U(CH_3)$_2$/DA, nearly 100% of the first pulse of ethylene was adsorbed and even in pulse 45, 30% was adsorbed. The total loss in ethylene expressed as C_2H_4/U was 42 in this pulse sequence. Th(CH_3)$_2$/DA gave nearly the same result. Th(neopentyl)$_2$/DA gave similar results, but we did not protract the experiment to such a large number of pulses. In an experiment with 8.0 μmol Cp'Th(bz)$_3$/DA, a hydrogenation run at -63°C led to 100% conversion, $\underline{N}_t$ > 3 sec^{-1}. Ten pulses of ethylene were then passed over the catalyst at 25°C. The adsorption of ethylene from pulse 10 was 98%. In general it made little difference whether polymerization runs followed activation directly or whether they followed hydrogenation runs. $\underline{N}_t$ per atom of M must initially have exceeded 0.1 sec^{-1} with the more active catalysts.

ThH_2/DA was less effective in polymerization. With 8.9 μmol (as monomer) ThH_2/DA conversions sank from 49 to 0% in pulses 1 to 14. Overall, C_2H_4/Th was 3.6. U(CH_3)Cl had a small activity, but Cp_3Th(n-butyl) had zero activity in pulse 1.

Carbon monoxide injected at 25°C led to partial inhibition of polymerization activity. For example, CO adsorption was measured on the U(CH_3)$_2$/DA described above after pulse 45. CO/U was 0.024. The conversion of a subsequent pulse of ethylene was 0.09.

VI. DISCUSSION

During the preparation of $U(CH_3)_2$/DA and $Th(CH_3)_2$/DA, roughly 10% of the CH_3 groups are converted to methane via three different processes. Presumably the same processes are involved with Th(neopentyl)$_2$/DA and Cp'Th(bz)$_3$/DA, but activation of ThH_2/DA leads to the liberation of H_2. All of these preparations lead to catalysts which are very active at -63°C. They are less active than Rh/SiO_2 if one compares Rh_s to total actinide. There are, of course, other very active hydrogenation catalysts, for example, metallic molybdenum on DA for which $\underline{N}_t$ per Mo_s is greater than 5.6 sec^{-1} at -46°C[12] and some supported dinuclear complexes of molybdenum.[13]

However, adsorption of carbon monoxide in the vicinity of CO/M = 0.03 poisons hydrogenation. Thus, per actual catalytically active site, $\underline{N}_t$ at -63°C is 30 sec^{-1} for $U(CH_3)_2$/DA and $\underline{N}_t$ for Cp'Th(bz)$_3$/DA must be greater. The exact ordering of the catalysts mentioned above is unclear at present since none of the rates of hydrogenation which have been measured on them has been demonstrated to be free of the effects of mass and heat transport.

The high activity which has required the use of low loadings of air-sensitive compounds and the small fraction of molecules of complex converted to active sites have made the supported actinide systems difficult to study. We do not yet know the kinetic orders in hydrogen and propylene. We do not know the precise nature of the sites, nor, therefore, the exact mechanisms of hydrogenation and polymerization. However, some things can be said.

The catalytically active sites must possess some special feature which permits rather strong reaction with carbon monoxide both with catalysts never exposed to hydrogen and from hydrogen on catalysts pretreated $\underline{H_2,100^o,1}$. Carbon monoxide may adsorb to form a formyl.[14]

Since thorium is confined to oxidation number +4, M is confined to +4 throughout the catalytic cycle. Therefore, oxidative addition of hydrogen as in the usual mechanism for hydrogenation on later transition metals is unlikely.[10] Similar problems appear in hydrogenations on rare earths, some complexes of early transition elements and main group elements. Perhaps a mechanism involving concerted steps[10] may be involved on actinide complexes/DA, Eq. 11. Both steps would need to have very low activation

$$\begin{matrix} H_2C{=}CHCH_3 \\ \vdots\ \ \vdots \\ M{-}H \end{matrix} \longrightarrow M{-}CH_2CH_2CH_3; \quad \begin{matrix} H{-}H \\ \vdots\ \ \vdots \\ M{-}CH_2CH_2CH_3 \end{matrix} \longrightarrow M{-}H + C_3H_8 \qquad (11)$$

energies so that $\underline{N}_t$ could be ~ 30 sec^{-1} at -63°C on $Cp'_2U(CH_3)_2$.

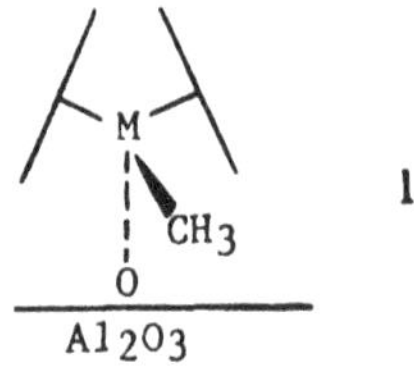

1

Protonolysis of $Cp'_2M(CH_3)_2$ must lead to a species like **1** where the distribution of the four bonds is roughly tetrahedral. In such a species the actinide atom would be severely sterically shielded by the two Cp' units shown as lines. Protonolysis, then, destroys the capacity of a molecule of complex to develop into a catalytic site. Accordingly, PDA and silica gel do not lead to active catalysts. In addition to the effect of shielding there are also electronic effects. Thus, in solution $Th(CH_3)(OCH_3)$ undergoes hydrogenolysis only 10^{-6} times as fast as $Th(CH_3)_2$. Shielding may explain why hydrogenolysis of $M{-}CH_3$ is incomplete compared with that in solution and why reaction of M-H with CH_3Cl is also incomplete. However, on this basis such shielded species should not be active sites for hydrogenation. If the "carbene" is associated with surface $Al^{3+}O^{2-}$, that species would also be severely shielded. Presumably, in the active site, one Cp' ring of a complex is adsorbed flat on the alumina so as to leave the atom of M exposed to reactants. $Cp'Th(bz)_3$ may lead to the most active catalyst because it is the least hindered.

Finally, why is Cp_3MR so inactive compared to complexes containing

$$\begin{matrix} CH_3 \\ / \\ M{-}CH_3 \\ | \\ \text{—}\!\!\!\!\text{—}CH_3 \end{matrix} \longrightarrow \begin{matrix} CH_3 \\ / \\ M \\ |\ \diagdown \\ \text{—}\!\!\!\!\text{—}CH_2 \end{matrix} + CH_4 \xrightarrow{H_2} \begin{matrix} CH_3 \\ / \\ M{-}H \\ | \\ \text{—}\!\!\!\!\text{—}CH_3 \end{matrix} \qquad (12)$$

M–H
$\diagdown$ CH_2 **2**

Cp' units? Conceivably during activation in helium the processes shown in Eq. 12 may occur to generate the active site, M-H. Alternatively, the middle species above may resist hydrogenolysis and the active M-H results from hydrogenolysis of the residual M-CH_3 to give as the active site, **2.** However, the M-CH_3 bond appears to be less reactive in Cp_3M-CH_3 in general than in $Cp'_2M(CH_3)_2$, for example in CO insertion.[15]

ACKNOWLEDGMENT

The experimental work described in this paper was performed by Ming-Yuan He from the Research Institute of Petroleum Processing, Beijing, People's Republic of China and Guoxing Xiong of the Dalian Institute of Chemical Physics, Dalian, People's Republic of China. We are indebted to P.J. Toscano, H.A. Stecher, and R.S. Sternal for the preparation of the complexes. This work was supported by the Division of Chemical Sciences of the Department of Energy under contract DEAC 02-81ER10980.

REFERENCES

1. D.C. Bailey and S.H. Langer, Chem. Rev., **81**, 109 (1981).
2. T.J. Marks, Science, **217**, 989 (1982).
3. T.J. Marks and R.D. Ernst in **Comprehensive Organometallic Chemistry** (G. Wilkinson, F.G.A. Stone and E.W. Abel, eds.), Pergamon Press, Oxford, 1982, Chapt. 21.
4. R.G. Bowman, R. Nakamura, P.J. Fagan, R.L. Burwell, Jr., and T.J. Marks, J. Chem. Soc., Chem. Commun., 257 (1981).
5. R.G. Bowman and R.L. Burwell, Jr., J. Catal., **63**, 463 (1980).
6. M.-Y. He, R.L. Burwell, Jr., and T.J. Marks, Organometallics, **2**, 566 (1983).
7. F.N. Tebbe and R.L. Harlow, J. Am. Chem. Soc., **102**, 6149 (1980).
8. K.C. Ott, J.B. Lee, and R.H. Grubbs, J. Am. Chem. Soc., **104**, 2942 (1982).

9. S.H. Pine, R. Zahler, D.A. Evans, and R.H. Grubbs, J. Am. Chem. Soc., **102**, 3270 (1980).

10. P.J. Fagan, J.M. Manriques, E.A. Maatta, A.M. Seyam, and T.J. Marks, J. Am. Chem. Soc., **103**, 6650 (1981).

11. P.H. Otero-Schipper, W.A. Wachter, J.B. Butt, R.L. Burwell, Jr., and J.B. Cohen, J. Catal., **50**, 494 (1977).

12. R.G. Bowman and R.L. Burwell, Jr., J. Catal., **63**, 463 (1980), Table I; J. Catal., **88**, 388 (1984).

13. Y. Sato, Y. Iwasawa, and H. Kuroda, Chem. Lett., 1101 (1982).

14. P.J. Fagan, K.G. Moloy, and T.J. Marks, J. Am. Chem. Soc., **103**, 6959 (1981).

15. D.A. Sonnenberger, E.A. Mintz, and T.J. Marks, J. Am. Chem. Soc., **106**, 3484 (1984).

13

Solvated Metal Atom Dispersed (SMAD) Catalysts
Highly Active Bimetallic Systems

Yuzo Imizu[a] and Kenneth J. Klabunde
Department of Chemistry
Kansas State University
Manhattan, KS 66506

ABSTRACT

New bimetallic catalytic materials have been prepared by the solvated metal atom dispersion (SMAD) procedure. A series of zero valent bimetallic composits Co-Cr/SiO_2, Co-Mn/SiO_2, and Co-Fe/SiO_2 were prepared and studied in catalytic test reactions and by H_2 chemisorption. A dramatic increase in activity for the Co/SiO_2 catalyst was found when Mn was incorporated into the Co particles; the increase being sufficient to create a catalyst active enough to hydrogenate 1-butene at -60°C in a diffusion controlled fashion. Selectivity/activity studies coupled with H_2 chemisorption results show that over half of the Co atoms are surface atoms, and that the Co clusters are probably affected by the Mn in an electronic fashion.

I. INTRODUCTION

The use of two or more metallic components in the preparation of heterogeneous catalysts has great potential for generating systems with

---

[a]On leave from Kitami Institute of Technology Kitami, Hokkaido 090, Japan.

increased activities, higher selectivities, and greater resistance to poisoning.[1] Indeed, the effect of adding a second metal to an active transition metal catalyst has received considerable study.[2] Sinfelt and coworkers[3] have worked extensively with bimetallic systems consisting of combinations of metals from Group VIII (Fe, Co, Ni families) and Group IB (Cu, Ag, Au). These materials showed considerably improved selectivities for a number of hydrocarbon reactions.[3] Some of these "bimetallic cluster catalysts" have been applied very successfully in commercial reforming of petroleum fractions.

Recently interest has been expressed in the preparation of "early transition metal - late transition metal" combinations, anticipating new and unusual carbon monoxide activation catalysts.[4] Hoever, there are considerable experimental problems involved in conventional preparative methods with such systems.

We have reported a new method of preparation of highly dispersed metallic catalysts by solvated metal atom dispersion (SMAD).[5,6,7] This method allows zero valent, ultra small metal particles to be dispersed on catalyst supports at room temperature or below, and no further reduction step is necessary.[6] The resulting pseudoorganometallic supported particles are completely new materials which have exhibited extremely high catalytic activities in a wide variety of reactions.[7]

Reported herein is an extension of the SMAD method where we have prepared bimetallic catalysts by simultaneous vaporization and solvation of two metals. This investigation has led to a new Co-Mn/SiO_2 catalyst that has shown profoundly improved catalytic activities compared with SMAD Co/SiO_2 or Mn/SiO_2 systems.[8] This dramatically improved activity is a new observation in bimetallic catalyst work. Normally selectivity is varied by addition of a second metal.[3,9]

In order to characterize our bimetallic systems (Cr-Co, Mn-Co, and Fe-Co/SiO_2), we carried out reaction rate studies and examined surface properties. Three test reactions were carried out: hydrogenation/isomerization of 1-butene, hydrogenation of 1,3-butadiene, and hydrogenation/hydrogenolysis of cyclopropane. Extensive hydrogen chemisorption studies were also carried out.

II. EXPERIMENTAL

A. Apparatus

Catalytic hydrocarbon reactions were carried out in a recirculation reactor with a volume of 496 ml. The products were analyzed periodically using an on-line gas chromatograph with a 1/4 in x 20 ft copper column packed with 30% propylene carbonate on 60/80 mesh chromosorb W (acid washed) thermostated at 0°C.[10]

The supported bimetallic catalysts were prepared using a stationary metal atom-vapor reactor as previously described.[11] For this work a special version was equipped with four water cooled copper electrodes, two W-Al_2O_3 metal vaporization crucibles, and two separate power supplies so that two metals could be vaporized simultaneously and independently.

Hydrogen chemisorption measurements were carried out in a conventional Pyrex glass volumetric adsorption apparatus having a volume of 32 ml when a 1 g sample was attached in a sample cell of 19 ml.[12]

B. Materials

The catalyst support was Ketjen SiO_2 R-10280 supplied by the Phillips Petroleum Company. This was ground and sieved to 20-40 mesh and had a surface area of 56 m^2/g and a pore volume of 1.06 ml/g. The metal components were supplied by Fisher (Fe), Matheson (Cr, Co), or Cerac, Inc. (Co). Toluene was used exclusively as the metal atom solvation/deposition medium, and was purified by refluxing over benzophenone ketyl followed by repeated freeze-pump-thaw degassing cycles.

The hydrocarbons were purified by passing through 4 Å molecular sieves at -78°C. Hydrogen was separated from traces of oxygen and water by passing through 4 Å molecular sieves at -78°C followed by freezing/condensing at -196°C into 13X molecular sieves. Just before use hydrogen was removed from the sieves under vacuum (only the initial portion was used).

C. Procedure

The SMAD bimetallic catalysts were prepared in a manner similar to that used in earlier single component SMAD catalysts.[7] Two metals were vaporized simultaneously (0.2 - 1 g of each) while 100 ml of toluene was inletted as a vapor. The three components immediately condensed on the liquid nitrogen cooled walls of the reaction vessel. After deposition, meltdown of the resulting matrix was permitted and the toluene-metal atom/cluster solution allowed to flow onto 20 g of SiO_2 (previously heat treated at 500°C for 3 hr in dry air, cooled and handled under pure nitrogen prior to and during transfer to the reaction chamber).

After preparation, a sample of the prepared catalyst (0.02 - 0.2 g) was placed in a Pyrex U-tube reactor with stopcocks on both ends (an inert atmosphere box was used for manipulations). The catalyst was outgassed at room temperature for about 30 min to $\leq 1 \times 10^{-5}$ Torr and used without further treatment.

The test reactions for 1-butene hydrogenation were carried out at -60°C, isomerization at -20°C, 1,3-butadiene hydrogenation at 0°C, and cyclopropane hydrogenation and hydrogenolysis at 0°C. Initial reaction rates were determined graphically from the slopes of product accumulation curves (as a function of reaction time). Unless otherwise specified, values were calculated based on the total number of Co atoms neglecting the contribution of the second metal.

Hydrogen adsorption isotherms were measured volumetrically at room temperature following outgassing. Reversible H_2 uptake was measured following the completion of the isotherm for total adsorption by outgassing at room temperature for 30 min and then measuring a second isotherm. The irreversible uptake was determined by taking the difference between the total gas uptake and the reversible uptake. The H_2 uptake values in Table I were determined at an equilibrium pressure of 100 Torr. The metal dispersions were calculated from the irreversible H_2 uptake data.

Elemental analyses were performed by Galbraith Laboratories, Inc. Samples were handled under inert atmosphere.

III. RESULTS

A. Hydrogen Chemisorption Over Co-Mn/SiO_2

Room temperature hydrogen adsorption isotherms are shown in Fig. 1 and 2. Irreversible H_2 uptake was only slightly pressure dependent. Since the total

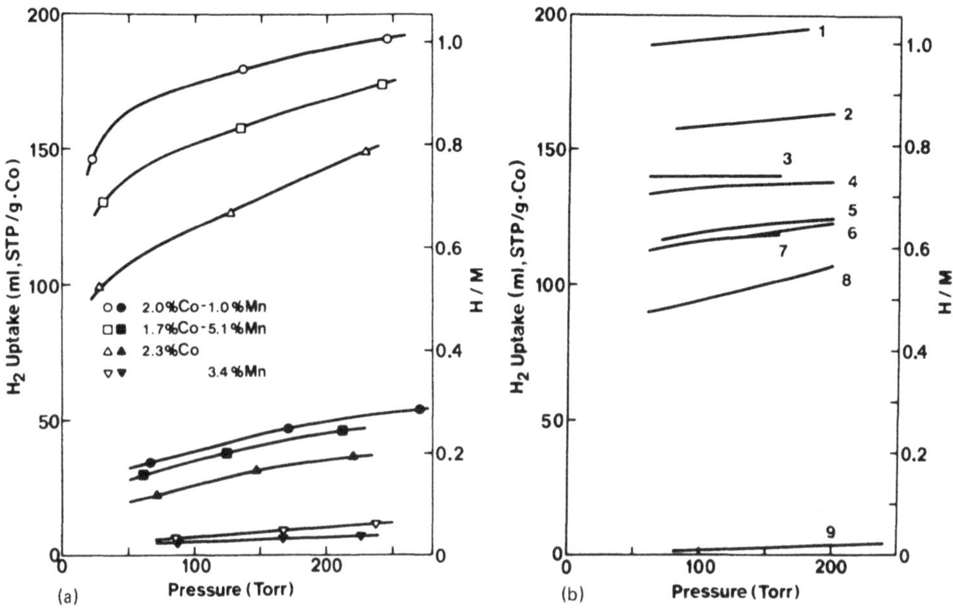

Fig. 1(a) Total and reversible H_2 adsorption at room temperature over bimetallic SMAD Co-Mn/SiO_2 Catalysts. Open symbol = Total; solid symbol = reversible.

Fig. 1(b) Irreversible H_2 adsorption at room temperature over bimetallic SMAD Co-Mn/SiO_2 catalysts.
1 = 3.6% Co - 0.8% Mn/SiO_2; 2 = 3.4% Co - 3.3% Mn/SiO_2;
3 = 3.3% Co - 2.6% Mn/SiO_2; 4 = 2.0% Co - 1.1% Mn/SiO_2;
5 = 3.2% Co - 2.5% Mn/SiO_2; 6 = 1.7% Co - 5.1% Mn/SiO_2;
7 = 4.1% Co - 0.1% Mn/SiO_2; 8 = 2.3% Co/SiO_2;
9 =3.4% Mn/SiO_2.

and the reversible H_2 uptake over Mn/SiO_2 were almost indistinguishable, and significantly smaller than that over Co-Mn/SiO_2, the data were obtained at an equilibrium pressure of 100 Torr (neglecting Mn atom contribution since Mn does not chemisorb H_2 effectively). The data are summarized in Table I. Dispersion (the fraction of surface exposed Co atoms) was calculated based on a 1:1 stoichiometry of hydrogen atoms to surface atoms.[13] Dispersion values exceeded 0.5 for Co-Mn/SiO_2 and Co/SiO_2.

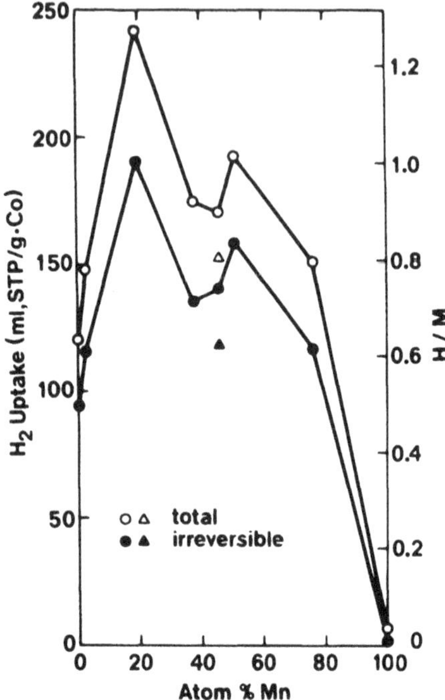

Fig. 2 Total and irreversible H_2 adsorption at room temperature over a bimetallic SMAD Co-Mn/SiO_2 catalyst as a function of composition. Triangular symbol = the catalyst was prepared by sequential co-deposition instead of tri-deposition.

Superdispersion was observed for 3.6% Co - 0.8% Mn/SiO_2 (about half the SiO_2 surface remained uncovered at this loading level). These results clearly demonstrate that the SMAD procedure leads to very highly dispersed metallic and bimetallic catalysts.

B. Elemental Analyses of Bimetallic SMAD Catalysts

Table II summarizes the C,H content of many of the catalysts. As with single component SMAD catalysts,[6,7] bimetallic systems incorporate carbonaceous fragments. The Cr/SiO_2 and Co-Cr/SiO_2 catalysts were quite rich in these fragments. The ratio of metal atoms to carbon atoms was

TABLE I

H_2 Uptake Over SMAD Co-Mn/SiO_2 Catalysts[a]

Catalyst[b]	at.% Mn	H_2 uptake[c] (ml, STP/g·Co) Total	Irreversible[d]	H/M[e]	Surface area[f] M^2/g·Co
2.3% Co	0	121	95	0.50	337
4.1% Co - 0.1% Mn	2.5	149	116	0.61	413
3.6% Co - 0.8% Mn	19.3	242	191	1.01	677
2.0% Co - 1.1% Mn	37.6	175	136	0.72	484
3.3% Co - 2.6% Mn	45.8	171	141	0.74	500
3.2% Co - 2.5% Mn[g]	46.0	154	119	0.63	423
3.4% Co - 3.3% Mn	51.0	193	159	0.84	565
1.7% Co - 5.1% Mn	76.3	152	117	0.61	416
3.4% Mn	100	7	2	—	—

[a]Measured at room temperature.

[b]SiO_2 support, wt, %.

[c]Data at an equilibrium pressure of 100 Torr neglecting Mn contribution.

[d]Obtained from the difference between the total and reversible H_2 uptakes.

[e]1:1 stoichiometry of H atom per surface Co atom was adopted.

[f]Data based on number of surface Co atoms = 1.51×10^{19} m^{-2} in Ref. 12, p. 296.

[g]Prepared by sequential co-deposition instead of tri-deposition (Mn - toluene first followed by Co-toluene).

consistent with the calculated ratio using the M/C values for the single metal catalysts, suggesting that carbonaceous incorporation is not affected by the intimate contact of the two metals.

C. Hydrogenation of 1-Butene

Product accumulation curves, determined as a function of reaction time are

TABLE II

Elemental Analyses of Bimetallic SMAD Catalysts

Catalyst[a]	at.%[b]	C(wt.%)	H(wt.%)	C/Mtot	C/Mtot[c]
2.6% Co - 1.6% Fe	39.4	0.65	0.17	0.74	1.09
1.5% Co - 3.4% Fe	70.5	0.84	0.19	0.81	0.82
4.1% Co - 0.1% Mn	2.5	1.10	0.21	1.28	1.40
3.3% Co - 2.6% Mn	45.8	0.86	0.15	0.69	0.97
3.2% Co - 2.5% Mn	46.0	0.77	0.15	0.64	0.97
1.7% Co - 5.1% Mn	76.3	0.53	0.17	0.36	0.66
3.6% Co - 0.7% Cr	18.1	1.18	0.18	1.32	1.69
2.0% Co - 2.2% Cr	55.5	1.98	0.31	2.16	2.21
0.6% Co - 3.7% Cr	87.5	1.90	0.31	1.94	2.66
2.3% Co		0.67	0.19	1.43	
4.7% Fe		0.57	0.07	0.56	
3.4% Mn		0.32	0.05	0.42	
3.8% Cr		2.52	0.41	2.84	
SiO_2[d]		0.051	0.04	——	
SiO_2[e]		0.12	0.05	——	

[a]SiO_2 support, wt. %.

[b]Atomic % of second metal.

[c]Calculated data based on data for monmetallic SMAD catalysts.

[d]Sample prepared by preheat treating at 500°C in dry air for 3h.

[e]Sample prepared by full SMAD procedure without metal vapor deposition.

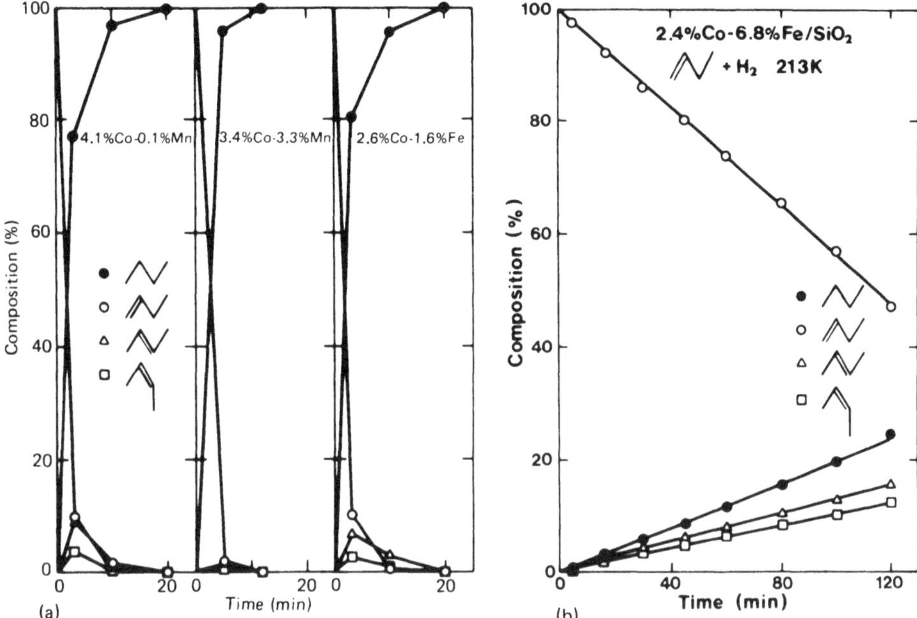

Fig. 3(a) Product accumulation curves as a function of reaction time at -60°C for 1-butene hydrogenation over bimetallic SMAD catalysts ($H_2/HC = 5/3$; $P_{tot} = 80$ Torr).

Fig. 3(b) Product accumulation curves as a function of reaction time at -60°C for 1-butene hydrogenation over a bimetallic SMAD Co-Fe/SiO_2 catalyst ($H_2/HC = 5/3$; $P_{tot} = 80$ Torr).

shown in Fig. 3. Initial reaction rates and selectivities calculated from these data are summarized in Table III. The most striking feature is the increase in the hydrogenation rate when Co is incorporated with Mn (note that Mn by itself is an inactive catalyst). Indeed, only 2.5 atom percent Mn increased the activity for hydrogenation by a factor of 10^2. Further increases were observed by continuing addition of Mn up to 51 atom percent. Since these catalysts were so active, even at -60°C (213 K), initial rates could not be determined accurately (essentially diffusion control). Lower catalytic reaction temperatures led to experimental difficulties due to the liquifaction of 1-butene. Despite these experimental difficulties, it is clear from Table

TABLE III

Initial Reaction Rates for 1-Butene Isomerization, 1-Butene Hydrogenation, 1,3-Butadiene Hydrogenation, and Cyclopropane Hydrogenation and Hydrogenolysis over Monometallic and Bimetallic SMAD Catalysts

		Initial Rates (molecules/M·atom·sec)[a]			
Catalyst[b]	at.%[c]	Isomerization of 1-butene[d]	Hydrogenation of 1-butene[e]	Hydrogenation of 1,3-butadiene[f]	Hydrogenation of cyclopropane[g]
2.6% Co - 1.6% Fe	39.4	0.017	0.44 —	0.14	
1.5% Co - 3.4% Fe	70.5	0.0096	0.0063 (0.00099)[i]	0.029	
2.4% Co - 6.8% Fe	74.9	0.00087	0.0016 (0.0017)[i]	0.0049	
4.1% Co - 0.1% Mn	2.5	0.030	0.57 —	0.30	0.0030 (0.00032)[j]
3.6% Co - 0.8% Mn	19.3	0.084	0.45 —		0.0056 (0.00057)[j]
2.0% Co - 1.1% Mn	37.6	0.025	1.1 —		0.0039 (0.00029)[j]
3.3% Co - 2.6% Mn	45.8	0.032	0.72 —		0.0037 (0.00033)[j]
3.2% Co - 2.5% Mn[h]	46.0	00.46	0.84 —		0.0056 (0.00060)[j]
3.4% Co - 3.3% Mn	51.0	0.025	0.60 —	0.099	0.0029 (0.00019)[j]
1.7% Co - 5.1% Mn	76.3	0.026	0.026 (0.018)[i]	0.0087	0.0057 (0.00055)[j]
3.6% Co - 0.7% Cr	18.1	0.0034	0.0051 (0.0034)[i]	0.011	
2.0% Co - 2.2% Cr	55.5	0.00022	0.0028 (0.0019)[i]	0.011	
0.6% Co - 3.7% Cr	87.5	0	0.0047 (0.0020)[i]	0.010	

2.3% Co	0.013	0.0058 (0.0060)[i]	0.024	0.00077 (0.000083)[j]
4.7% Fe	0.00053	0.00022 (0.0013)[i]	0.0052	
3.4% Mn	0	0.000013 (0.000029)[i]	0.00030	0
3.8% Cr	0	0 (0)	0	0

[a]Data for Co-Fe and Co-Cr were based on 100% Co dispersion neglecting second metal contribution. Data for Fe, Cr, and Mn were based on 100% dispersion of each metal.

[b]SiO_2 support, weight %.

[c]Atomic % of second metal.

[d]P_{HC} = 30 Torr, without H_2, -20°C.

[e]H_2/HC = 1.67, P_{tot} = 80 Torr, -60°C.

[f]H_2/HC = 2, P_{tot} = 60 Torr, 0°C.

[g]H_2/HC = 3, P_{tot} = 120 Torr, 0°C.

[h]Prepared by sequential co-deposition instead of tri-deposition (Mn-toluene first followed by Co-toluene).

[i]Data for accompanying isomerization.

[j]Data for accompanying hydrogenation.

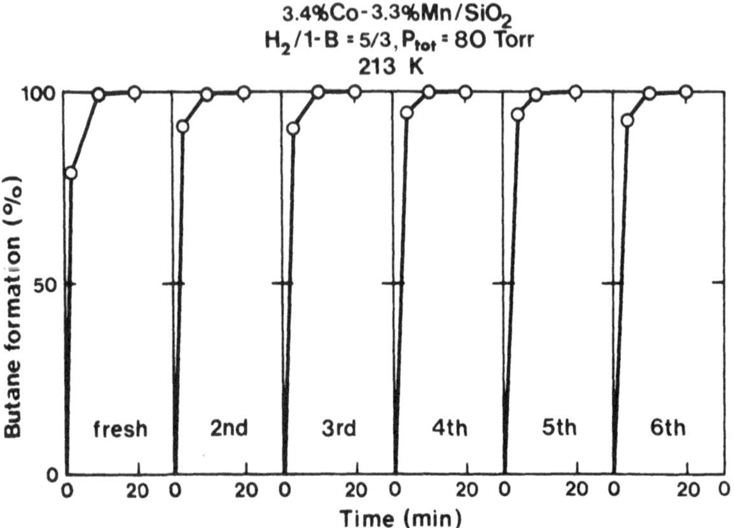

Fig. 4 Repetitive 1-butene hydrogenation experiments over a bimetallic SMAD Co-Mn/SiO_2 catalyst. 1-B/H_2 = 0.56, P_{tot} = 98 Torr on 6th experiment.

III that the activity of Mn/SiO_2 is 10^2 lower than that of Co/SiO_2 and at least 10^4 lower than that of Co-Mn/SiO_2.

These bimetallic SMAD catalysts exhibited long life times as demonstrated by repetitive 1-butene hydrogenation (Fig. 4). After six repetitive experiments in which a total of 0.54 mmole of 1-butene was converted over 9.7 μmole of Co active sites on the surface, no sign of deactivation was observed. Thus, hydrogenation sites turned over more than 560 times without any deactivation.

The Co-Cr and Co-Fe/SiO_2 catalysts were also examined. Both Cr/SiO_2 and Fe/SiO_2 were less active than Co/SiO_2. Addition of Fe to Co/SiO_2 caused an increase in activity, but large amounts of Fe were detrimental. In the case of Cr, all loadings were ineffectual on the Co activity.

D. Hydrogenation of 1,3-Butadiene

Initial reaction rates calculated from the rate of disappearance of 1,3-

TABLE IV

Selectivities for 1,3-Butadiene Hydrogenation over Bimetallic SMAD Catalysts[a]

Catalyst[b]	Conversion %	Composition of butenes, % 1	trans	cis	Selectivity[c]
2.6% Co - 1.6% Fe	37.4	43.0	35.9	21.1	0.41
1.5% Co - 3.4% Fe	19.6	67.2	16.3	16.5	0.96
2.4% Co - 6.8% Fe	23.4	66.5	14.7	18.8	0.96
4.1% Co - 0.1% Mn	25.1	39.1	38.8	22.1	0.57
3.4% Co - 3.3% Mn	18.0	57.0	25.7	17.3	0.56
1.7% Co - 5.1% Mn	22.8	68.8	17.4	13.8	0.95
3.6% Co - 0.7% Cr	16.0	66.0	18.8	15.2	0.96
2.0% Co - 2.2% Cr	24.6	55.7	27.4	16.9	0.94
0.6% Co - 3.7% Cr	13.2	50.3	31.6	18.1	0.95
2.3% Co	22.8	67.8	17.3	14.9	0.98
4.7% Fe	28.5	67.2	18.0	14.8	0.97
3.4% Mn	17.0	75.6	13.9	10.5	1
3.8% Cr	inactive	——	——	——	——

[a]Reaction conditions, H_2/HC = 2, P_{tot} = 60 Torr, 0°C.

[b]SiO_2 support, wt. %.

[c]Ratios of butenes to (butenes + butane).

butadiene are included in Table III while selectivities are included in Table IV. Rates of reactions were similar to those found with 1-butene. Less active 1-butene hydrogenation catalysts behaved as selective catalysts for 1,3-butadiene conversion to butenes (one double bond hydrogenated). However, highly active 1-butene hydrogenation catalysts, such as 3.4% Co - 3.3% Mn/SiO_2, rapidly converted both double bonds in the diene, and were not selective to butenes. However, this was a transient phenomenon. After

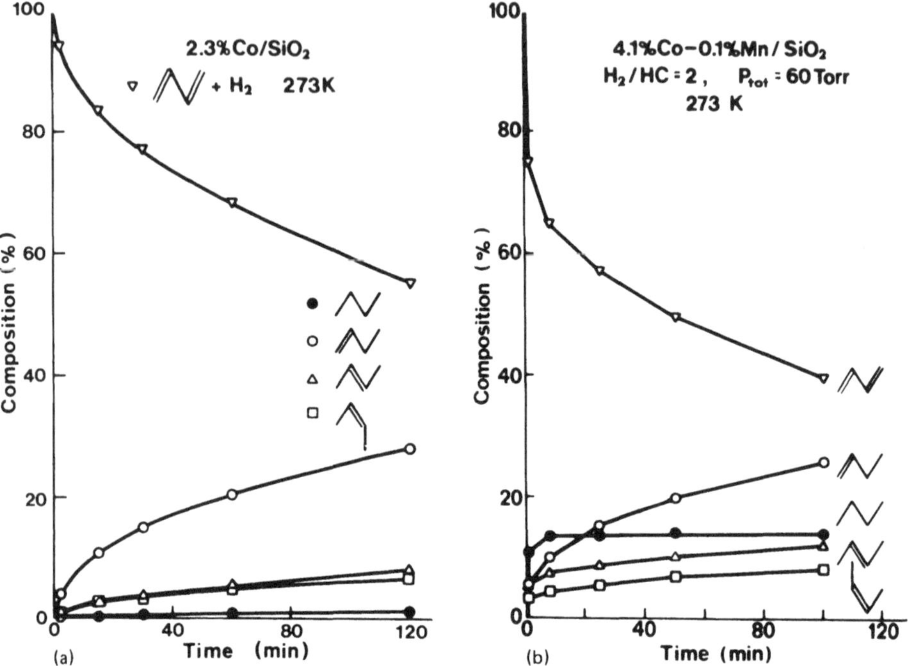

Fig. 5(a) Product accumulation curves as a function of reaction time at 0°C for 1,3-butadiene hydrogenation over a SMAD catalyst ($H_2/HC = 2$; $P_{tot} = 60$ Torr).

Fig. 5(b) Product accumulation as a function of reaction time at 0°C for 1,3-butadiene hydrogenation over a bimetallic SMAD catalyst.

a certain reaction time, one double bond conversion was observed (Fig. 5 and 6).

E. Cyclopropane Conversion

Following a 10-20 minute induction time, cyclopropane underwent hydrogenation and hydrogenolysis at 24°C, Eq. 1. Subsequent experiments

$$\text{cyclo-}(CH_2)_3 + H_2 \xrightarrow{\text{hydrogenation}} CH_3CH_2CH_3 \qquad (1)$$

$$\xrightarrow{\text{hydrogenolysis}} CH_4 + CH_3CH_3$$

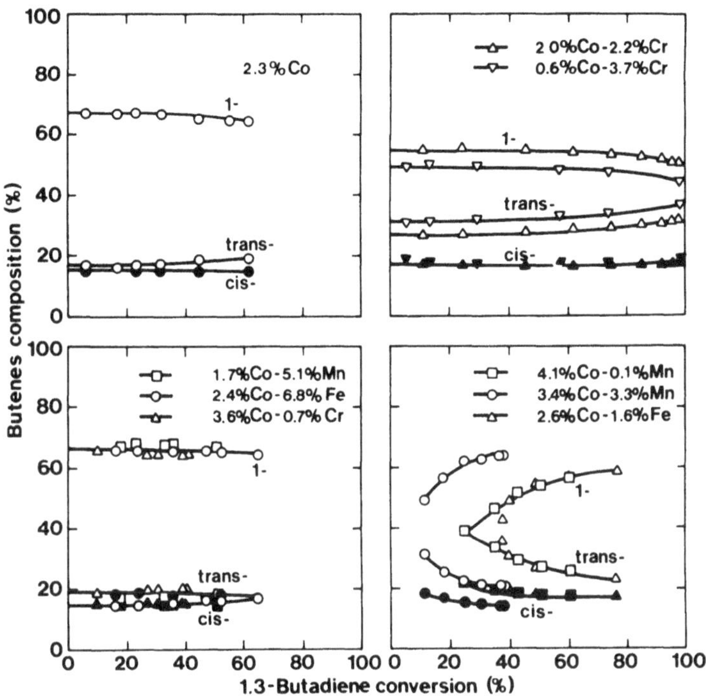

Fig. 6 Integrated product selectivities of 1,3-butadiene hydrogenation as a function of hydrogenation conversion.

with the same catalyst sample did not require an induction period (Fig. 7). Initial rates are shown on Table III. Fig. 8 shows initial rates vs composition of Mn. The interesting feature is that cyclopropane conversion rates for specific catalysts were very similar for both hydrogenation and hydrogenolysis. The most active catalyst was 3.6% Co - 0.8% Mn/SiO_2, which was seven times more active than Co/SiO_2. The Mn/SiO_2 catalyst showed no activity. The hydrogenation/hydrogenolysis ratio did not very significantly with Mn loading, and stayed about 10.

F. Comparison of Catalytic Activities with Other Metal Catalysts

Davis and Somorjai[14] collected and evaluated kinetic data for hydrocarbon reactions over metal catalysts. Comparisons of these data with our SMAD

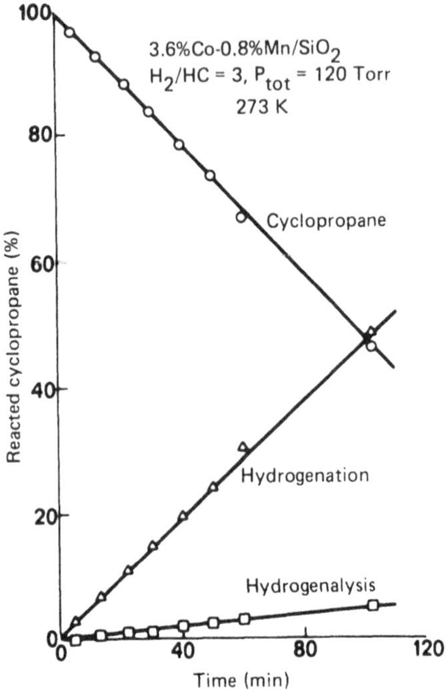

Fig. 7 Product accumulation curves as a function of reaction time at 0°C for cyclopropane conversion over a bimetallic SMAD catalyst.

catalyst data are shown in Table V. The reaction rates at certain temperatures were calculated by using activation energies of 8 kcal/mole and 12 kcal/mole for 1-butene hydrogenation and cyclopropane conversion, respectively. Surprisingly, our SMAD Co based bimetallic catalysts show higher activities for 1-butene hydrogenation than Pt(223) single crystals. Similarly, commercial Co catalysts for cyclopropane conversion were a factor of 50 less active than our Co-Mn/SiO_2 SMAD system.

G. Isomerization of 1-Butene

The isomerization of 1-butene obeyed the first order rate equation (Eq. 2)

$$2.3 \log (X_e - X) = k_t + 2.3 \log X_e \tag{2}$$

where X and X_e represent conversion at time t and at equilibrium,

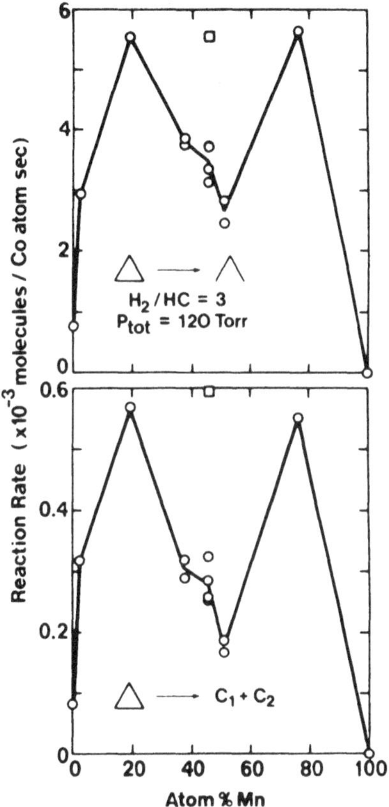

Fig. 8 Manganese composition dependence of the initial rates of cyclopropane conversion at 0°C. Square symbol = the catalyst prepared by sequential co-deposition instead of tri-deposition.

respectively (Fig. 9). This equation holds rigorously for a reactant going to a single product, the reaction being first order in both directions.[15] Initial reaction rates were calculated from the first order reaction rate constants and are included in Table III.

These SMAD bimetallic catalysts catalyzed the 1-butene isomerization reaction in the absence of H_2, in sharp contrast to most conventional catalysts.[16] (This is apparently due to the presence of carbonaceous groupings and M-H species in the SMAD particle.) Addition of the second

TABLE V

Comparison of the SMAD Co-Mn/SiO_2 Catalyst with Precious Metal Catalysts in the Reactions of 1-Butene and Cyclopropane[a]

Catalyst	T(°d)	Rate (molecules/cm^2·sec)
	1-Butene hydrogenation	
3.6% Co - 0.8% Mn/SiO_2[b]	-60	>6.8×10^{14} [c]
3.6% Co - 0.8% Mn/SiO_2[b]	25	>1.5×10^{17} [c,d]
Pt (223)	25	1.3×10^{16}
	Cyclopropane hydrogenation	
35% Co/kieselguhr	25	1×10^{12}
3.6% Co - 0.8% Mn/SiO_2[b]	25	4×10^{13} [c,e]
0.5 - 5% Pt/Al_2O_3	25	7×10^{13}
	Cyclopropane hydrogenolysis	
35% Co/kieselguhr	25	5×10^{11}
3.6% Co - 0.8% Mn/SiO_2[b]	25	6×10^{12}, [c,e]
5% Rh/C		8×10^{12}

[a]The data for Pt and Rh are from Ref. 14.

[b]Prepared by SMAD procedure.

[c]Data based on number of surface Co atoms = 1.51×10^{19} m^{-2} in Ref. 12, p. 296.

[d]Data adjusted from -60°C by using E_a = 8 kcal.

[e]Data adjusted from 0°C by using E_a = 12 kcal.

metal showed different effects. Any amount of Mn and small amounts of Fe increased the activity of the Co/SiO_2 system. However, Cr was detrimental. The compositional data has been plotted in triangular form in Fig. 10. The superposition of the points for Co/SiO_2 and the various Co-M/SiO_2 catalysts indicates that selectivities were not affected by the second metal.

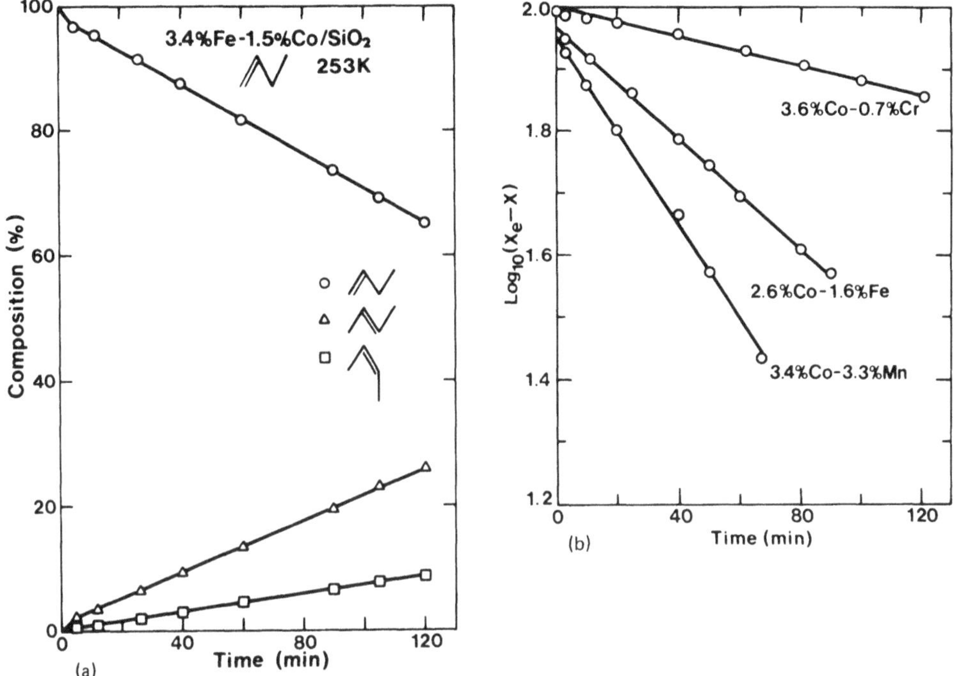

Fig. 9(a) Product accumulation curves as a function of reaction time at -20°C for 1-butene isomerization over a bimetallic SMAD catalyst (P_{req} = 30 Torr).

Fig. 9(b) First-order plots for 1-butene isomerization at -20°C over bimetallic catalysts.

Hydrogenation of 1-butene accompanied isomerization. Selectivity features are shown in Fig. 11. Addition of Fe to the Co/SiO_2 system favored isomerization while Cr addition favored hydrogenation.

IV. DISCUSSION

A. The SMAD Bimetallic Process

These results, demonstrating the remarkable catalytic activities of bimetallic SMAD catalysts, along with H_2 chemisorption data demonstrating extremely high dispersions, support our earlier reports on SMAD

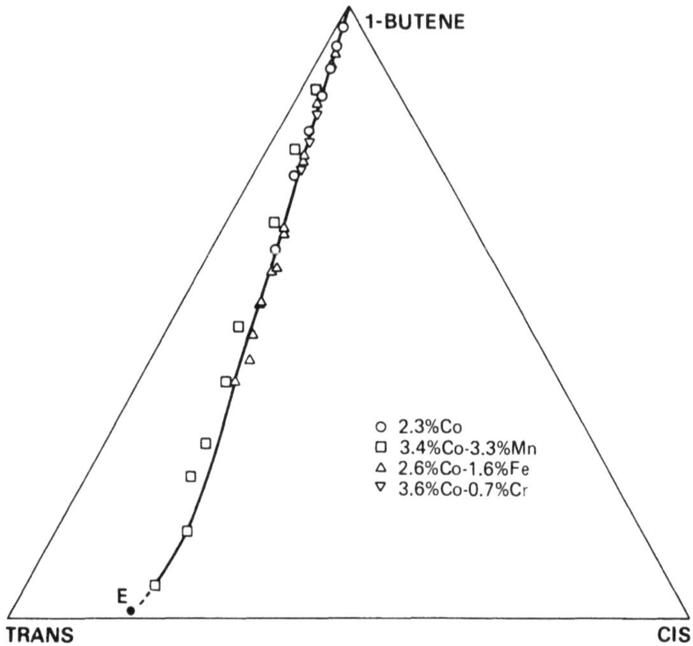

Fig. 10 Mole fraction of products of 1-butene isomerization at -20°C over bimetallic SMAD catalysts.

catalysts.[6,7] These bimetallic systems are even more unusual however, since the second metal has a significant effect on the catalytically active metal.

A closer look at the SMAD process is needed in order to try to understand these effects.

The metals employed here all vaporize monatomically, and are solvated by excess toluene. Upon warming and meltdown, some clustering (nucleation) occurs, but the cluster growth process competes with a reaction channel where the growing clusters react with the host solvent.[6,17] This way carbonaceous groups are incorporated into the small clusters as they form and attach themselves to the support surface. The resultant pseudo organometallic particles are truly new materials and exhibit novel properties.[6,7] The carbonaceous groups appear to have a beneficial effect on activities/selectivities, and also may help in binding of the particle to the

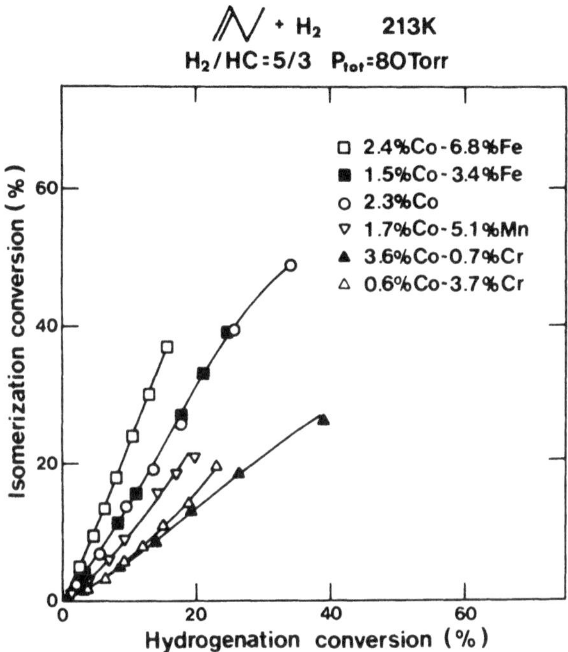

Fig. 11. Isomerization rate vs. hydrogenation rate plots for the reaction of 1-butene and H_2 at -60°C over bimetallic SMAD catalysts.

support.[7c] This beneficial effect of the carbonaceous materials contradicts popular current thought that carbonaceous groups are detrimental. One should note that Somorjai and coworkers have also observed such effects in their single crystal metal work (ethylidyne on Pt and Rh surfaces participated in ethylene hydrogenation).[18,19]

The SMAD process is illustrated in Scheme 1.

In the case of bimetallic SMAD systems the decomposition sequence of the arene complexes may play an important role in controlling the geometrical arrangement of the final metal particles. It is known that bis(toluene)chromium is stable over 100°C,[20] whereas toluene solvated iron decomposes at -79°C.[21] Generally, the ease of decomposition is believed to be $(toluene)_nCo > Fe > Mn > Cr$.[20] As a consequence it would be expected that Co clustering would precede the second metal, and that the second metal

-10-

M + toluene $\xrightarrow{-196°C}$

M' + toluene $\xrightarrow{-196°C}$

-96°C warm and melt

$M_xM'_y$

about -50°C

CH_3 $M_xM'_y$ R H

up to room temp SiO_2

CH_2 M_xM_y R H CH

SiO_2

Mainly C_1 fragments make up the carbonaceous groupings

Scheme I

would deposit on the outside of the Co clusters. It should be noted that a sequential (not simultaneous) co-deposition of metals also provided an active catalyst with similar properties as the normally prepared simultaneous deposition (Table III and Fig. 2). This result indicates that the critical part of the bimetallic SMAD process is not the metal atom deposition, but the cluster growth/reaction/support deposition <u>after</u> meltdown and intimate mixing of the solvated atoms and clusters.

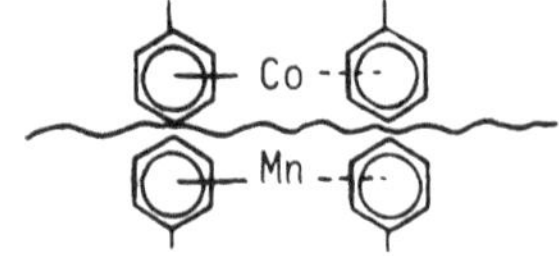

(A layering experiment where toluene/Co was deposited on top of the toluene/Mn matrix, followed by meltdown and contact with SiO_2)

Further work with XPS, EXAFS, and other surface spectroscopic studies are needed in order to characterize these bimetallic SMAD clusters and catalysts.

B. Catalytic Properties of the Bimetallic SMAD Materials

Two concepts must be considered when trying to rationalize the dramatic increase in activity of a Co/SiO_2 catalyst when Mn is added; (1) ensemble/structural effects (2) electronic/ligand effects.[2c,2b,22,23]

For some reactions an increase in catalytic activity could be due to a change in selectivity (a side reaction is inhibited).[2e] For example the Au-Pt alloy was reported to show this behavior in the isomerization of n-hexane.[24] However, such activity increases are not as large as we observe for the Mn effect on Co (10^2 increase in rate even at -60°C for hydrogenation).

Ponec and coworkers[25] have reported that a suppression of hydrogenolysis and an enhancement of hydrogenation of cyclopropane is found for Cu-Ni vs Ni. The decreased hydrogenolysis activity of Ni when Cu was added was explained by surface Cu atoms breaking up the types of arrays of Ni atoms required for hydrogenolysis (ensemble/structural effect). Our work with cyclopropane shows no such change in selectivity when inactive Mn is added to active Co (as compared to Ponec adding inactive Cu to active Ni). So we conclude our results may not be due to ensemble/structural effects.

Similarly, it has been reported that Cu addition to Ni significantly decreased hydrogenation activity for 1,3-butadiene, and the 1-butene/2-butene ratio tended to increase with increasing Cu content.[26] Again, our results with $Co\text{-}Mn/SiO_2$ are quite different; that is, increased activity and little dependence of the 1-butene/2-butene ratio on Mn content. This provides more circumstantial evidence that we are not dealing with an ensemble/structural effect, although further work is clearly warranted on this point.

Hydrogen chemisorption studies showed that more than half of the Co atoms were exposed, and the extent of dispersion did not depend on Mn content. It is also worthy to note only about half of the SiO_2 was covered with Co, according to these data. Furthermore, cyclopropane selectivity

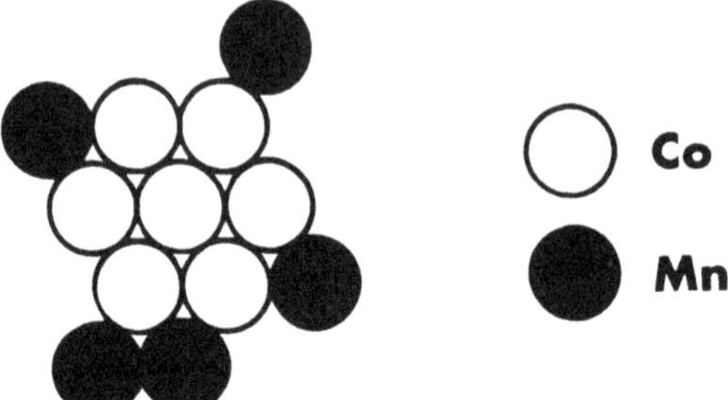

Fig. 12. Schematic surface arrangement of Co and Mn atoms (Top view).

features suggest that Mn atoms do not break up necessary arrays of active Co atoms. Also, butadiene activities/selectivity did not support an ensemble effect.

Based on these considerations and the SMAD procedure, where Co is probably deposited first, we propose a working model where Co atoms are clustered together with Mn atoms (and carbonaceous groups) attached on the outside (Fig. 12). In this way, direct interaction of Mn and Co is possible, arrays of Co atoms could still be available, and the presence of the Mn could have an electronic/ligand effect on the Co particle.

It should again be emphasized that these SMAD materials are new, and direct comparisons with pure metallic alloys perhaps not justified. Thus, these conclusions must be tentative.

V. CONCLUSIONS

Solvated metal atom dispersion of two metals simultaneously or sequentially, followed by exposure of the metal atom solutions to SiO_2, leads to highly dispersed remarkably active bimetallic catalysts. An unusual activating effort of Mn on Co was observed, where a 10^2 increase in activation for hydrogenation was observed. Some Co-Mn/SiO_2 catalysts were so active that hydrogenation of 1-butene took place at -60°C at a diffusion controlled rate. Such activities greatly exceeded those reported for Pt(223) single crystals as well as commercial Co catalysts. Other metals, such as Cr and Fe had less dramatic effects on Co activities.

Studies of selectivities/activities in hydrogenation, isomerization, and hydrogenolysis, and H_2 chemisorption studies led to the tentative conclusion that Mn exhibited an electronic effect on Co, thus enhancing its activity.

ACKNOWLEDGEMENTS

Support of the National Science Foundation and the Phillips Petroleum Company is acknowledged with gratitude.

REFERENCES

1. (a) B.C. Gates, J.R. Katzer, and G.C.A. Schuit, **Chemistry of Catalytic Processes**; McGraw-Hill, New York, 1979; (b) C.N. Satterfield, **Heterogeneous Catalysis in Practice**; McGraw-Hill, New York, 1980.
2. (a) J.H. Sinfelt, Catal. Rev., **9**, 147 (1973); (b) V. Ponec, Catal. Rev., **11**, 41 (1975); (c) W.M.H. Sachtler and R.A. Santen, Adv. Catal., **26**, 69 (1977); (d) D.A. Dowden, In **Catalysis**, (C. Kemball, ed.) (Specialist Periodical Report); The Chemical Society, London, 1978, Vol. 2, Chapter 1; (e) V. Ponec, Adv. Catal., **32**, 149 (1983).
3. (a) J.H. Sinfelt, J.L. Carter, and D.J.C. Yates, J. Catal., **24**, 283 (1972); (b) J.H. Sinfelt, Acc. Chem. Res., **10**, 15 (1977); (c) J.H. Sinfelt, Science, **195**, 641 (1977); (d) J.H. Sinfelt, **Bimetallic Catalysts**; John Wiley & Sons, New York, 1983.
4. (a) G.C. Demitras and E.L. Muetterties, J. Am. Chem. Soc., **99**, 2796 (1977); (b) C.P. Casey, R.M. Bullock and F. Nief, J. Am. Chem. Soc., **105**, 7574 (1983).
5. (a) K.J. Klabunde, H.F. Efner, L. Satek and W. Donley, J. Organomet. Chem., **71**, 309 (1974); (b) K.J. Klabunde, H.F. Efner, T.O. Murdock, and R. Roppel, J. Am. Chem. Soc., **98**, 1021 (1976).
6. K.J. Klabunde and Y. Tanaka, J. Mol. Catal., **21**, 57 (1983).
7. (a) K.J. Klabunde, S.C. Davis, H. Hattori and Y. Tanaka, J. Catal., **54**, 254 (1978); (b) K. Matsuo and K.J. Klabunde, J. Catal., **73**, 216 (1982); (c) K. Matsuo and K.J. Klabunde, J. Org. Chem., **47**, 843 (1982).
8. K.J. Klabunde and Y. Imizu, J. Am. Chem. Soc., **106**, 2721 (1984).
9. A.L. Dent and M. Lin, in **Hydrocarbon Synthesis from Carbon Monoxide and Hydrogen**, ACS Advan. Chem. Ser. **178** (E.L. Kugler, F.W. Steffgen, ed., American Chemical Society, Washington, D.C., 1979, p. 47-63.

10. J.A. McKenna, Jr. and J.A. Idleman, Anal. Chem., **31**, 2000 (1959).
11. K.J. Klabunde, P.L. Timms, S. Ittel and P.S. Skell, Inorg. Syn., **19**, 59 (1979).
12. J.R. Anderson, **Structure of Metallic Catalysts**, Academic Press, New York, 1975, Chapter 6.
13. R.C. Reuel and C.H. Bartholomew, J. Catal., **85**, 63 (1984).
14. (a) S.M. Davis and G.A. Somorjai, **Chem. Phys. Solid Surf Hetrog. Catal.**, (D.A. King and D.P. Woodruff, eds.), Elsevier, Amsterdam, 1982, Vol. IV, p. 217; (b) G.A. Somorjai, **Chemistry in Two Dimensions: Surfaces**, Cornell University Press, Ithaca, 1981, p. 414.
15. J.W. Hightower and W.K. Hall, J. Am. Chem. Soc., **89**, 778 (1967).
16. J.E. Germain, **Catalytic Conversion of Hydrocarbons**, Academic Press, London, New York, 1969.
17. (a) K.J. Klabunde and S.C. Davis, J. Am. Chem. Soc., **100**, 5973 (1978); (b) S.C. Davis, S. Severson and K.J. Klabunde, J. Am. Chem. Soc., **103**, 3024 (1981).
18. G.A. Somorjai, Proc. Welch Foundation Conf. XXV, Heterogeneous Catal., Hilton Hotel, Houston, TX, Nov. 9-11, 1981, p. 83, p. 107.
19. B.E. Bent, B.E. Koel, F. Zaera, and G.A. Somorjai, "Abstracts of Papers", 187th ACS National Meeting, Division of Colloid and Surface Chemistry, St. Louis, Miss., Apr. 8-13, 1984; American Chemical Society: Washington, D.C.
20. P.S. Skell, D.L. Williams-Smith, and M.J. McGlinchey, J. Am. Chem. Soc., **95**, 3337 (1973).
21. K.J. Klabunde, **Chemistry of Free Atoms and Particles**, Academic Press, New York, 1980, Chapters 4 and 5.
22. M. Boudart, Adv. Catal., **20**, 153 (1969).
23. J.A.R. Van Veen, J.F. Van Baar, C.J. Kroese, J.G.F. Coolegem, N. deWit, and H.A. Colijn, Ber. Bunsenges. Phys. Chem., **85**, 693 (1981).
24. (a) J.W.A. Sachtler and G.A. Somorjai, J. Catal., **81**, 77 (1983); (b) G.A. Somorjai and S.M. Davis, Chemtech, **13**, 502 (1983).
25. J.M. Beelen, V. Ponec, and W.M.H. Sachtler, J. Catal., **28**, 376 (1973).
26. P.F. Carr and J.K.A. Clarke, J. Chem. Soc. A, 985-987 (1971).

14

Amphora Catalysts

D.R. Herrington
The Standard Oil Company (Ohio)
4440 Warrensville Center Road
Cleveland, Ohio 44128

ABSTRACT

Unique fixed bed catalysts consisting of hollow spheres with a hole connecting the hollow center to the exterior are termed "amphora" catalysts. With an effective particle thickness 0.25X the overall particle diameter and a high surface-to-volume ratio, such catalysts offer distinct advantages for reactions which are heat and/or mass transfer limited. The amphora shape is produced by subjecting slurry droplets of active catalyst and/or catalyst precursors to controlled drying conditions. A possible mechanism for this physical phenomenon will be discussed. Because the formation of the amphora shape is not related to any specific chemical composition, amphora catalysts have a wide range of applicability. Data will be presented to show the performance advantages of amphora catalysts for a number of potential applications such as hydrocarbon oxidations, the Claus process, catalytic reforming, and hydrotreating. These data will compare identical catalyst formulations prepared in the amphora shape _vs_ conventional shapes such as solid spheres, tablets, and extrudates. The hydrotreating examples include both vapor phase and trickle bed conditions, leading to some conclusions as to the effect of liquid hold-up in the interior of the sphere.

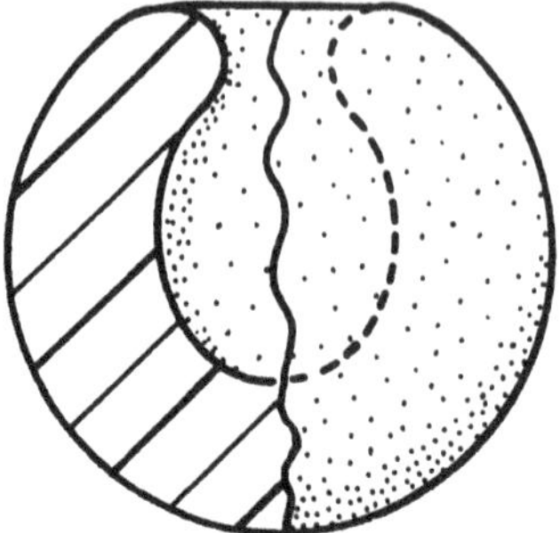

Fig. 1. Cross-sectional drawing of a representative particle.

I. INTRODUCTION

In the late 1950's when Sohio was developing its acrylonitrile technology, J.L. Callahan[1] was working on a spray-drying process for making fluid bed catalyst. When viewed under a microscope, certain samples of the spray-dried catalyst contained particles which had unusual shapes. These particles had channels leading from the outside to a hollow center (Fig. 1). Callahan recognized that this unusual shape could offer unique advantages when scaled up to a size suitable for use in fixed bed applications.

Mass and heat transfer limitations are often important considerations in heterogeneous catalysis. Chemical reactions that are diffusion-limited benefit from the use of smaller catalyst particles because of their higher ratio of surface area to volume and their shorter internal diffusion pathlength. By minimizing residence time of product molecules within the catalyst particle, secondary reactions can be minimized and selectivity increased. In addition, exothermic reactions benefit from small catalyst particles because heat build-up in the catalyst interior is minimized and the higher surface-to-volume ratio enhances heat transfer to the bulk medium. Unfortunately, there is a finite limit to the size of catalyst particles for fixed bed applications. Decreasing particle diameter decreases void volume, so as particles get smaller, undue pressure drop will be encountered in the reactor. This consideration has limited the practical diameter of fixed-bed catalyst particles to $\geq$1.5 mm for most applications.

The new Callahan particles offered a unique combination of advantages. The novel shape had a higher surface-to-volume ratio than solid spheres, extrudates, and even some shaped extrudates, and the spherical particles would pack more efficiently than tablets or extrudates. Most importantly, since the effective particle thickness was 0.25X that of the overall particle diameter, a bed of 3.0 mm diameter particles would have the performance characteristics of 0.75 mm solid spheres.

Callahan named the new particles amphorae after the ancient Greek urns that were used to carry cargo on trading ships. He and a number of co-workers began to develop techniques for making amphora-shaped fixed bed catalysts in various compositions for use in a variety of heat and/or mass transfer-limited reactions.

II. RESULTS AND DISCUSSION

A. Preparation of Amphora Catalysts

Slurries of active catalyst and/or catalyst precursors were prepared and droplets of these slurries were distributed onto beds of powdered bed materials. The bed materials served to cushion the slurry droplets and maintain their sphericity during drying, which was effected by overhead heat lamps. The most critical parameters for successful amphora formation were found to be (a) the rheology of the slurry and (b) the establishment of a differential rate of drying.

The most plausible mechanism for amphora formation, developed by G.K. Meloy, is a variation of one of the classic mechanisms of the drying of a sphere in spray-drying processes. In that mechanism, a rigid shell is initially formed on the surface of the sphere so that further drying occurs by migration of liquid from the sphere interior to the surface. During this migration, solid particles are carried with the liquid. Because the rate of liquid migration and evaporation is faster than the rate of diffusion of the solid particles back into the interior, a hollow sphere is formed.

Meloy hypothesized that the variation of this mechanism that results in amphora formation is one of unequal evaporation rates over the surface of the spherical particle. This is attained by a unidirectional air flow over the slurry droplets as they are drying. The air flow speeds evaporation on one side of the exposed portion of the droplet, making that portion rigid but leaving an area of greater fluidity on the other side of the droplet capable

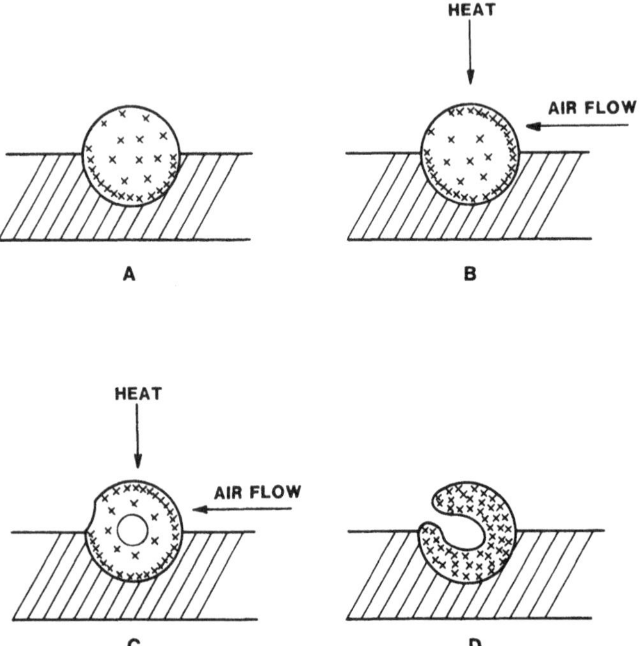

Fig. 2. Mechanism of amphora formation.

of plastic flow. As drying continues and natural shrinkage occurs, the particle first forms a hollow center and then, to relieve surface stresses and minimize surface energy, the area of greater fluidity shrinks into the particle and the hole is "pulled".

The process is depicted in Fig. 2. In Fig. 2A the droplet has just penetrated the bed. The portion of the droplet in contact with the bed quickly becomes rigid due to liquid transfer to the bed by capillary action. Fig. 2B shows application of the directional air flow (which may also be a directional application of heat), which increases the evaporation rate on the side of the droplet facing the source of air flow or heat, causing that portion of the droplet to become rigid but leaving a surface of lower solids content and greater plasticity on the opposite side. In Fig. 2C migration of liquid and suspended solids from the interior of the sphere leaves a void; but liquid and suspended solids also migrate from the more fluid area of the surface to the areas of the sphere that are drier and have greater evaporation rates.

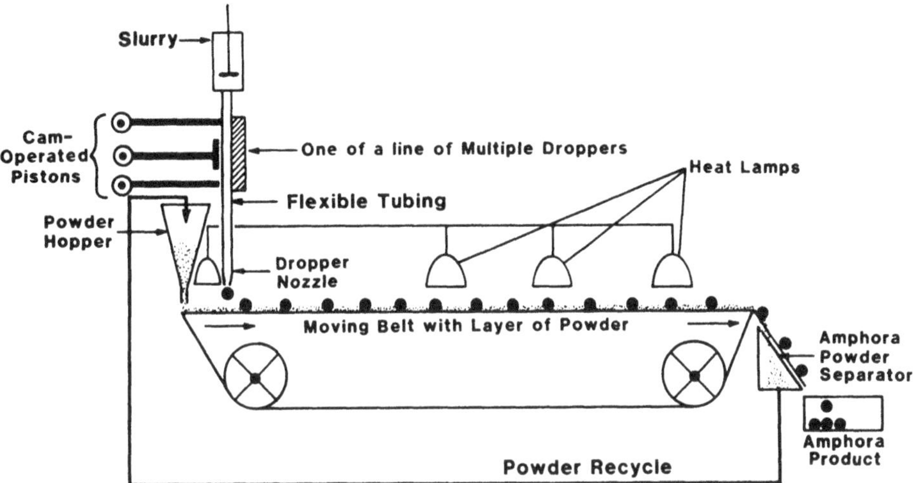

Fig. 3. Apparatus for continuous manufacture of amphora.

This relieves surface stresses building up as shrinkage occurs, which might otherwise result in cracked particles. This is seen as a dimple forming on the surface of the sphere and is the onset of the amphora hole formation. In Fig. 2D migration and shrinkage continue to form the amphora hole.

Amphorae were prepared in a variety of diameters ranging from 1.5 mm to 6.0 mm simply by varying the size of the slurry-dropping nozzle and the density of the bed material. Many different bed materials were used: sand, clay, alumina, carbon, sugar, cellulose, diatomaceous earth, even full range spray-dried and calcined catalysts. In addition to these hydrophilic bed materials, hydrophobic bed materials such as fluorinated graphite, powdered Teflon, or porous materials impregnated with hydrophobic fluids were quite useful when dealing with aqueous slurries. With hydrophobic bed materials, the region of slowest drying is that portion of the slurry droplet in contact with the bed, and the amphora hole forms on the bottom of the particle. This eliminates the need for a unidirectional air flow or directional heat source.

Because of the formation of the amphora shape is a natural phenomenon that occurs during drying, amphora particles can be formed from a variety of compositions.[2] Amphora-shaped alumina and/or silica supports, zeolites, and complex metal oxide compositions such as oxidation or ammoxidation catalysts have been prepared. The physical properties of the particles (pore volume, bulk density, crush strength, etc.) can be readily varied by adjusting the solids content of the slurries or by adding modifying agents and binders to the slurries. Highly porous solid spheres can also be fabricated by the slurry-dropping technique, these spheres having macropore diameters unattainable in tablets or extrudates.

To demonstrate that amphorae could be produced on a continuous basis, the apparatus[3] shown in Fig. 3 was developed. A moving belt passes under a feed hopper where it picks up a layer of bed material. The bed material is preheated and next passes under a row of dropping nozzles. Slurry is pumped through these nozzles by the action of a series of cam-operated pistons on flexible tubing leading from the slurry reservoir to the nozzles. The belt carrying the slurry droplets then passes under a bank of heat lamps where the amphorae form and dry sufficiently to permit them to be handled. At the end of the table, the amphorae are screened off and the powdered bed material is recycled. Using this device, liter quantities of amphora catalyst were prepared for testing in laboratory and pilot plant reactors.

B. Performance Evaluations

To demonstrate that amphora catalysts could outperform conventionally shaped catalysts in reactions limited by heat and/or mass transfer, a number of applications were selected to test the versatility of the new particle shape. The list included oxidation, ammoxidation, oxydehydrogenation, hydrotreating, reforming, and Claus chemistry. A few of these are discussed below to illustrate the performance advantages of amphora catalysts.

1. Oxidation reactions

Hydrocarbon oxidation reactions represent an area in which both activity and selectivity can be affected by proper control of heat and mass transfer within catalyst particles. Table I shows the results of oxidizing propylene to acrolein/acrylic acid using a Sohio catalyst formulation that was fabricated

TABLE I

Oxidation of Propylene to Acrolein/Acrylic Acid (Amphorae vs. Tableted Catalyst)

		Corrected % per pass converion			
Catalyst form	Temp. (°C)	Acrylic acid	Acrolein	Total Conversion	% Useful
5 mm X 2.8 mm Tablets	310	4.8	74.8	81.7	79.6
5 mm X 2.8 mm Tablets	329	7.7	82.6	92.9	90.3
5 mm X 2.8 mm Tablets	338	8.8	82.0	93.6	90.8
3.1 mm Diameter amphora	310	7.1	84.3	93.9	91.4
3.1 mm Diameter amphora	327	10.8	83.1	96.8	93.9
3.1 mm Diameter amphora	338	13.1	82.6	99.5	95.7

as conventional tablets and in the amphora shape. Activity and selectivity have both been enhanced by using amphora-shaped catalyst. Table II compares the performance of amphorae and solid spheres for the oxidation of o-xylene to phthalic anhydride. Here also, an identical catalyst formulation was used in both cases, the only difference being the shape of the particles. The improvement in selectivity at higher temperatures with the amphora-shaped catalyst is striking.

Amphorae and tablets were prepared from the same catalyst formulation and used for the oxydehydrogenation of 2-butene. The data are

TABLE II

Oxidation of o-Xylene to Phthalic Anhydride (Amphorae vs. Solid Spheres)

			Corrected % per pass conversion		
Catalyst form	Temp. (°C)	Exotherm (°C)	Phthalic Anhydride	CO	CO_2
Amphora	353	18	67.8	3.3	11.4
Spherical	348	20	71.2	7.3	21.5
Amphora	368	20	78.8	5.3	14.1
Spherical	361	26	63.8	9.5	26.7
Amphora	381	22	77.7	5.9	16.4
Spherical	376	40	53.3	12.9	33.9

TABLE III

Oxydehydrogenation of 2-Butene to Butadiene (Amphorae vs. Tablets)[a]

Catalyst form	Temp. (°C)	Corrected % per pass converson			
		Butadiene	Maleic anhydride	Total conversion	% Useful
Tablet	302	40.4	1.1	44.5	41.5
Amphora	302	48.2	1.5	53.5	49.7
Tablet	329	69.5	3.4	78.5	72.9
Amphora	329	72.6	6.7	88.0	79.3

[a] C_4/air = 1/16; contact time = 3.3 s; pressure = atmospheric.

contained in Table III. In this case, the more efficient heat and mass transfer provided by the amphora shape have imparted the dual benefits of improved selectivity and activity.

2. Claus chemistry

An alumina catalyst was fabricated into 1.5 mm diameter extrudates and 1.5 mm diameter amphorae and evaluated for the conversion

$$2H_2S + SO_2 \longrightarrow 3S + 2H_2O.$$

This "Claus chemistry" is extremely sensitive to both heat and mass transfer effects. The data in Table IV show a clear advantage for amphorae in this application. The extrudate shape has given a conversion quite typical of commercial catalysts in a single stage, which illustrates the need for

TABLE IV

Claus Chemistry (232°C) (Amphorae vs. Extrudate)

Catalyst form	H_2S conversion, %	Maximum catalytic conversion (thermodynamic limit)
Extrudate	89.0	95.5
Amphora	95.3	95.5

[a] Feed stream = 17.4 vol. % H_2S, 8.7 vol. % SO_2.

TABLE V

Reforming Mid-continent Naptha (Amphorae vs. Extrudate)

Catalyst form	Temp. (°C)	Time on stream (h)	RON of product
Extrudate	480	3	87.6
Amphora	480	3	99.5
Extrudate	510	6	96.6
Amphora	510	6	103.7

multiple reactors to achieve acceptable conversions. The amphora catalyst, on the other hand, is operating at the thermodynamic limit of the reaction under these conditions. Such performance in a commercial process could conceivably eliminate the need for at least one reactor.

3. Reforming

Catalytic reforming is yet another application for amphora catalysts. Identical reforming catalyst compositions were prepared in amphorae and extrudate shapes and used for the reforming of mid-continent naphtha. The data in Table V illustrate the improvements in research octane number that are made possible with amphora-shaped catalyst.

4. Hydrotreating

The area of catalysis that has seen perhaps the greatest activity with regard to varying catalyst shape in order to overcome diffusion limitations is hydrotreating. A variety of catalyst shapes have appeared in the literature over the years. Spherical catalysts have been used in place of extrudates or tablets because of their higher surface-to-volume ratio and better packing efficiency. A number of novel extruded catalysts have been developed, such as multilobed extrudates.[4,5]

The relative efficiencies of a number of shaped hydrotreating catalysts have been compared by de Bruijn et al.[6] In that study, the relative activities of the shaped catalysts were correlated with their particle size, Lp

TABLE VI

Effect of Catalyst Particle Shape on Surface-to-Volume Ratio[a]

Shape	Size	Lp, mm	S/V ratio, mm^{-1}
Cylinder	1.5 mm	0.322	3.1
Sphere	1.5 mm	0.250	4.0
Trilobe	1.5 mm	0.234	4.3
Quadrilobe	1.5 mm	0.189	5.3
Amphora	1.5 mm	0.168	6.0
Crushed	20-60 mesh	0.055	18.2

[1]All but the amphora data is taken from reference 6.

(defined as the volume of the particle divided by its geometric surface area). Table VI indicates how the particle shape can influence Lp and the surface-to-volume ratio (1/Lp) and illustrates the high S/V ratio of amphora particles. Fig. 4, adapted from reference 6, shows the correlation of Lp with the relative weight activity (RWA) of various shaped catalysts for resid hydrotreating. This experimental relationship predicts that an amphora-shaped catalyst of similar composition and porosity, with its Lp of 0.168, would have a relative weight activity surpassed only by ground and screened catalyst.

Alumina supports were prepared in the form of 1.5 mm diameter extrudates and 1.5 mm diameter amphorae and impregnated with solutions of cobalt or nickel nitrate and ammonium heptamolybdate to achieve standard hydrotreating catalyst compositions. These were evaluated with various petroleum feedstocks ranging from kerosene to vacuum gas oil. The results of the kerosene treating are shown in Table VII. These data are perhaps the best example of the remarkable performance capabilities of amphora catalysts. At a liquid hourly space velocity of 2.0, the extruded catalyst removed 92% of the sulfur in the kerosene feed. This performance was matched and even slightly surpassed by the amphora catalyst at 2 1/2 times the throughput. Since the two catalysts had virtually the same chemical

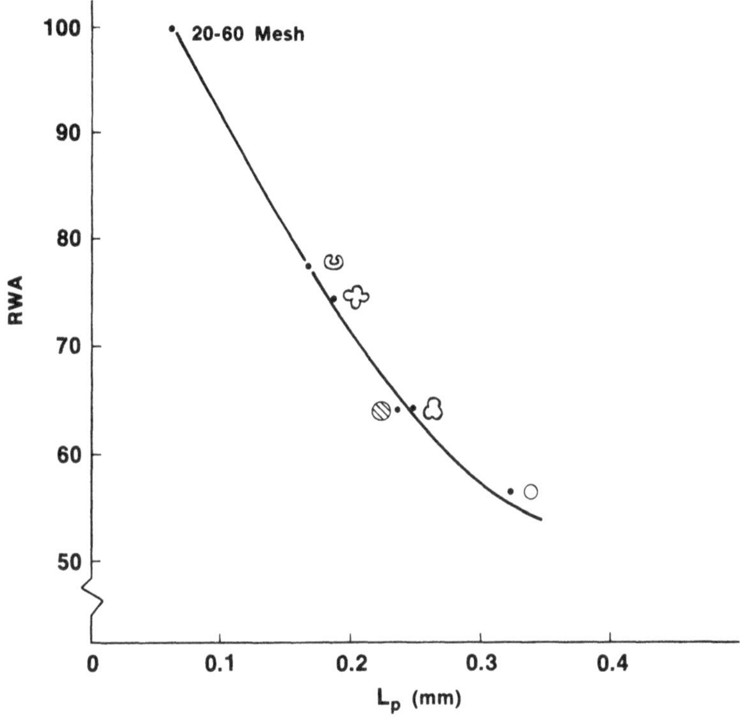

Fig. 4. Relative weight activity as a function of particle size in resid hydrotreating (adapted from reference 6).

TABLE VII

Hydrodesulfurization of Straight Run Kerosene at 316°C (Amphorae vs. Extrudate)

	Sulfur removal (%)	
LHSV (h^{-1})	Amphora	Extrudate
2	94.1	92.0
3	96.9	88.6
4	94.1	83.0
5	92.8	78.2
6	88.9	73.1

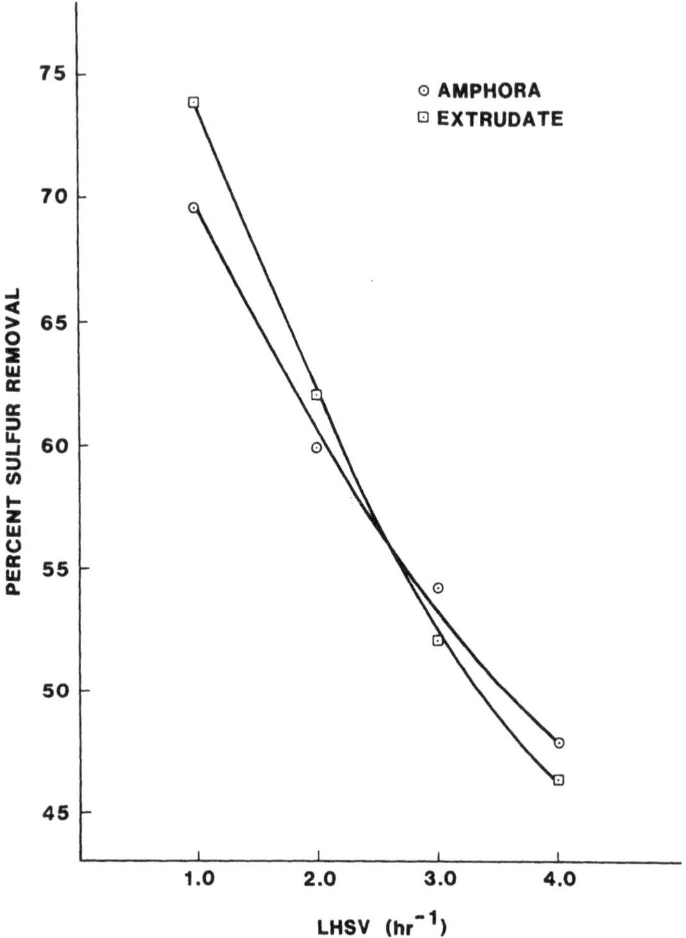

Fig. 5. Crossover effect observed during the hydrodesulfurization of light Iranian vacuum gas oil (b.p. range 213-580°C, 1.42 wt. % S) at 371°C, 34 atm.

composition and physical properties, this difference must be attributed to the amphora shape.

When a vacuum gas oil was used as the hydrotreating feedstock, an interesting phenomenon was encountered. For the first time, an amphora catalyst was being tested under trickle bed conditions, since only a small fraction of the vacuum gas oil is vaporized under reaction conditions. At low

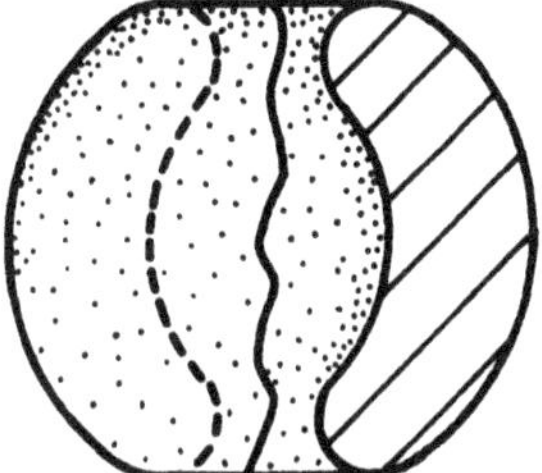

Fig. 6. Cross-sectional drawing of a typical Amphora II particle.

throughputs, the amphora catalyst fell below the extrudates in performance (Fig. 5). We attribute this to a "teacup effect" caused by stagnant pools of liquid accumulating in the interior of those amphora particles whose holes were oriented upward in the downflow reactor. However, at LHSVs higher than 2.6, the amphora curve crosses over and exceeds that of the extruded catalyst. At these higher throughputs, the exchange of liquid in the amphora interior is apparently efficient enough to prevent stagnation and permit the full use of each particle. The mass velocity in the laboratory reactor at the crossover point (LHSV 2.6) was approximately 75 $lb/ft^2/hr$, which fortuitously is well below typical mass velocities in commercial reactors for the hydrotreating of these types of feedstocks ($\sim$5000 $lb/ft^2/hr$).

Although the "teacup effect" did not seem to present a problem in commercial trickle bed beactors, a new amphora particle was developed to eliminate any potential problems in these applications. By varying the conditions under which the slurry droplets were dried, amphora particles having a second hole located 180° from the first opening (Fig. 6) could be formed.[7] These new particles were designated Amphora II.

The vacuum gas oil hydrotreating experiments were repeated comparing Amphora II particles with amphora and extruded catalysts having virtually the same chemical composition and physical properties. The results of these tests are illustrated in Fig. 7. The Amphora II particles not only eliminated the "teacup effect" but also outperformed both the amphora snd extrudate catalyst shapes over the entire range of throughputs. This is attributed to a more efficient gas/liquid flow through the Amphora II catalyst bed.

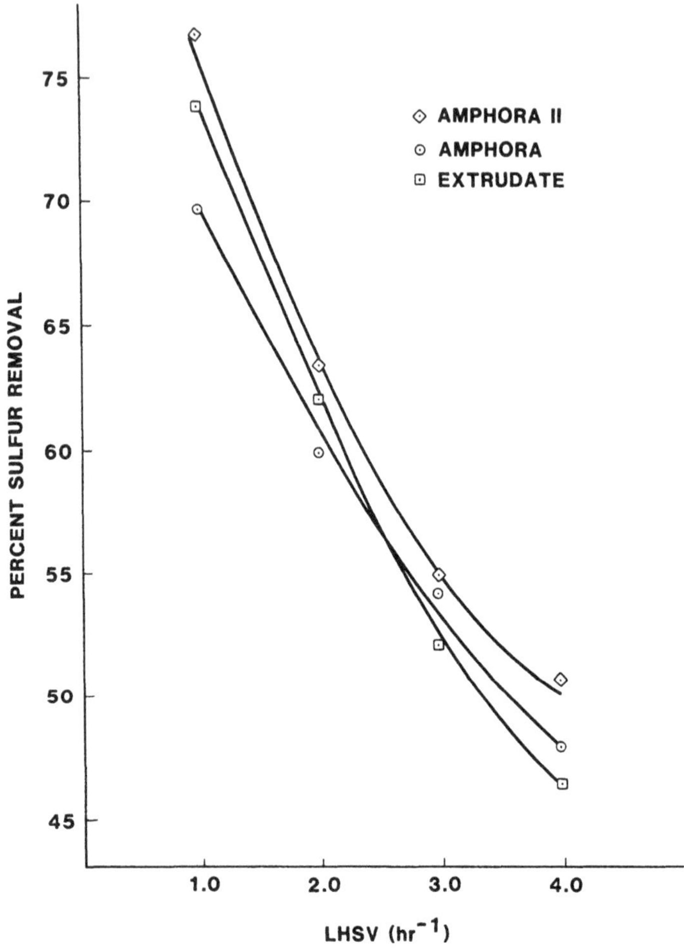

Fig. 7. Beneficial effect of Amphora II in the hydrodesulfurization of light Iranian vacuum gas oil.

III. CONCLUSIONS

The experimental results show that amphora-shaped catalysts are capable of outperforming conventionally shaped catalysts in virtually any fixed bed application where heat and mass transfer effects are significant. The greater the heat or mass transfer limitations, the greater are the potential benefits. Since the formation of the amphora particle shape is a physical phenomenon independent of chemical composition, amphora catalysts have potential for application in widely diverse areas such as chemical processing, petroleum/synthetic fuels refining, and auto exhaust catalysts.

ACKNOWLEDGEMENTS

The author is grateful to The Standard Oil Company (Ohio) for permission to present this work. In addition to those colleagues specifically mentioned or referenced herein, the contributions of J.P. Bartek, L.P. Crosbie, D.G. Farrington, R.K. Grasselli, H.F. Hardman, O.A. Kiikka, D.B. Terrill, and R.J. Zagata are gratefully acknowledged.

REFERENCES

1. J.L. Callahan, U.S. Patent No. 3,044,965 (1962).
2. J.L. Callahan, A.F. Miller, and W.G. Shaw, U.S. Patent No. 3,966,639 (1976).
3. J.L. Callahan, A.F. Miller, and W.G. Shaw, U.S. Patent No. 3,848,033 (1974).
4. R.B. Jacobs and G.B. Holkstra, U.S. Patent No. 3,764,565 (1973).
5. W.R. Gustafson, U.S. Patent No. 3,990,964 (1976).
6. A. deBruijn, I. Naka, and J.W.M. Sonnemans, Ind. Eng. Chem. Proc. Res. Dev., **20**, 40 (1981).
7. D.R. Herrington and A.P. Schwerko, U.S. Patent No. 4,170,569 (1979).

PART IV

CATALYTIC OXIDATION

15

Industrial Uses of Catalytic Oxidation and the Direct Oxidation of Olefins to Glycols

R.G. Austin,[+] R.C. Michaelson and R.S. Myers
Exxon Chemical Company
4 Pearl Court
Allendale, NJ 07401

ABSTRACT

The mechanism and utility of several industrially important homogenously catalyzed oxidation processes will be discussed. Special emphasis will be given to the Oxirane process for the production of olefin oxides, the Mid Century process for the synthesis of terephthalic acid, and the Halcon process for the synthesis of ethylene glycol.

Recent work at Exxon Chemical Company aimed at the direct production of vicinal glycols by the oxidation of olefins will also be discussed. A highly selective oxidation of olefins with molecular oxygen has been discovered. This oxidation of olefins to glycols utilizes a binary catalyst system composed of an osmium and a copper salt and shows quantitative uptake of oxygen. The effect of copper, osmium, halide, and oxygen levels will be presented.

In addition, glycols can be produced from olefins by oxidation with alkyl hydroperoxides using an osmium catalyst and an alkali metal promoter.

[+]Present Address: Exxon Chemical Company
Chemical Technology Center
P.O. Box 5200
Baytown, Texas 77522

This essentially quantitative reaction produces a valuable coproduct as well as the vicinal glycol. The kinetics and overall yields of this selective oxidation will be presented.

I. INTRODUCTION

Major applications of homogeneous catalysis in the chemical industry has been the subject of many recent reviews and symposia.[1] The demonstration of high selectivity and economic efficiency for a variety of transformations catalyzed by soluble transition metal compounds have led to increasing interest and application to industrial processes.

Well known processes in the area of oxidation, namely, the Oxirane propylene oxide, Mid-Century terephthalic acid, and the Halcon ethylene glycol route have received much attention. Mechanistic studies, in general, have followed commercialization and many oxidation processes are not understood even now. The following discussion will be centered on proposed mechanisms of these three processes and will fit nicely into our recent discoveries.

The Mid-Century/Amoco process for oxidation of p-xylene to terephthalic acid uses air as the oxidant, a mixture of cobalt, manganese and bromide salts as the catalyst system, and acetic acid as the solvent.[1-4] The mechanism is basically a hydrogen abstraction by bromine atoms, or other odd electron species such as ROO•, R•, or RO• or possibly even dioxygen complexes. Because of the relative stability of the benzyl radical, this abstraction is much easier than for alkanes in general. The mechanism for this liquid-phase oxidation in acetic acid via activation by bromine is shown simply in Eq. 1 and 2. A proposed bromine cycle is shown for this

$$CH_3C_6H_4CH_3 + Br^\bullet \longrightarrow {}^\bullet CH_2C_6H_4CH_3 \qquad (1)$$

$$ {}^\bullet CH_2C_6H_4CH_3 + O_2 \longrightarrow {}^\bullet OOCH_2C_6H_4CH_3 \qquad (2)$$

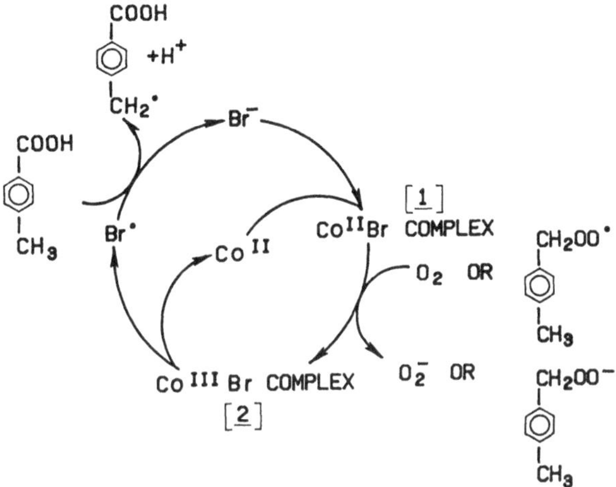

Scheme 1

oxidation in Scheme 1. Complexation of bromide to cobalt (II) may facilitate electron transfer to molecular oxygen to form a cobalt (III) bromide complex. Cobalt (III) can abstract an electron from a bromide ligand to yield a bromine atom which then can easily abstract a hydrogen atom from a methyl group to continue the process and complete the catalyst cycle.[1]

One of the major applications of homogeneous catalysis is the Oxirane process for production of propylene oxide,[5] based on the catalytic transfer of oxygen from an alkyl hydroperoxide to an olefin catalyzed by a soluble molybdenum containing salt. This selective epoxidation of olefins with tert-butyl hydroperoxide was reported by Indictor and Brill[6] in 1965. The reaction with propylene, as shown in Eq. 3, also produces a valuable

$$CH_3{-}CH{=}CH_2 + (CH_3)_3COOH \xrightarrow{Mo} CH_3{-}\underset{\diagdown O \diagup}{CH{-}CH_2} + (CH_3)_3COH \qquad (3)$$

coproduct, tert-butyl alcohol.

The mechanism of oxygen transfer to olefins from hydroperoxides has been studied extensively over the past few years.[2,7] The Sharpless mechanism is based on a labeling study in which an olefin was oxidized by a molybdenum compound which contained both an oxo and peroxy function. Only the peroxy oxygen was transferred to the olefin.[8] A mechanism

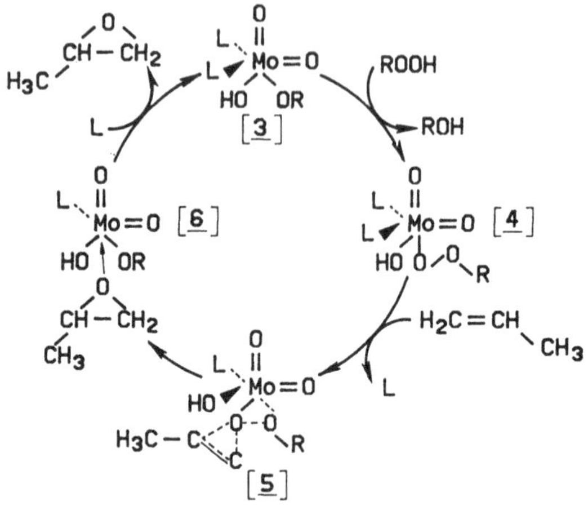

Scheme 2

consistent with this study is shown in Scheme 2. The key feature of this mechanism involves coordination of the alkylperoxy group through the distal oxygen as shown in Eq. 4. Despite other related studies on this complex mechanism many uncertainties still exist.[1]

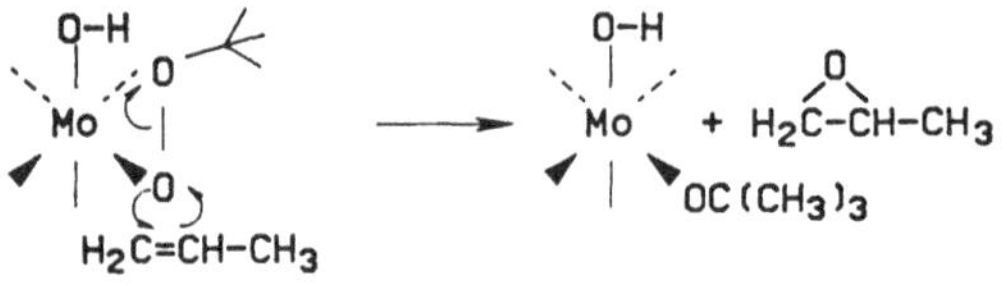

(4)

According to the mechanism proposed by Sharpless, an alkyl molybdate compound **3** forms an alkylperoxomolybdate **4** through exchange with alkyl hydroperoxide. This complex then coordinates with an olefin which probably displaces a neutral ligand such as ROH. The intermediate **5** is produced and the Mo-bound oxygen is transferred to the olefin. This transfer appears to form the epoxide which remains coordinated to the metal as shown in **6**. The cycle is completed when the epoxide is displaced by either ROH or another molecule of alkyl hydroperoxide.

Another process which has not been commercialized is a homogeneously catalyzed route to ethylene glycol. The Halcon process produces a mixture of monohydroxyacetates and diacetates that are

$$H_2C{=}CH_2 + 2HOAc + \frac{1}{2}O_2 \xrightarrow{Te,\ Br^-} AcO{-}CH_2{-}CH_2{-}OAc + H_2O \tag{5}$$

hydrolyzed selectively to form the glycol. The first step is summarized in Eq. 5. The catalyst system consists of an acetic acid solution of TeO_2 and HBr. The system gives high selectivity (> 95%) to ethylene diacetate and ethylene monohydroxyacetate.[1]

The mechanism of this oxidation is not clearly understood. The oxidation couple of the Te/Br system is analogous to the Wacker acetaldehyde synthesis. A proposed mechanism based on a sequence of conventional reactions is shown in Scheme 3. The oxymetallation by Te(IV) of ethylene in the presence of acetic acid leads ultimately to the diacetate. The intermolecular or intramolecular reductive elimination of intermediate **8** could give the mixture of products found (e.g. hydroxyacetates, bromoacetates).

With these homogeneous catalytic oxidations in mind, the remainder of the paper will be centered on some recent results obtained in our laboratories directed toward the synthesis of glycols using molecular oxygen or tert-butyl hydroperoxide as the oxidant.

II. RESULTS AND DISCUSSION

We have discovered a selective single step reaction to produce glycols[9] according to Eq. 6. A detailed study of this reaction using propylene as the olefin has been carried out. High selectivity (> 99%) to glycol has been

H_2O $TeBr_4$ [7] $H_2C{=}CH_2$

$2HBr + \frac{1}{2}O_2$

$TeBr_2$ [10] Br_3Te–CH_2CH_2–Br [8]

AcO OAc +HBr HOAc

HOAc Br_3Te–CH_2CH_2–OAc [9] HBr

Scheme 3

$$CH_3CH = CH_2 + \frac{1}{2} O_2 + H_2O \xrightarrow[\text{PROMOTER}]{\text{CAT/COCAT}} CH_3\underset{\displaystyle |}{\overset{\text{OH}}{C}}H-\overset{\text{OH}}{C}H_2 \tag{6}$$

achieved under moderate conditions (e.g. $\geq$100°C, <1000 psi). Remarkably, no over oxidation products such as ketols or acids are formed in this reaction.

Osmium tetroxide catalyzed oxidation of olefins in the presence of oxygen has previously been reported by ICI[10] and Celanese.[13] Limitations of high pH and the production of oxalic acid and CO_2 in the case of ICI are noteworthy.[10,12] Similar pH restrictions and low rates are found in the Celanese work.[10,11,13]

Our binary catalyst system is comprised of a noble metal and a first row transition element. Osmium, introduced into the system as the tetroxide or the trihalide and a copper co-catalyst (e.g. $CuCl_2$, $CuBr_2$) are typically employed. Halides are also efficient co-catalysts for this reaction. Polar organic solvents miscible with water are suitable for this homogeneous reaction. In addition, the presence of a promoter has an accelerating effect. The overall stoichiometry, Eq. 6, for this facile oxidation shows quantitative uptake of oxygen. An initial screening program pointed out a number of key features related to this system.

No significant reaction was observed under our conditions in the absence of osmium or transition metal co-catalyst (e.g. copper). It is

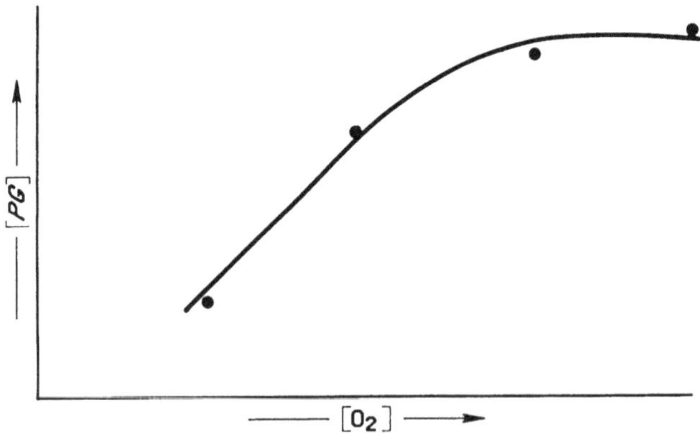

Fig. 1. PG production versus O_2 concentration.

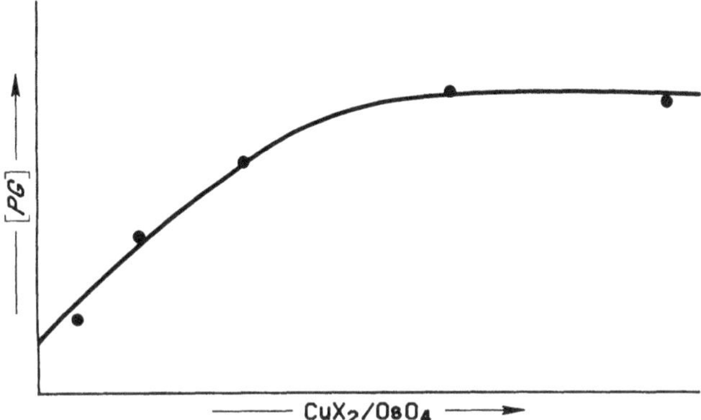

Fig. 2. PG production versus Cu/Os ratio.

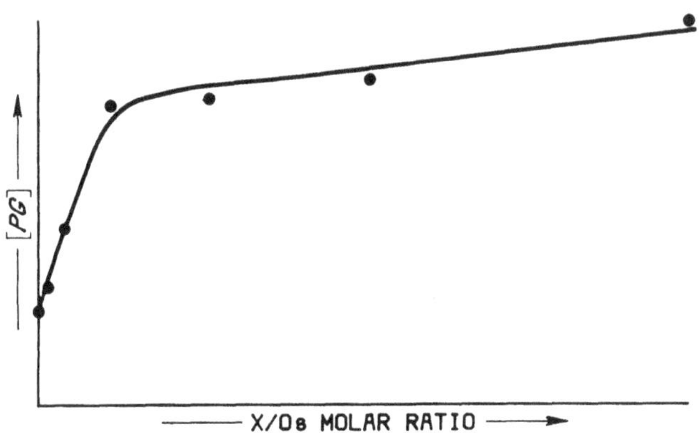

Fig. 3. PG productions versus X/Os molar ratio.

noteworthy that non-volatile osmium sources (e.g., $OsBr_3$) may be used as well. Other important observations include first order kinetics on osmium and comparable rates for other olefins such as ethylene and 1-butene.

Fig. 1 shows the effect of glycol production on increasing oxygen concentration at constant reaction time. The maximum rate is obtained before the upper explosion limit is exceeded. The effect of the transition metal co-catalyst to osmium molar ratio on propylene glycol production can be seen in Fig. 2. The reaction becomes zero order in copper above a specific copper to osmium ratio.

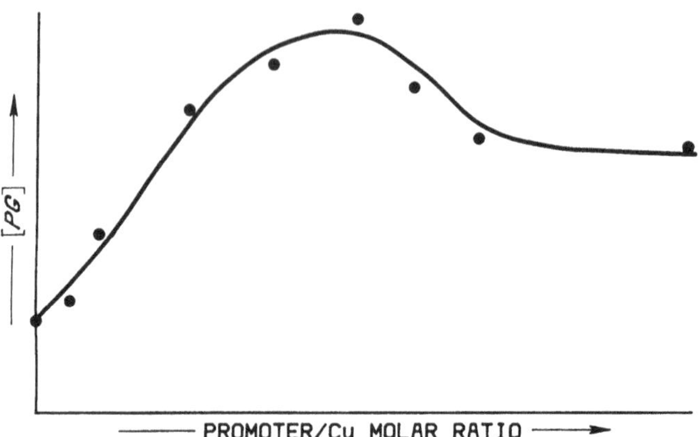

Fig. 4. PG production versus promoter/Cu molar ratio.

TABLE I

Propylene Conversions at Elevated Temperature with Multiple O_2 Additions

TIME	TEMP °C	O_2 CONV.	PROPYLENE CONV.
60 MIN.	T	89%	X
60 MIN.	T+10°	100%	2X
60 MIN.	T+20°	100%	3X

- O_2 ADDED IN SEVERAL INCREMENTS THROUGHOUT THE REACTION
- UEL WAS NOT EXCEEDED
- SELECTIVITY ON PROPYLENE ~ 100%

Similarly, the molar ratio of halide to osmium is a critical parameter. Fig. 3 illustrates that the reaction becomes essentially zero order in halide above a specific halide to osmium ratio.

For the maximum rate, a narrow range of promoter to copper ratio should be employed as is seen from Fig. 4. The rate of reaction increases steadily until a specific promoter to copper ratio is achieved and then declines afterward.

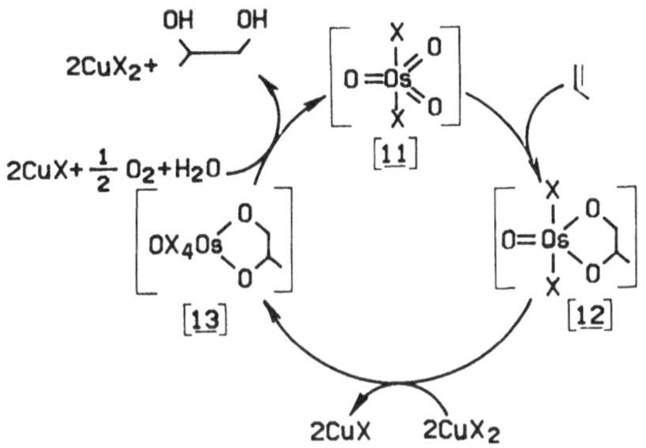

Scheme 4

Each of these factors is important to the overall selectivity and conversions of this hydroxylation reaction. High olefin conversions are obtained in this hydroxylation reaction. Quantitative oxygen conversion is also achieved at various temperatures as shown in Table I. No loss in selectivity to glycol is observed at these high olefin conversions. Generally, reactions of this nature are run at relatively low hydrocarbon conversion in order to obtain the desired selectivity and to avoid overoxidation products.

A proposed mechanism consistent with the available data is summarized in Scheme 4. An oxyhalide osmium(VIII) intermediate **11** complexes with a mole of olefin to generate a reduced osmate(VI) ester **12**. The oxidation of this osmium(VI) species to the corresponding osmium(VIII) complex **13** is aided by CuX_2. This complex is hydrolyzed to produce the glycol and the CuX_2 is regenerated by oxidation of the Cu(I) species by molecular oxygen. The presence of CuX_2 along with the promoter may be the key to avoiding the over oxidation products normally found in these reactions. Studies of this interesting and novel catalyst system are continuing.

Another reaction we have investigated is the oxidation of olefins with organic hydroperoxides. Our novel catalyst system consists of an osmium source (e.g. OsO_4, OsX_3) and an alkali metal halide.[14] Overall batch results for a typical reaction for formation of ethylene glycol are shown in Table II. High yields of ethylene glycol are obtained. Tert-butyl alcohol is the coproduct when tert-butyl hydroperoxide is used as the oxidant. The

TABLE II

TBHP Oxidation of Olefin

$$(CH_3)_3C\text{-}OOH + CH_2{=}CH_2 + H_2O \xrightarrow[\text{2) NaX}]{\text{1) OsO}_4} (CH_3)_3C\text{-}OH + \underset{OH}{CH_2}\text{-}\underset{OH}{CH_2}$$

TYPICAL YIELDS BASED ON TBHP

GLYCOL	> 80
TBA	95
ACETONE	
METHANOL	
DTBP	

EXCELLENT YIELDS BASED ON OLEFIN (>98%)

TABLE III

Kinetic Data

REACTANT	REACTION ORDERS
Os	1±0.1
C_2H_4	0.55±0.1
TBuOOH	0
NaX	0
H_2O	0.5

excellent selectivity to glycol (>98%) for this facile oxidation is also noteworthy. The major by-products consisting of acetone and methanol arise from the decomposition of the hydroperoxide.

The kinetics of the reaction have been studied under typical batch conditions as can be seen from Table III. The reaction was determined to be first order in osmium, independent of TBHP and halide, and 0.5 order in water and ethylene.

A proposed mechanism consistent with the kinetic data and the stereospecificity is outlined in Scheme 5. The osmium(VIII) oxy-halide

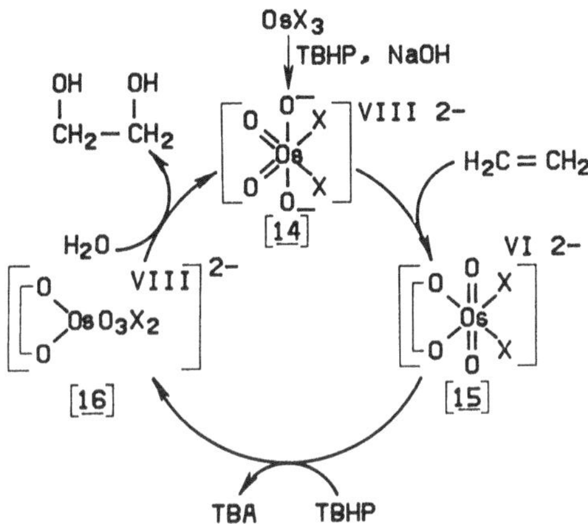

Scheme 5

species **14** is reductively complexed with olefin to generate the Os(VI) osmate ester **15**. Whether the oxidative hydrolysis of the osmate(VI) ester **15** is sequential as shown in this scheme or concerted is not conjectured. Nevertheless, the ester intermediate is oxidized to an osmium(VIII) oxyhalide **16** and hydrolyzed to produce the glycol and continue the cycle. The oxidation of intermediates **15** by TBHP also produces the co-product, TBA.

III. SUMMARY

Homogeneous catalysis in the oxidation of olefins continues to be an active area of research. Mechanistic studies of known oxidation processes has led to a better understanding of oxygen transfer whether using molecular oxygen or organic hydroperoxides as primary oxidants.

The discovery of new reactions such as our novel selective vicinal hydroxylation of olefins using the Os/Cu/promoter catalyst systems will initiate further studies and potential commercializations. An understanding of many of these complex metal-halogen catalysts systems is still far from complete.

REFERENCES

1. G.W. Parshall, J. Mol. Catal., 4, 243 (1978).

2. R.A. Sheldon and J.K. Kochi, **Metal-Catalyzed Oxidations of Organic Compounds**, Academic Press, New York, 1981.
3. D.S.S. Ravens, Trans. Faraday Soc., **55**, 1768 (1959).
4. Y. Kamiya, J. Catal., **33**, 480 (1974).
5. R. Landau, Hydrocarbon Process., 46 (1967).
6. N. Indictor and W.F. Brill, J. Org. Chem., **30**, 2074 (1965).
7. R.A. Sheldon, J. Mol. Catal., **7**, 197 (1980).
8. A.O. Chong and K.B. Sharpless, J. Org. Chem., **42**, 1587 (1977).
9. R.G. Austin and R.C. Michaelson, U.S. Patent 4,390,739 (1983).
10. M. Schroder, Chem. Rev., **80**, 187 (1980).
11. J.F. Cairns and H.L. Roberts, J. Chem. Soc., C, 640 (1918).
12. J. Perichon, S. Palms, and R. Bovet, Bull. Soc. Chim. Fr., 982 (1963).
13. Celanese Corporation, British Patent 1028940, (1966).
14. R.G. Austin and R.C. Michaelson, U.S. Patent 4,314,088 (1982), U.S. Patent 4,393, 253 (1983), European Patent Applications 77,202 (1983).

16

Mechanistic Studies of the Selective Oxygen Oxidation of Sulfides to Sulfoxides Catalyzed by Dihalo Ruthenium (II) Complexes

Dennis P. Riley+

The Procter & Gamble Company

Miami Valley Laboratories

Cincinnati, Ohio 45247

ABSTRACT

The selective and facile molecular oxygen oxidation of dialkyl sulfides to their sulfoxides can be effectively catalyzed by neutral ruthenium(II) complexes of the type cis or trans $RuX_2(DMSO)_4$. The results of kinetic studies for two of these catalysts, cis $RuCl_2(DMSO)_4$ and trans $RuBr_2(DMSO)_4$, show that these catalytic oxidations are first-order with respect to total catalyst concentration, less than first-order in oxygen concentration, zero-order in the sulfide substrate, and approximately first-order in alcohol. These and other observations plus ^{18}O-labeling studies support a mechanism involving oxidation of a"Ru(II)" species to give an oxidized metal species, probably "Ru(IV)" and peroxide. The active sulfide oxidant is peroxide, whose concentration is approximated by the steady-state assumption. The reductant of the oxidized metal is the solvent alcohol, thus completing the catalytic cycle. The actual catalyst is generated in situ by reaction of the $RuX_2(DMSO)_4$ complex with excess sulfide. While many species form in situ, the actual catalyst has been

+Present Address: Monsanto Co., 800 N. Lindbergh Blvd., St. Louis, Missouri 63167

tentatively identified as the all trans-RuX_2(sulfoxide)$_2$(thioether)$_2$ complex by independent syntheses and electrochemical studies.

I. INTRODUCTION

While the selective oxidation of thioethers to their sulfoxides is potentially very useful, a thorough review of the literature reveals that there is only one example of the use of metal-based catalysts to promote this oxidation. The catalysts described in the literature also suffer from poor selectivities and very slow rates.[1] The paucity of examples of such homogeneous metal ion catalyzed oxidations is somewhat surprising in view of the large number of complexes known to catalyze the oxygen oxidation of tertiary phosphines to phosphine oxides. Using such catalyzed phosphine oxidations as models for sulfide oxidation, we screened all of the complexes known to catalyze phosphine oxidations[2-4] or to form metal-O_2 adducts.[2,5] None of these complexes afforded any activity for the oxidation of thioethers.

It is known that the dioxygen oxidation of phosphines to phosphine oxides catalyzed with low-valent metal complexes, such as $Pt^0(PPh_3)_4$, generates peroxide as the active oxidant.[3a] These catalysts form inner-sphere peroxide complexes (O_2 adducts). Thus, for an oxidation to occur the substrate must be able to displace O_2^{2-}. We have found that sulfides do not react with such complexes, consequently, they are apparently not able to displace coordinated peroxide. For this reason we believe that to achieve effective dioxygen catalysis with substrates, such as sulfides, that are apparently poorer ligands than phosphines it would be essential to use metal complexes that can undergo outer-sphere electron transfer to give free peroxide in solution. We chose to investigate ruthenium(II) complexes as potential catalysts, since there are a number of Ru(II) compounds known to undergo outer-sphere oxidation with molecular oxygen to yield peroxide.[6-8]

In separate studies involving the oxygen oxidation of tertiary amines, we had discovered that catalytic chemistry could be realized using the oxygen oxidation of ruthenium(II) to produce peroxide.[9] Based on this reasoning we believed that with a suitable ligand environment about ruthenium(II) we could generate an active catalyst for sulfide oxidations. After screening a large number of ruthenium(II) complexes, we found that the <u>cis</u> $RuCl_2(DMSO)_4$ (DMSO = dimethylsulfoxide) complex is an effective catalyst for the oxygen driven conversion of thioethers to sulfoxides, giving both a high selectivity for sulfoxide over sulfone (sulfoxide/sulfone $\geq$25) and much faster rates than observed with any catalyst reported in the literature.[10] Our attempts to improve this catalyst by synthesizing other structural analogues revealed that the anion plays an important role in determining the activity of the $RuX_2(DMSO)_4$ complexes as catalysts. Not only is it necessary to have two anions coordinated to the metal but the identity of the anions is very important with catalytic activity increasing in the order $I^- \ll Cl^- < SnCl_3^- < SCN^- < Br^-$.[11,12] Further synthetic attempts to increase catalytic activity revealed that the presence of phosphie ligands bound to the ruthenium(II) catalyst greatly influenced the catalytic activity[13] which increased in the order $RuX_2(PR_3)_3 \ll RuX_2(DMSO)_2(PR_3)_2 < RuX_2(DMSO)_n(PR_3)$, n = 2 or 3. Also the nature of the phosphine itself plays a major role in the activity of these catalysts since in complexes of stoichiometry $RuX_2(DMSO)_n(PR_3)$ the activity increases in the order $P(OR)_3 < PPh_3 \approx AsPh_3 < PBu_3$.[13]

The kinetic and mechanistic studies presented here reveal that the mechanism of action of these catalysts proceeds by a common mechanism. In addition, synthetic and electrochemical studies, aimed at elucidating the structure of the active catalyst species generated in solution with the $RuX_2(DMSO)_4$ complexes are described.

II. EXPERIMENTAL

The experimental procedures[14] and syntheses of complexes have been detailed elsewhere.[11,13] In general, catalytic pressure reactions were run in a Griffin-Worden pressure vessel fitted with sample lines under an oxygen atmosphere and shaken in a thermostatically controlled oil bath.[14] These reaction mixtures were sampled periodically and the samples were analyzed directly using a Hewlett-Packard 5880A gas chromatograph with a flame ionization detector.

All cyclic voltammograms were measured in 0.10 <u>M</u> tetra-n-butylammonium tetrafluoroborate in methylene chloride. The methylene chloride was dried by passing through two columns of dry alumina and then distilling over CaH_2 under dry N_2. The supporting electrolyte was twice recrystallized from ethyl acetate-hexane. A single compartment, three electrode cell was used. It had a platinum working electrode, platinum wire auxiliary electrode, and a Ag/AgCl reference electrode in H_2O separated from the CH_2Cl_2 by a ceramic frit. The voltammograms were measured with a PAR 173 potentiostat and a PAR universal programmer and recorded on an oscilloscope.

III. RESULTS AND DISCUSSION

A. Reaction Mechanism

The complexes listed in Table I are all catalysts for the selective aerial oxidation of sulfides to sulfoxides in alcohol solvents. While the rates vary with the different oxidation catalysts, the oxidation reaction has identical overall kinetics; namely, the reaction rates exhibit zero-order sulfide substrate dependence (see Fig. 1 for a typical reaction profile), first-order

TABLE I

Turnover Numbers of Several Ru(II)-Based Catalysts for the Molecular Oxygen Oxidation of Decyl Methyl Sulfide[a]

Complex	T.N. (Hr^{-1})
cis $RuCl_2(DMSO)_4$	6
trans $RuBr_2(DMSO)_4$	19
$RuCl_2(PPh_3)_3$	0.5
$RuCl_2(DMSO)_2(PPh_3)$	9.5
$RuBr_2(DMSO)_3(PPh_3)$	22
$RuBr_2(DMSO)_3(PBu_3)$	24
$RuBr_2(DMSO)_3(POBu_3)$	11
$RuBr_2(DMSO)_2(Diphos)$	5.5
$RuBr_2(DMSO)_3(AsPh_3)$	22.5

[a]T = 100°C, O_2 Pressure = 100 psi, Solv = MeOH.

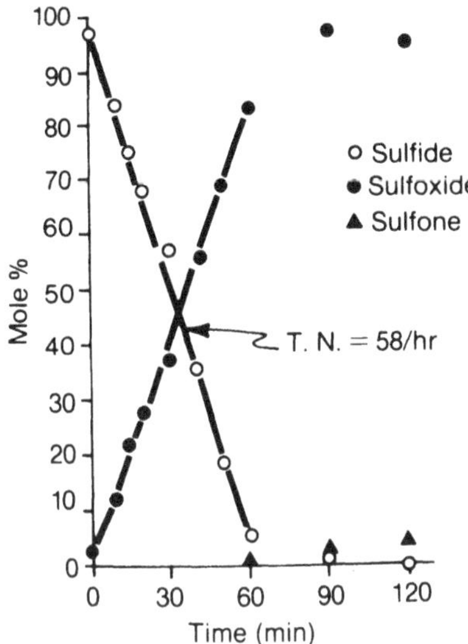

Fig. 1. A reaction profile for the molecular oxygen oxidation (200 psi) of decyl methyl sulfide (0.2 M) at 105°C catalyzed by $RuBr_2(DMSO)_3(PPh)_3$ (2.85 mmol) in methanol: (O) sulfide, (●) sulfoxide, (Δ) sulfone.

dependence in total catalyst concentration, and variable but less than first-order oxygen dependence (Fig. 2 for a plot of V_{Obs} versus P_{O_2} for two of these catalysts). As can be seen in Fig. 2 the dependence of the rate on oxygen pressure approaches first-order at lower pressures and approaches zero-order at higher oxygen pressures.

In order to clarify the mechanistic possibilities several other studies were carried out. Addition of oxygen-18 labeled H_2O to these reactions coupled with GC-mass spectral analyses of reaction aliquots revealed that no oxygen-18 containing sulfoxide is formed during the reaction. This eliminates one possible route (Eq. 1) for reducing the oxidized metal during the catalytic cycle that was postulated by earlier workers using $RuCl_2(PPh_3)_3$.[1]

$$2Ru^{III}SR_2 + H_2O \longrightarrow 2Ru^{II} + \overset{\overset{O}{\uparrow}}{S}R_2 + 2H^+ \qquad (1)$$

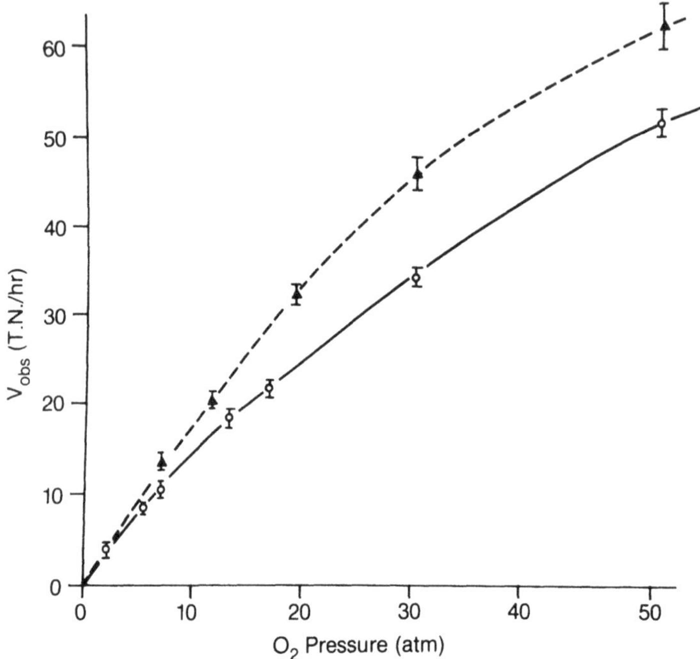

Fig. 2. Plot of V_{Obs} (T.N./hr) versus P_{O_2} for the O_2 oxidation of decyl methyl sulfide (0.15 M) in methanol with $[Ru]_{tot} = 3.0 \times 10^{-3}$ M: (Δ) *cis* $RuCl_2(DMSO)_4$ at 108°C and (O) *trans* $RuBr_2(DMSO)_4$ at 95°C.

Additional studies were carried out in which the fate of the alcohol solvent was monitored during the course of the reaction. In all cases some amount of solvent was oxidized (e.g. isopropanol to acetone). The quantity oxidized equaled the amount of sulfoxide produced throughout the reaction. These two facts support the conclusion that the reductant of the oxidized metal is the alcohol solvent.

Based on these studies and the known fact that ruthenium(II) complexes can react with oxygen in an outer-sphere electron transfer to produce peroxide, we believe that these catalytic thioether oxidations proceed via the mechanism shown in Eqs. 2-4. The valence of the oxidized metal in these reactions could either be

$$Ru(II) + O_2 \underset{k_{-1}}{\overset{k_1}{\rightleftharpoons}} "Ru^{ox}" + O_2^{2-} \quad (2)$$

$$SR_2 + H_2O_2 + ROH \xrightarrow{k_2} \overset{O}{\overset{\uparrow}{S}}R_2 + H_2O + ROH \quad (3)$$

$$"Ru^{ox}" + R_1R_2CHOH \xrightarrow{k_3} Ru(II) + R_1R_2C{=}O + 2H^+ \quad (4)$$

ruthenium(III) or ruthenium(IV). It is likely that the oxidant of the alcohol solvent is not ruthenium(III). This is supported by the high selectivity for sulfoxide and by the lack of radical autoxidation products. Since the oxidation of alcohols by ruthenium(III) produces alcohol radicals[15] and since thioethers are very susceptible to free-radical initiated autoxidations,[16] the involvement of Ru(III) in this chemistry can be eliminated. But this still does not eliminate the possibility that a Ru(II) species is oxidized to a Ru(III) species which undergoes disproportionation to yield a Ru(IV) species and Ru(II) [8,17,18] (Eq. 5). This also requires that superoxide

$$2\ "Ru(III)" \rightleftharpoons "Ru(II)" + "Ru(IV)" \quad (5)$$

be produced by the oxidation of a Ru(II) complex with oxygen. Since superoxide will not itself oxidize thioethers to sulfoxides,[19] this Ru(III) route requires that either superoxide oxidizes another Ru(II) complex (Eq. 6) or that superoxide dismutates (Eq. 7). Since HO_2 is a more active oxidant and also undergoes dismutation (Eq. 7) more readily than superoxide anion,[7]

$$"Ru(II)" + O_2^{\dot{-}} \longrightarrow Ru(III) + O_2^{2-} \quad (6)$$

$$2\ O_2^{\dot{-}} \longrightarrow O_2 + O_2^{2-} \quad (7)$$

added protons would be expected to increase the observed reaction rate. But added acid (HSO_3CF_3) has no effect on the rates or product distributions.[11] As a consequence, the intermediacy of a Ru(III) species in this chemistry is unlikely. This conclusion means that the direct oxidation of a Ru(II) complex with O_2 will produce peroxide and Ru(IV) directly.

The kinetic studies also support this interpretation. An integrated rate expression can be derived from Eqs. 2-4, where Ru^{ox} = Ru(IV). We have shown elsewhere[10,11,13] that this rate expression is as shown in Eq. 8, where a steady-state approximation is applied to the peroxide concentration and where $[Ru]_t = [Ru(IV)] + [Ru(II)]$. This rate law is in excellent agreement

$$V = \frac{k_1 k_3 [Ru]_t [O_2] [ROH]}{k_1 [O_2] + k_3 [ROH]} \tag{8}$$

with the observed kinetics. It correctly predicts both the first-order dependence on the total Ru added and the zero-order substrate dependence. Significantly, it predicts a variable and less than unity oxygen dependence, as observed. This expression implies that a plot of $[Ru]_t/V$ versus $[O_2]^{-1}$ should be linear with a slope equal to $1/k_1$ and an intercept of $1/(k_3[ROH])$. Such plots over O_2 pressures from 50-800 psi are indeed linear for all of the catalysts we have investigated (corr. coeff. > 0.98). The values of k_1 and k_3 are listed in Table II for three of these catalysts and reveal that the reduction of the alcohol (k_3) is the inherently slow step in this process. However, since the solvent, methanol, is about 25.0 M and the O_2 concentration at 100 psi is about 0.05 M[19] and since k_3 is never less than 50 times smaller than k_1, the denominator of Eq. 8 is dominated by the $k_3[ROH]$ term. As a consequence, at lower O_2 pressures and high alcohol concentrations the observed rate of thioether conversion is given by the approximation shown by Eq. 9. Thus at O_2 pressures of

$$V = \frac{-d[SR_2]}{dt} = k_1 [Ru]_t [O_2] \tag{9}$$

TABLE II

Values of the Rate Constants k_1 and k_3 Obtained from Linear Plots of $[Ru]_t/V_{obs}$ Versus $[O_2]^{-1}$

Complex	Temp. (°C)	$k_1 (M^{-1}sec^{-1})$	$k_3 (M^{-1}sec^{-1})$
cis $RuCl_2(DMSO)_4$	108	6.3×10^{-2}	2.1×10^{-3}
trans $RuBr_2(DMSO)_4$	95	5.3×10^{-2}	1.1×10^{-3}
trans $RuBr_2(DMSO)_3(PPh_3)$	95	6.3×10^{-2}	3.8×10^{-3}

about 100 psi or less, Eq. 9 will be valid and the rate-determining step is the oxidation of the metal with oxygen.

The effect of temperature on the reaction rate was studied for the three catalysts, cis-$RuCl_2(DMSO)_4$, trans-$RuBr_2(DMSO)_4$, and trans-$RuBr_2$-$(DMSO)_3(PPh_3)$ in methanol at 100 psi O_2 pressure. In each case linear Arrhenius plots were obtained over at least 40°K ranges. From the slopes and intercepts of these straight-line plots the activation energies and entropies were calculated for these catalysts: cis $RuCl_2(DMSO)_4$, E_a = 22.3 kcal/mole and $\Delta S^{\neq}$ = 8.6 eu; trans $RuBr_2(DMSO)_4$, E_a = 23.7 kcal/mole and $\Delta S^{\neq}$ = -2.6 eu; and trans $RuBr_2(DMSO)_3(PPh_3)$, E_a = 21.9 kcal/mole and $\Delta S^{\neq}$ = -3.5 eu.

These values reveal that the activation energy of the oxygen reaction with Ru(II) is about twenty kcal/mole which is consistent with a bimolecular rate-determining step. Even more important is the low activation entropy values of these catalyst systems. These values are consistent with an outer-sphere electron-transfer,[6] and they support the original contention that to achieve dioxygen catalysis with poor ligands, it is necessary to use metal complexes that can undergo outer-sphere electron transfer to give free peroxide in solution.

B. Catalyst Identity

In Table III is shown a compilation of observed rates of oxidation of different thioether substrates with $RuX_2(DMSO)_4$ catalysts. The striking feature of these data is the profound effect that the thioether has on the rate. The mechanism described above (Eqs. 2-4), predicts that thioether substrate should have no effect on the reaction rate, because the reaction is essentially a catalytic alcohol oxidation sequence and the substrate thioether simply traps the peroxide as it forms. This indicates that the thioether substrate must react with the $RuX_2(DMSO)_n(PR_3)_x$ (where X=O or 1) complexes to generate the active catalyst complex(es) in solution (Eq. 10).

$$RuX_2(DMSO)_4 + \text{xs-}SR_2 \rightleftharpoons RuX_2(DMSO)_{4-x}(SR_2)_x + DMSO \qquad (10)$$

To better understand this selective catalytic system, the nature of the species formed was studied in situ.[19] Reactions were carried out at 100°C

TABLE III

The Observed Rates of Molecular Oxygen Oxidation of Different Thioether Substrates at 105°C and 100 PSI O_2 in Methanol

Substrate	Turnover Number (mole sub/mole cat hr^{-1}) cis $RuCl_2(DMSO)_4$	trans $RuBr_2(DMSO)_4$
dimethyl sulfide	7.2	24
tetrahydrothiophene	4.0	9.5
decyl methyl sulfide	8.5	27
diethyl sulfide	52	121
di-n-butyl sulfide	78	172
di-isobutyl sulfide		131
di-tert-butyl sulfide	0	0
thioanisole ($PhSCH_3$)	4	8

under argon in methanol in a pressure vessel with a 20-50 fold excess of thioether over Ru. The reactions were followed by TLC on silica gel. In only a few minutes a distribution of different complexes forms and remains unchanged even after an hour. These solutions were subjected to medium pressure (50-70 psi) column chromatography under argon using silica gel and 97% CH_2Cl_2/3% MeOH as the eluting solvent. Table IV shows the results of several such reactions in which various substrates were utilized. The description of the synthesis and characterization of these products are described elsewhere.[13,19] The salient feature of these results is the profound effect the steric bulk of the thioether has on the relative yield and the stoichiometry of the complexes formed. The bulkiest thioether, di-t-butyl sulfide gives only one detectable product; the monokis(thioether) complex $RuX_2(DMSO)_3(SR_2)$. In contrast the least bulky ligand, tetrahydrothiophene,[20] gives predominantly a trans-dihalotetrakis(sulfide) complex, trans $RuBr_2(SR_2)_4$, although other species are present in low levels. As the steric bulk increases from tetrahydrothiophene to dimethyl

TABLE IV

Distribution of Species Isolated and Characterized from the Reaction of an Excess of Thioether with a $RuX_2(DMSO)_4$ Complex in Methanol under Argon

Catalyst	Sulfide	Products (% Yield)
cis $RuCl_2(DMSO)_4$	SMe_2	$RuCl_2(SMe_2)_4$ (50%)[a]
		$RuCl_2(SMe_2)_3(DMSO)$ (33%)[a]
		$RuCl_2(Me_2S)_2(DMSO)_2$ (12%)[b]
trans $RuBr_2(DMSO)_4$	THT[c]	85% $RuBr_2(S(CH_2)_4)_4$[a]
trans $RuBr_2(DMSO)_4$	SMe_2	Same distribution as chloro
trans $RuBr_2(DMSO)_4$	SEt_2	$RuBr_2(Et_2S)_3(DMSO)$ (28%)[a]
		$RuBr_2(Et_2S)_2(DMSO)_2$ (65%)[b]
trans $RuBr_2(DMSO)_4$	t-Bu_2S	$RuBr_2(DMSO)_3$(t-Bu_2S) (90%)

[a]trans dihalo complex from far IR and ^1H NMR.
[b]2:1 mixture of the trans, cis, cis- and trans, trans, trans-complexes.
[c]tetrahydrothiophene

sulfide the amount of tetrasubstituted tetrakis(thioether) complex decreases and other trans-dihalo complexes such as the tris(thioether) and the two trans-dihalo isomeric bis(thioether) complexes become increasingly abundant. An even bulkier ligand, diethyl sulfide, does not afford any of the tetrakis(thioether) complex but gives roughly equimolar amounts of the three complexes, trans $RuBr_2(SEt_2)_3(DMSO)$, trans-dihalo cis, cis $RuBr_2(SEt_2)_2(DMSO)_2$, and all trans-$RuBr_2(SEt_2)_2(DMSO)_2$.

These distribution results permit the making of some qualitative statements concerning the likelihood that a specific structure is a catalytically active species in these oxidations. First, a monokis(thioether) complex can be ruled out with some certainty, since no activity is observed in the di-t-butyl sulfide system and since this is the only complex formed in this system. If complexes of the stoichiometry trans $RuX_2(SR_2)_4$ were the active species in these systems, then the greatest activity should be observed with tetrahydrothiophene or dimethyl sulfide as a substrate.

TABLE V

Reversible Half-Wave Potentials ($E_{1/2}$) for One-Electron Oxidations of Ru(II) Complexes of the Stoichiometry $RuX_2(DMSO)_{4-n}(SR_2)_n$

Complex	$E_{1/2}V$ (versus Ag/Ag^+ in CH_2Cl_2)
cis $RuCl_2(DMSO)_4$	+1.5[a]
trans $RuBr_2(DMSO)_4$	+1.5[a]
trans $RuBr_2(DMSO)_3(t\text{-}Bu_2S)$	+1.34
trans Chloro-$RuCl_2(DMSO)_2(Me_2S)_2$	+0.75 and +1.04[a]
trans Bromo-$RuBr_2(DMSO)_2(Et_2S)_2$	+0.79 and +1.01
trans $RuCl_2(DMSO)(Me_2S)_3$	+0.85
trans $RuCl_2(Me_2S)_4$	+0.61
trans $RuBr_2(Me_2S)_4$	+0.65
trans $RuBr_2(THT)_4$[b]	+0.59
trans $RuCl_2(THT)_4$[b]	+0.57

[a]Irreversible

[b]Tetrahydrothiophene

Consequently, these complexes can also be ruled out as the catalyst candidate. The complexes of stoichiometry $RuX_2(SR_2)_3(DMSO)$ also can be eliminated as catalyst candidates. If this stoichiometry afforded an active catalyst, then similar rates would be expected with dimethyl sulfide or diethyl sulfide since this type of complex is present in about the same proportion in the two systems. This then leaves the two trans-dihalo bis(thioether)bis(sulfoxide) complexes as the most likely candidates for the structure of the catalyst in these systems. In fact, the relative abundance of the 2,2,2-isomers is about five times as great with diethyl sulfide as with dimethyl sulfide. This five-to-one ratio is about the same as the rate ratio observed with the trans $RuBr_2(DMSO)_4$ catalyst for the oxidation of diethyl sulfide and dimethyl sulfide. The question this conclusion leaves is which or both of these two isomers is the catalyst and can this conclusion be tested independently?

To obtain a better understanding of the redox properties of these various complexes, their electrochemistry was studied using standard cyclic voltammetry. In Table V are listed the half-wave potentials for the complexes trans-dihalo $RuX_2(SR_2)_n(DMSO)_{4-n}$. As can be seen the sulfoxide ligand appears to stabilize the ruthenium(II) to oxidation very effectively. Apparently the S-bound sulfoxide ligand is either an excellent π-acceptor or weak σ-donating ligand (or both) when coordinated to low-spin Ru(II). Each replacement of a sulfoxide ligand with a thioether ligand decreases the redox potential by about 0.2-0.3V. A break in this trend is observed with the unusually low redox potential exhibited by one of the 2,2,2-complexes. Since the separation of this mixture of the two isomers with stoichiometry trans-dihalo $RuX_2(SR_2)_2(DMSO)_2$ was not possible a model complex[19] with a bidentate ligand, 3,6-dithiooctane was synthesized and characterized. This complex has the trans-dihalo cis,cis $RuX_2(SR_2)_2(DMSO)_2$ structure and it possesses a reversible oxidation wave at 1.05V. This indicates that the structure of the 2,2,2-isomer which possesses the anomalously low oxidation potential is the all trans $RuX_2(SR_2)_2(sulfoxide)_2$ complex.

The relative ease of oxidation of the all trans-isomer indicates that it is indeed unusual electronically and is a likely candidate for an oxidation catalyst. In addition, this complex also has a ligand environment about it that favors a lower oxidation state, namely, there are two sulfoxide ligands bound to ruthenium. From the electrochemical data the sulfoxide ligands would definitely be poor ligands for stabilizing a ruthenium(IV) complex; i.e., good π-acceptor, poor σ-donor. This leads to the conclusion that the all trans 2,2,2-Ru(II) complex is the catalyst species in these reactions because it alone possesses the proper combination of electronic factors. This complex can be oxidized by oxygen because of its geometry which affords an easier to oxidize Ru(II) ion. Further, its stoichiometry is such that the ligand environment is a destabilizing one in the oxidized Ru(IV) complex; consequently, the oxidized complex can be reduced by the alcohol solvent.

Additional studies are in progress in this area. We hope to be able to synthesize an all trans 2,2,2-Ru(II) complex with multidentate ligands about the Ru(II) center. In this way we hope to make a very active catalyst and at the same time control the stereochemistry about the Ru(II) so that the intimate mechanistic details can be better studied.

REFERENCES

1. M.A. Ledlie, K.G. Allum, V.I. Howell, and R.G. Pitkethly, J. Chem. Soc., Perkin I, 1734 (1976).
2. R.A. Sheldon and J.K. Kochi, in **Metal-Catalyzed Oxidations of Organic Compounds**, Academic Press, New York, 1981, pp 72-119.
3. (a) A. Sen and J. Halpern, J. Am. Chem. Soc., **99**, 8337 (1977) (see references therein). (b) R. Barral, C. Bocard, I.S. deRoch, and L. Sajus, Kinet. Katal., **14**, 164 (1973). (c) B.S. Tovrog, S.E. Diamond, and F.J. Mares, J. Am. Chem. Soc., **101**, 270 (1979).
4. R. Barral, C. Bocard, and I.S. deRoch Tetrahedron Lett., 1963 (1972).
5. S. Otsuka, A. Nakamura, Y. Tatsumo, and M. Miki, J. Am. Chem. Soc., **94**, 3761 (1972).
6. J.R. Pladziewicz, T. Meyer, J.A. Broomhead, and H. Taube, Inorg. Chem., **12**, 639 (1973).
7. D.M. Stanbury, O. Haas, and H. Taube, Inorg. Chem., **19**, 518 (1980).
8. B.S. Tovrog, S.E. Diamond, and F.J. Mares, J. Am. Chem. Soc., **101**, 5067 (1979).
9. D.P. Riley, J. Chem. Soc., Chem. Commun., 1530 (1983).
10. D.P. Riley, Inorg. Chem., **22**, 1965 (1983).
11. D.P. Riley and R.E. Shumate, J. Am. Chem. Soc., **106**, 3179 (1984).
12. J.D. Oliver and D.P. Riley, Inorg. Chem., **23**, 156 (1984).
13. D.P. Riley, Inorganica Chim. Acta, (In press).
14. R.E. Shumate and D.P. Riley, J. Chem. Educ., (In press).
15. M.S. Thompson and T.J. Meyer, J. Am. Chem. Soc., **104**, 4106 (1982).
16. (a) J.A. Howard and S. Korcek, Can. J. Chem., **49**, 2178 (1971). (b) L. Bateman, J.I. Cunneen, and J. Ford, J. Chem. Soc., 3056 (1956).
17. DeF.P. Rudd and H. Taube, Inorg. Chem., **10**, 1543 (1971).
18. R.R. Gagne and D.N. Marks, Inorg. Chem., **23**, 65 (1984).
19. Unpublished results from our laboratories.
20. C.A. Tolman, Chem. Revs. **77**, 313 (1977).

17

Asymmetric Epoxidation

Bryant E. Rossiter
Hoffman LaRoche, Inc.
Nutley, NJ 07110

ABSTRACT

In 1980, Katsuki and Sharpless reported the first practical method for asymmetric epoxidation of allylic alcohols which is highly enantioselective for a broad range of substrates. Because of its high efficiency, the ready availability of reagents and the ease of execution as well as the proven utility of epoxides as synthetic intermediates, asymmetric expoxidation has been widely embraced by organic chemists. These same characteristics have given this reaction a high potential for industrial acceptance.

Recently several variations of this reaction have been developed which include the kinetic resolutions of allylic alcohols, 1,2-amino alcohols, hydroxysulfides, and propargylic alcohols and the asymmetric formation of chlorodiols from allylic alcohols. These and other aspects of the Sharpless epoxidation will be discussed.

I. INTRODUCTION

Over the past several years, the discovery and development of means to perform catalytic asymmetric chemical transformations has increased dramatically.[1] These reactions are not only promoting the understanding of the subtle principles involved in asymmetric transformations, but are also

Asymmetric Epoxidation of Primary Allylic Alcohols

D-(-)-diethyl tartrate

$Ti(OPr\text{-}\underline{i})_4$, TBHP

CH_2Cl_2, -20°C

L-(+)-diethyl tartrate

Scheme 1

significantly changing the way chemists make stereochemically complex organic substrates. Indeed, some of these reactions have found industrial use.

II. RESULTS AND DISCUSSION

One of the most versatile and interesting of these is the Sharpless epoxidation (Scheme 1).[2-4] In this reaction, pro-chiral allylic alcohols are epoxidized enantioselectively with tert-butyl hydroperoxide (TBHP) and a titanium(IV) alkoxide catalyst bound to a chiral dialkyl tartrate ligand. It is similar in nature to other metal-catalyzed epoxidations in which a transition metal in its d^0 oxidation state catalyzes the epoxidation of olefins and allylic alcohols with alkyl hydroperoxides (Scheme 2).[3,5] It has been proposed[5b] that epoxidation occurs first by assembling a titanium-TBHP-allylic alcohol-dialkyl tartrate complex, **2**, followed by formation of the intermediate peroxo complex, **3**, which then transfers its distal peroxo oxygen to the bound allylic alcohol. This complex exchanges its tert-butoxide and epoxyalkoxide ligands for TBHP and allylic alcohol and the process begins anew. The actual catalytic species is believed to be a 2:2 titanium(IV)-tartrate dimer based on the kinetics of the reaction, measurements by differential vapor phase osmometry (CH_2Cl_2) and by Rayleigh light scattering (cyclohexane) as well as electron impact mass

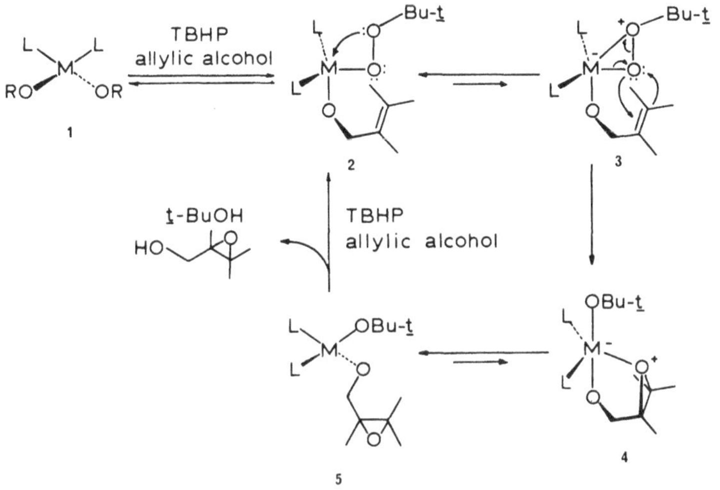

Scheme 2

spectrometry.[3] One possible structure for the catalyst is represented in Fig. 1. This is analogous to the structure found for the complex of vandium(IV) and tartaric acid, $Na_4[(VO)_2(tart)_2]$[6] and is consistent with IR and NMR data.[3] While the exact origin of the high enantioselectivity of this reaction has not been completely elucidated, it is believed to be strongly related to the stereoelectronic alignment of the olefin and the peroxo ligand in complex **3** (Scheme 2) as they approach one another and eventually react.[3,5b] The peroxo ligand, rigidly bound in a stereochemically well-defined manner, will prefer to react with one face or the other of the bound allylic alcohol leading to enantioselective epoxidation.

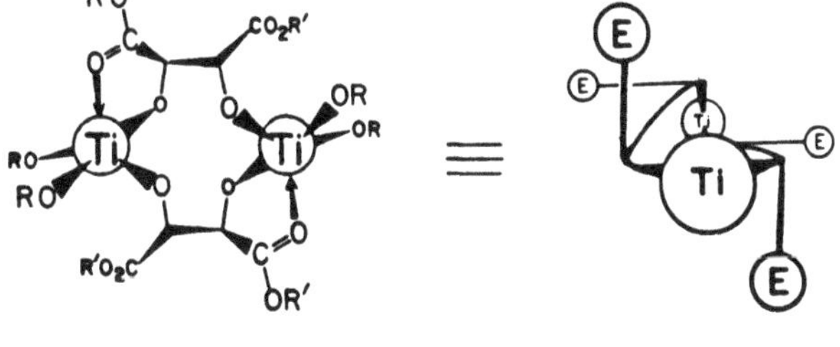

Fig. 1

TABLE I

Asymmetric Epoxidation of Prochiral Allylic Alcohols with $Ti(OiPr)_4$, (+)-DET, and TBHP

	substrate	epoxide	% e.e.	yield	ref.
1	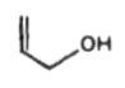	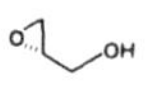	95	15	3b
2	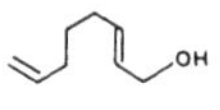	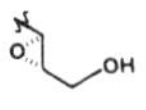	95	80	18
3	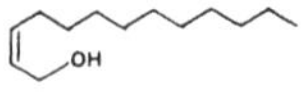	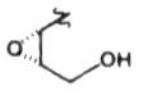	90	82	2
4	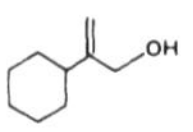	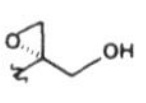	95	81	2
5	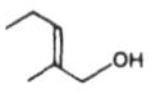	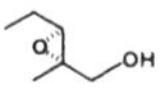	95	79	18
6*	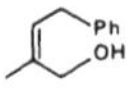	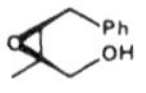	91	90	17
7	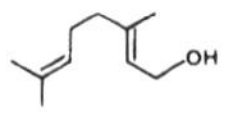	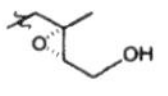	95	77	2
8	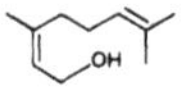	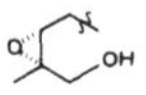	94	79	2
9*	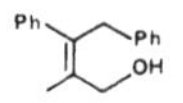	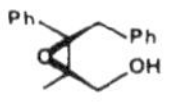	94	90	17

*epoxidation w/ (-)-DET

The reaction is unusually efficient and predictable. The epoxyalcohol products are generally obtained in >90% e.e. and often >95% e.e. Table I shows a representative sample of substrates successfully epoxidized by this method. In addition, for most cases, the absolute configuration of the resulting epoxide is constant for a given enantiomer of a tartrate ester regardless of the substitution pattern of the allylic alcohol. As shown in Scheme 1, when L-(+)-tartrate is used the allylic alcohol, as drawn, tends to be epoxidized from the underside whereas when a D-(-)-tartrate is used it is

TABLE II
Allylic Alcohol Expoxidations

	SUBSTRATE T	Ti (OR)$_4$/ (+)-tartrate	% e. e.	Reaction Time	Major Product Enantiomer
10	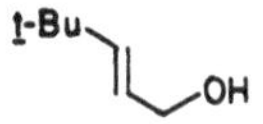 t-Bu, OH	Ti(OiPr)$_4$/DET	95	96 h	2S
11	t-Bu, OH	Ti(OtBu)$_4$/DET	85	15 h	2S*
12	t-Bu, OH	Ti(OtBu)$_4$/DET	60	15 h	2S*
13	t-Bu, OH	Ti(OiPr)$_4$/DET	25	192 h	2S
		Ti(OtBu)$_4$/DET	25	192 h	2S

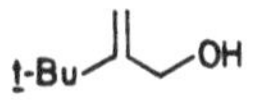

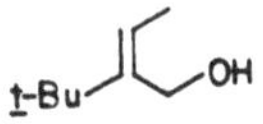

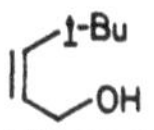

epoxidized from the top. For prochiral allylic alcohols there are no known exceptions to this selectivity rule and only a few exceptions for chiral allylic alcohols (see below). The efficiency of the reaction is remarkably insensitive to pre-existing chirality in the starting material in contrast to what is observed with many enzymes or other asymmetric reactions which are often highly sensitive to substrate structure. These features, along with the ease with which the reaction is performed and the overall utility of 2,3-epoxy alcohols in general, has made this reaction an extremely useful one among synthetic chemists concerned with stereochemically complex substrates.[4,7]

To better understand the power and scope of this reaction it is instructive to consider those situations in which it performs somewhat inefficiently. As shown in Table II,[3a,8] an allylic alcohol substituted in the *trans* position with the bulky *tert*-butyl substituent is capable of being epoxidized with high enantioselectivity. Substitution in the *cis* position causes a slight drop in the enantioselectivity as does substitution of a bulky group in the 2 position (entries 11 and 12). A dramatic drop in enantioselectivity occurs when the olefin is substituted with a bulky group in

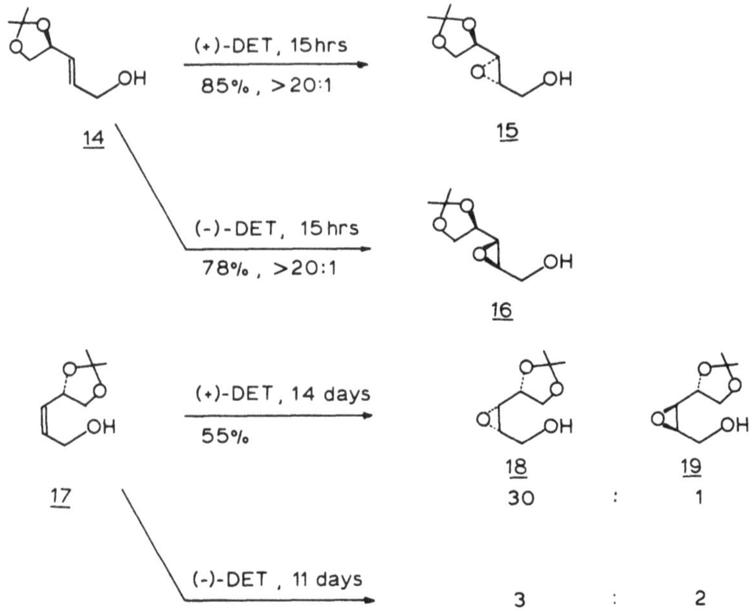

Scheme 3

the cis position (entry 13). There is also a noticeable drop in the rate of reaction for Z-allylic alcohols.

A similar trend is seen with chiral allylic alcohols such as **14** and **17** (Scheme 3).[9] The trans allylic alcohol reacts with both (+)-and (-)-diethyltartate (DET) at a reasonable rate and with high selectivity. The cis isomer is epoxidized slowly with high selectivity when (+)-DET is used and with poor selectivity when (-)-DET is used. The latter case is one of the few exceptions known to the powerful selectivity rule governing this reaction. Thus, in general, the selectivity for this reaction tends to be very good except for substrates with bulky substituents in the cis position.

Occasionally one encounters situations in which the epoxy alcohol product is unstable to the reaction conditions. In the presence of various nucleophiles such as TBHP and iso-propanol as well as internal nucleophiles and the Lewis acidic Ti(IV) alkoxides, one finds the products opening to form diols (Scheme 4).[3,10] This not only destroys the product but also creates catalyst inhibiting diols. One way of partially preventing this is to use

Formation of Catalyst Inhibitors

Ti Nu: → Ti Nu

Scheme 4

titanium(IV) tetra-t-butoxide. This substitutes the relatively non-nucleophilic t-butoxide for i-propoxide. Sharpless et. al.[11] found that the use of this form of titanium(IV) enabled them to substantially increase the yield of the epoxide **21** from 15% with $Ti(OiPr)_4$ to 51% (Scheme 5).

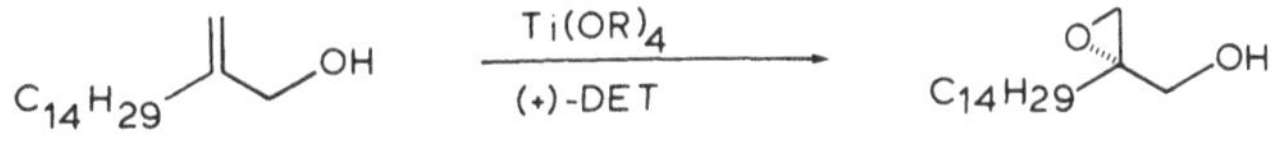

$C_{14}H_{29}$ OH (20) —$Ti(OR)_4$, (+)-DET→ $C_{14}H_{29}$ O OH (21)

R = iPr 15%
= tBu 51%

Scheme 5

In some cases the epoxy alcohol is too reactive. Even though the starting material was completely consumed, none of the desired product was isolated when highly p-methoxycinnamyl alcohol was epoxidized. The less electron rich and consequently less reactive p-nitro and p-chlorocinnamyl alcohols[10] have been epoxidized in very good yield. Interestingly 3,4-dimethoxy-α-methylcinnamyl alcohol has been epoxidized in good yield (Scheme 6).[12] The α-methyl group apparently helps to stabilize the epoxy

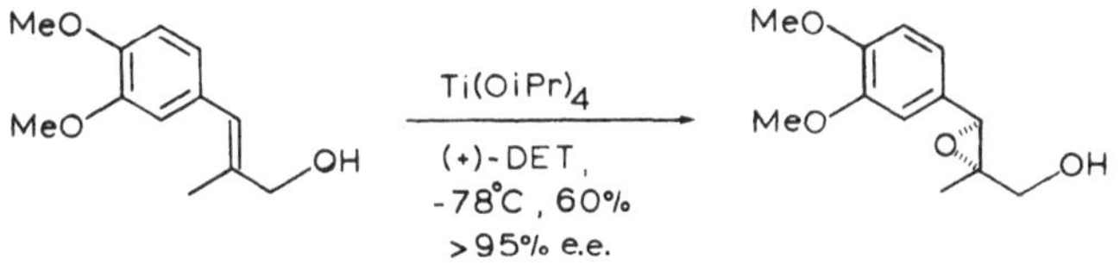

MeO MeO OH (22) —$Ti(OiPr)_4$, (+)-DET, -78°C, 60%, >95% e.e.→ MeO MeO O OH (23)

Scheme 6

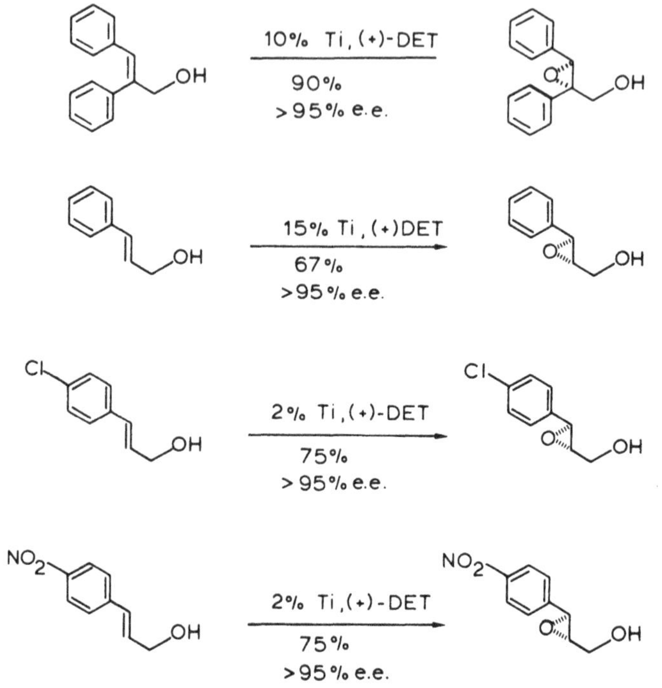
OH
10% Ti, (+)-DET
90%
>95% e.e.
OH
OH
15% Ti, (+)DET
67%
>95% e.e.
OH
Cl
OH
2% Ti, (+)-DET
75%
>95% e.e.
Cl
OH
NO2
OH
2% Ti, (+)-DET
75%
>95% e.e.
NO2
OH

Scheme 7

alcohol against nucleophilic attack.

Even though this reaction is catalytic in nature, it has generally been run with stoichiometric amounts of reagent. There have been instances when the catalytic nature of the reaction has been clearly manifested. The most dramatic examples are those involving cinnamyl alcohols. As one can see from Scheme 7, good yields and enantioselectivities can be obtained when the reaction is run catalytically, using as little as 2% of the reagent. With highly refined conditions, it seems likely that lower catalyst loadings could be used with any of these substrates. While these reactions look promising in terms of increasing the efficiencies of these reactions, cinnamyl alcohols in general tend to be good epoxidation substrates. It would therefore be interesting to consider the catalytic epoxidation of a less cooperative substrate. Recently, the reaction of trans 2-hexen-2-ol was developed using 54 mole % catalyst (Scheme 8).[13] As shown, the yield and %

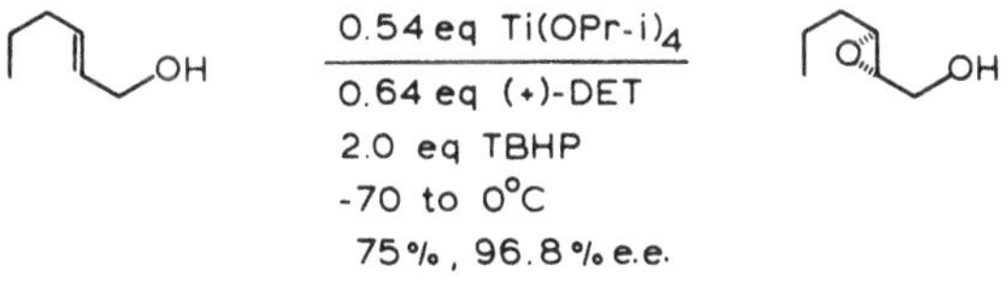

0.54 eq Ti(OPr-i)$_4$
0.64 eq (+)-DET
2.0 eq TBHP
-70 to 0°C
75%, 96.8% e.e.

Scheme 8

e.e. are both quite good. It has also been demonstrated that even less of the catalyst (as low as 10%) could be used although at that level the reaction rate was slow and 100% conversion of the starting material was difficult to attain. These results clearly indicate that for many substrates, the use of 1 equivalent of reagent is wasteful and a more economical epoxidation can be attained with less reagent. This also helps in terms of simplifying an often tedious work-up procedure.[4,10]

While running the reaction homogeneously with catalytic quantities of reagent is desirable, it would be even more desirable to use a heterogeneous catalyst. This has been attempted by Farrall, Alexis and Trecanter[14] who formed a polymer bound tartrate ester (Scheme 9). Using this polymer in the presence of Ti(O$\underline{i}$Pr)$_4$ and TBHP they effected the epoxidation of

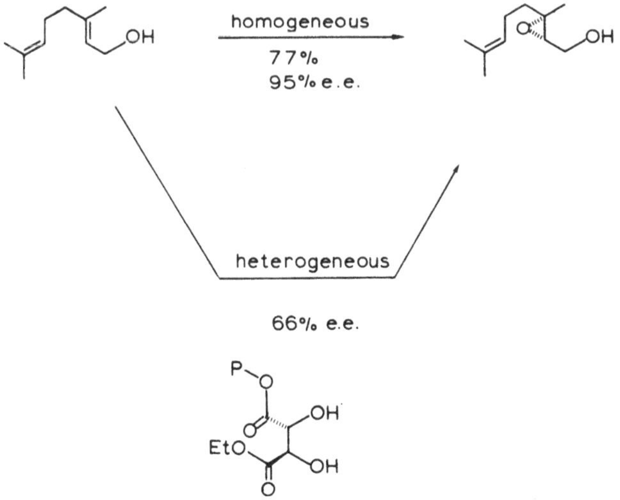

homogeneous
77%
95% e.e.

heterogeneous

66% e.e.

P–O, OH, EtO, OH, O

Scheme 9

geraniol with e.e.'s as high as 66%. The polymer could be recycled although loss of activity accompanied recycling. Although not as enantioselective as the homogeneous reagent, it represents an interesting first attempt in this area.

III. SUMMARY

The Sharpless epoxidation has the potential to work well for many substrates although sometimes less than optimum results are realized by those attempting to carry out this reaction. To get the highest possible enantioselectivity, two major points should be kept in mind.

(1) The correct ratio of Ti(IV) to tartrate ester must be used. For asymmetric epoxidations this is a ratio of 1: $\geqslant$ 1.1.[15]

(2) More subtly, all reagents should be carefully dried. It has been shown that small amounts of water in the reaction mixture will cause a drop in the enantiomeric excess of the products.[16] For instance, asymmetric epoxidation of α-phenylcinnamyl alcohol normally occurs with 99% enantioselectivity (Scheme 10). When one equivalent of H_2O was added to the reaction mixture, the enantioselectivity of the reaction dropped to 49%.

This reaction has been used to make a wide variety of stereochemically complex substrates including insect pheromones, pharmaceuticals and sugars.[4,7,17] One of the first uses of the reaction as well as its first industrial use was in the synthesis of (+) Disparlure, the sex attractant of the male Gypsy moth.[17] In this synthesis (Scheme 11) the cis-allylic alcohol is epoxidized in good yield and with reasonably good enantioselectivity. The epoxy alcohol could be recrystallized to near enantiomeric purity. Oxidation with Collins' reagent followed by Whittig olefination and hydrogenation with Wilkinsons' catalyst gave the desired

"dry" 99% e.e.
1 eq H_2O 48% e.e.

Scheme 10

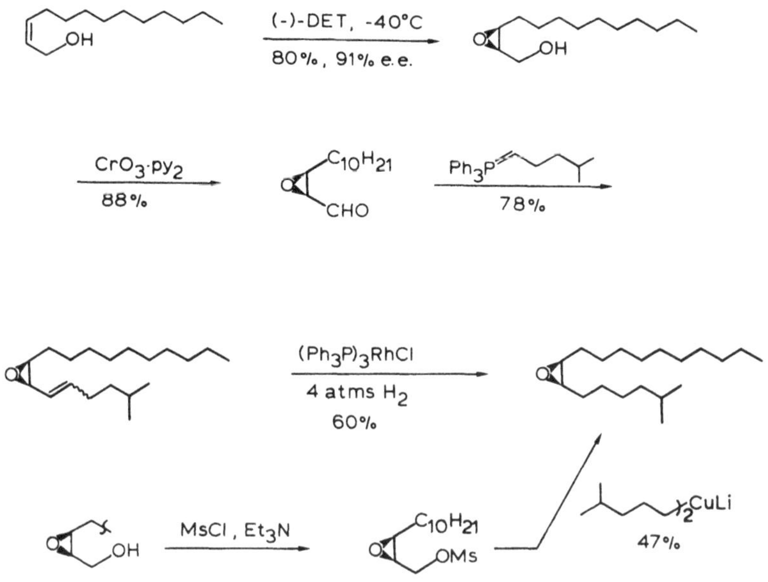

(-)-DET, -40°C

80%, 91% e.e.

$CrO_3 \cdot py_2$

88%

$C_{10}H_{21}$

CHO

Ph_3P

78%

$(Ph_3P)_3RhCl$

4 atms H_2

60%

MsCl, Et_3N

$C_{10}H_{21}$

OMs

CuLi

47%

Scheme 11

product. Alternately the epoxy mesylate could be obtained in high yield and reacted with the shown cuprate reagent to give the desired product.

In addition to the asymmetric epoxidation of prochiral allylic alcohols has been the asymmetric epoxidation/kinetic resolution of chiral, especially secondary, allylic alcohols (Scheme 12).[15]

Recently some new discoveries have shown that the Ti(IV)-tartrate-TBHP system is capable of enantioselective oxidation with substrates other than allylic alcohols. Using a 2:1 ratio of Ti(IV)-tartrate it was shown that 1,2-amino alcohols could often be efficiently kinetically resolved (Scheme 13).[19]

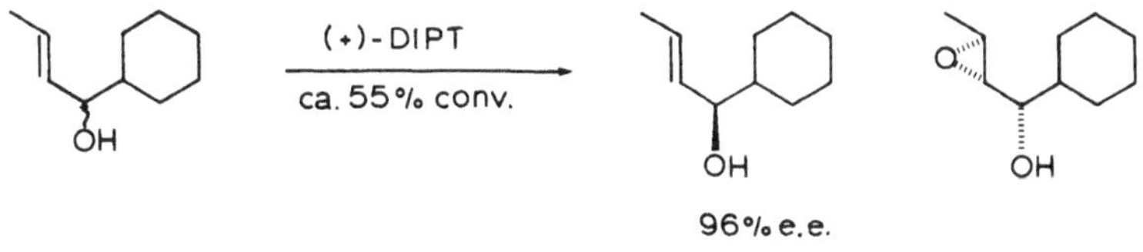

(+)-DIPT

ca. 55% conv.

OH

96% e.e.

Scheme 12

Ph–CH(OH)–CH$_2$–N(pyrrolidine) → 1) 1.2(+)-DIPT, 2.0 Ti(OiPr)$_4$; 2) 0.6 TBHP → Ph–CH(OH)–CH$_2$–N(pyrrolidine) 37% 95% e.e. + N-oxide 59% 63% e.e.

Scheme 13

It has also been shown that using a 2:1 ratio of $TiCl_2(O\underline{i}Pr)_2$, DET that allylic alcohols can be oxidized to chlorodiols with reasonably good enantioselectivity (Scheme 14).[11] These substrates, when treated with base give the epoxy alcohol which has an absolute configuration opposite that which one would obtain from normal asymmetric epoxidation. The intermediate in this reaction is most likely the epoxy alcohol.

$C_{14}H_{29}$ allylic alcohol → 2 $TiCl_2(OiPr)_2$, 1 (+)-DET, TBHP, 0°C → chlorodiol 68% e.e.

Scheme 14

As is evident, the Sharpless epoxidation has been a fruitful area of research in asymmetric synthesis. No doubt it will continue to be so for some time to come.

ACKNOWLEDGEMENTS

I wish to express gratitude to Professor K. Barry Sharpless and Mr. M.G. Finn for their help in preparing this manuscript.

REFERENCES

1. J.D. Morrison, **Asymmetric Synthesis**, Academic Press, New York, Vol. V, in press.
2. T. Katzuki and K.B. Sharpless, J. Am. Chem. Soc., **102**, 5974 (1980), T. Katzuki and K.B. Sharpless, European Patent 0-046-033, (1981).
3. For more complete discussions of the mechanism see:
(a) M.G. Finn and K.B. Sharpless in **Asymmetric Synthesis**, (J.D. Morrison, ed.) Academic Press, New York, Vol. V, in press; (b) K.B.

Sharpless, S.S. Woodard, and M.G. Finn, Pure Appl. Chem., **55**, 1823 (1983); (c) S.S. Woodard, Ph.D. Dissertation, Stanford University, Stanford, CA (1981).

4. B.E. Rossiter, in **Assymmetric Synthesis**, (J.D. Morrison, ed.) Academic Press, New York, Vol V, in press.
5. (a) R.A. Sheldon and J.K. Kochi, **Metal Catalyzed Oxidations of Organic Compounds**, Academic Press, New York, 1981 Chapter 3; (b) K.B. Sharpless and T.R. Verhoeven, Aldrichimica Acta, **12**, 63 (1979).
6. R.E. Tapscott and G.L. Robbins, Inorg. Chem., **15**, 154 (1976); (b) R.E. Tapscott, R.L. Belford, and I.C. Paul, Coord. Chem. Rev., **4**, 323 (1969); (c) S.K. Hahs, R.B. Ortega, R.E. Tapscott, C.F. Campana, and B. Morosin, Inorg. Chem., **21**, 664 (1982).
7. C.H. Behrens and K.B. Sharpless, Aldrichimica Acta, **16**, 67 (1983).
8. M.J. Schweiter, S.M. Dissertation, Massachusetts Institute of Technology, Cambridge, MA (1984).
9. (a) T. Katsuki, A.W.M. Lee, P. Ma, V.S. Martin, S. Masamune, K.B. Sharpless, D. Tuddenham and F.J. Walker, J. Org. Chem., **47**, 1378 (1982); (b) N. Minami, S.S. Ko, and Y. Kishi, J. Am. Chem. Soc., **104**, 1109 (1982).
10. D. Tuddenham and K.B. Sharpless, unpublished results.
11. L.D.-L. Lu, R.A. Johnson, M.G. Finn and K.B. Sharpless, J. Org. Chem., **48**, 728 (1984).
12. K.B. Sharpless, personal communication.
13. J.G. Hill, K.B. Sharpless, C.M. Exon, and R. Regenye, Org. Syn., submitted for publication.
14. M.J. Farrall, M. Alexis, and M. Trecarten, Nouv. J. Chim., **7**, 449 (1983).
15. V.S. Martin, S.S. Woodard, T. Katsuki, Y. Yamada, M. Ikeda, and K.B. Sharpless, J. Am. Chem. Soc., **103**, 6237 (1981).
16. J.G. Hill, B.E. Rossiter, and K.B. Sharpless, J. Org. Chem., **48**, 3607 (1983).
17. K.B. Sharpless, C.H. Behrens, T. Katsuki, A.W.M. Lee, V.S. Martin, M. Takatani, S.M. Viti, F.J. Walker, and S.S. Woodard, Pure Appl. Chem., **55**, 589 (1983).

18. (a) B.E. Rossiter, T. Katsuki, and K.B. Sharpless, J. Am. Chem. Soc., **103**, 464 (1981); (b) B.E. Rossiter, Ph.D. Dissertation, Stanford University, Stanford, CA (1981).
19. S. Miyano, L.D.-L. Lu, S. Viti, and K.B. Sharpless, J. Org. Chem., **48**, 3608 (1983).

18

C-C Double Bond Cleavage by Molecular Oxygen

Helmut Bönnemann, Werner Brijoux, Norbert Pingel
and Dieter M.M. Rohe
Max-Planck-Institut für Kohlenforschung
P.O. Box 01 13 25, D - 4330 Mülheim a.d. Ruhr

ABSTRACT

The conversion of alkenes by molecular oxygen to give carbonyl compounds through C=C double bond cleavage is catalyzed by acetylacetonato- or cyclopentadienyl rhodium complexes, allyl rhodium systems, or $(Ph_3P)_3RhCl$. For example, 2,3-dimethyl-2-butene is transformed into acetone, butadiene into acrylaldehyde, while styrene gives benzaldehyde. The reaction requires enhanced concentrations of oxygen in solution. This can be achieved by using compressed air or a loop reactor at normal pressure.

I. INTRODUCTION

Although the dioxygen molecule has a diradical ground state, molecular oxygen reacts rather slowly with organic substrates at ambient temperatures. This kinetic stability of the free triplet O_2 may be overcome by interaction with transition metal complexes. In MO terms, the metal-O_2 interaction can be interpreted in terms of successive one-electron transfers from the metal into the partially vacant antibonding π^* orbitals.[1] This results in the stepwise formation of superoxide and peroxide ions which can coordinate with the metal giving superoxo- and peroxo-complexes.[2] The interatomic O-O distance is elongated considerably and as a consequence, dissociation of the O-O bond becomes easier on going from left to right in Fig. lb. The sp^2 lone pair orbitals (n_O) are oriented similarly to the four hydrogen atoms in ethylene (Fig. la). Thus it might be expected that a lone pair of the oxygen molecule could interact with a transition metal to give an end-on complex or that the double bond could interact to give a side-on complex.

Since Vaska[3] reported the first reversible O_2 side-on complex in 1963 numerous examples have been found for the principle types of metal-dioxygen attachment (see Fig. 2). When the electronegativity of the halide is decreased, the Ir-O_2 bond strength obviously increases so that the side-on dioxygen is fixed irreversibly in the iodide complex. Apparently, the cobalt and iron complexes shown belong to the end-on type, and have a metal-O-O angle of around 120°. The so called "picket fence" Fe(porphyrin)-O_2 complex, serves as an excellent model for the natural oxygen carriers,[4] since the ligand prevents the intermolecular oxidation of the Fe(II)O_2 species. In contrast, superoxo cobalt(III) complexes rapidly react with a Co(II) giving the μ-peroxo-type of a binuclear complex. (Co-O-O bond angle = 110.8°.)[5] The O-O bond distance is 1.469 A, very close to that of a peroxide ion.

II. RESULTS AND DISCUSSION

Having observed that certain monovalent metal complexes in the cobalt triad promote the oxygenation of olefinic substrates, we turned our attention to the homogeneous oxidation of olefins using molecular oxygen. A variety of Co(I), Rh(I) and Ir(I) complexes are known to afford mononuculear peroxo complexes in which the oxygen is coordinated side-on to the three-valent metal center. In this report we focus on the role of rhodium catalysts in homogeneous olefin oxidations.

Read, et al.[8] first reported the co-oxygenation of 1-alkenes and triphenylphosphine by molecular oxygen catalyzed by an O_2-complex derived from $(Ph_3P)_3RhCl$. This reaction gives mainly methylketones and equimolar amounts of triphenylphosphine oxide. Mimoun[9] used $RhCl_3{\cdot}3H_2O$: $Cu(ClO_4)$

(a)
σ bonding MO
sp² lone pair orbitals
(n_o)
π bonding MO
1 π_u (p_x)
π* anti bonding MO
1 π*_g (p_x)
(b)
O₂
+e⁻
O₂⁻
+e⁻
O₂²⁻
+2e⁻
2O²⁻
Type
Dioxygen
Superoxide
Peroxide
Bond order
2.0
1.5
1.0
0
O–O Å
1.21
1.33
1.49

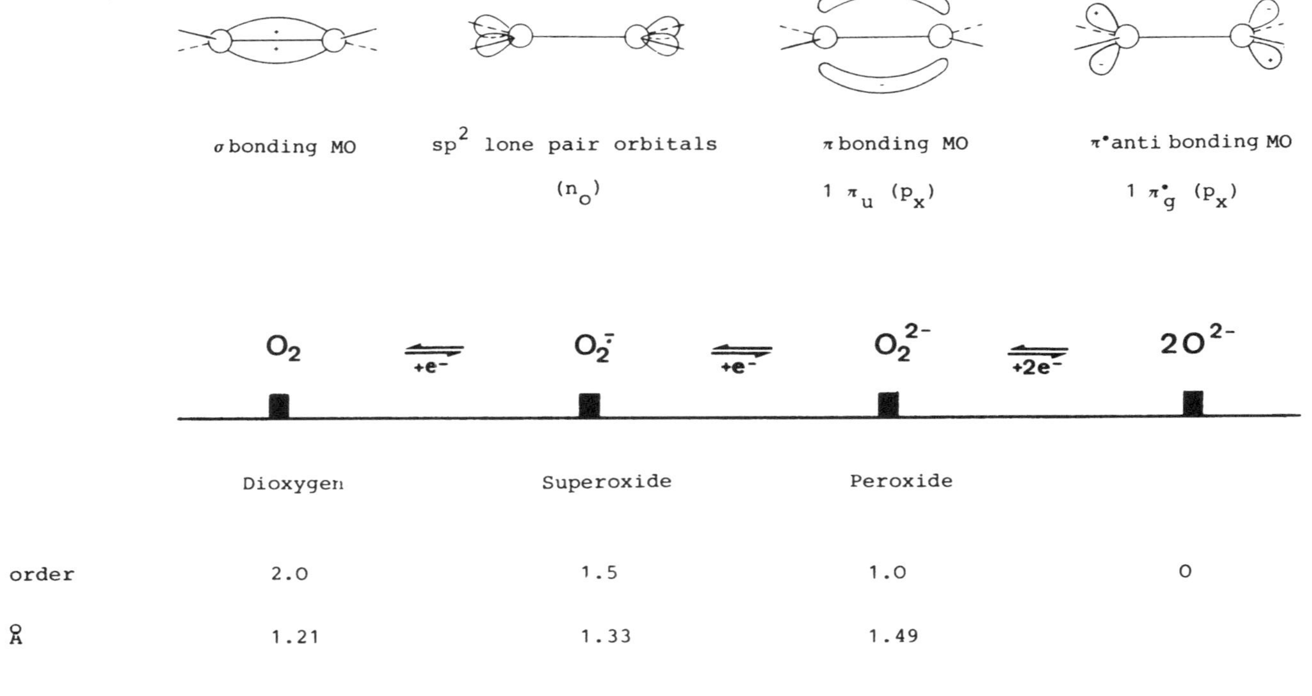

Fig. 1.

Compound	Type	Geometry	O–O (Å)	Ref.
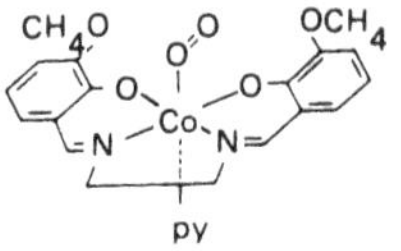 CH_4O, O_2, OCH_4, Co, N, py	reversible O_2 complex	Co-O-O	1.26	[6]
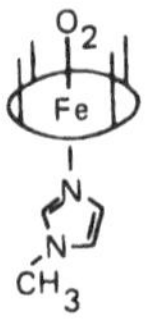 O_2, Fe, N, CH_3	reversible O_2 complex	Fe-O-O	1.24	[4]
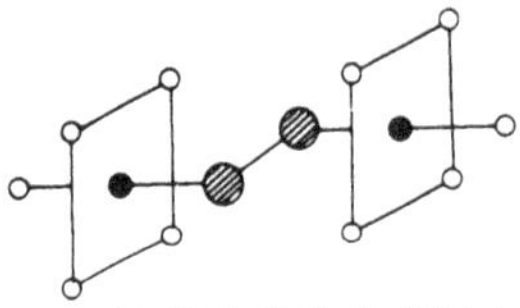 $[(NH_3)_5CoO_2Co(NH_3)_5]^{4+}$	μ peroxo complex	Co-O-O-Co	1.47	[5]
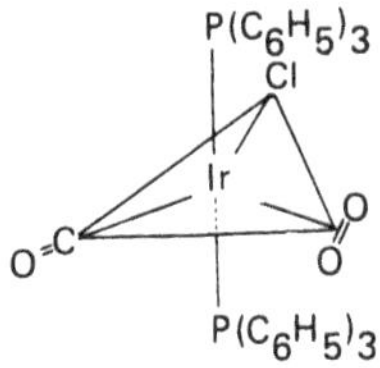 $P(C_6H_5)_3$, Cl, Ir, O=C, O, O, $P(C_6H_5)_3$	reversible O_2 complex	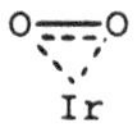 O–O, Ir	1.30	[3]
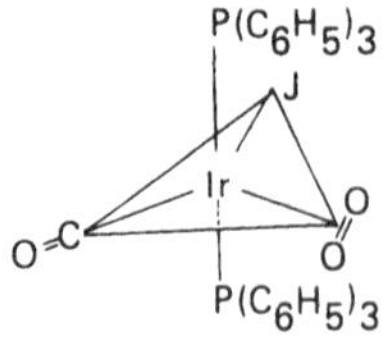 $P(C_6H_5)_3$, J, Ir, O=C, O, O, $P(C_6H_5)_3$	irreversible O_2 complex	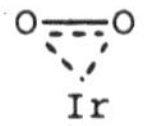 O–O, Ir	1.51	[7]

Fig. 2. Principal types of metal dioxygen complexes

$$M^{I} + O_2 \rightleftharpoons M^{II}\text{—O—O}\cdot \rightleftharpoons M^{III}\langle^{O}_{O}|$$

Superoxo Peroxo

M = Co, Rh, Ir

Fig. 3.

as the catalyst and suggested a mechanism similar to that shown in Fig. 4. It appears significant that excess phosphine acts as an acceptor for one of the oxygen atoms. We explored the catalytic oxygenation in the absence of phosphine.[10] When air was bubbled through a toluene/styrene solution of acetylacetonatorhodium at 80°, the only reaction observed was the formation of acetophenone (Fig. 5). However, when the same reaction was carried out in an autoclave charged with dry compressed air at 80° and 50 bar, the oxidative cleavage of the alkene double bond became the predominant pathway of the Rh-catalyzed reaction (Fig. 5). Under these conditions turnover numbers of ∿60 mole benzaldehyde per mole of catalyst were found with only traces of acetophenone present. Interestingly, no benzoic acid, which could be produced by autoxidation of the aldehyde, was detected. Furthermore, a free radical reaction has been excluded by

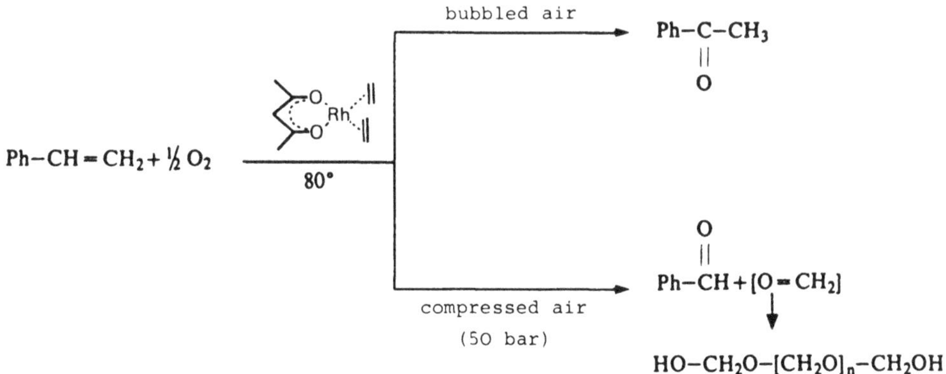

Fig. 4.

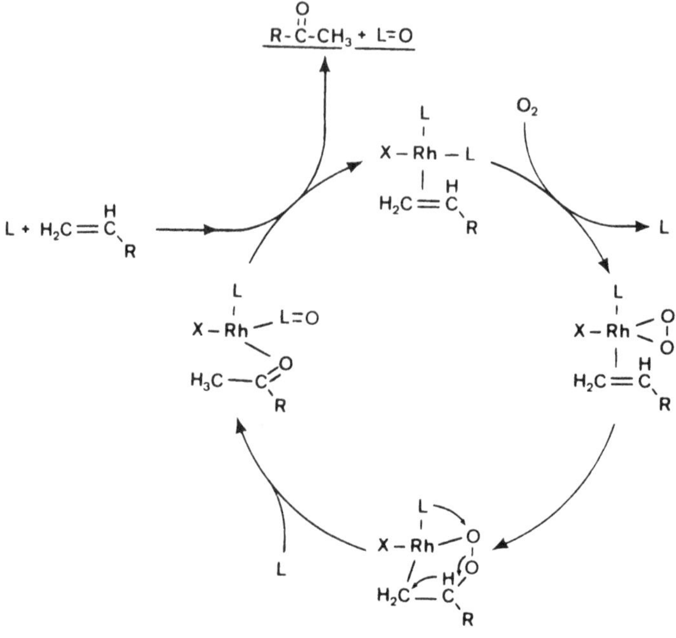

$$R{-}CH{=}CH_2 + PPh_3 + O_2 \xrightarrow{(Ph_3P)_3RhX} R{-}\overset{\overset{\displaystyle O}{\|}}{C}{-}CH_3 + Ph_3P{=}O$$

$L = PPh_3$; $X = Cl, CN, OCN, SCN$

Fig. 5.

showing that the addition of a radical scavenger had no influence on the course of the reaction. Formaldehyde, the second product of the catalytic C-C bond cleavage, was identified in the form of polyoxymethylene.

The analogous reaction of α-methylstyrene with compressed air in the presence of acetylacetonatorhodium (Fig. 6) yielded acetophenone and polyoxymethylene along with traces of the oxirane. Thus, when the reaction was carried out with a higher oxygen concentration in solution, the main products were, again, derived from the catalytic C-C bond fission and oxidation by molecular oxygen. To answer the question of how the

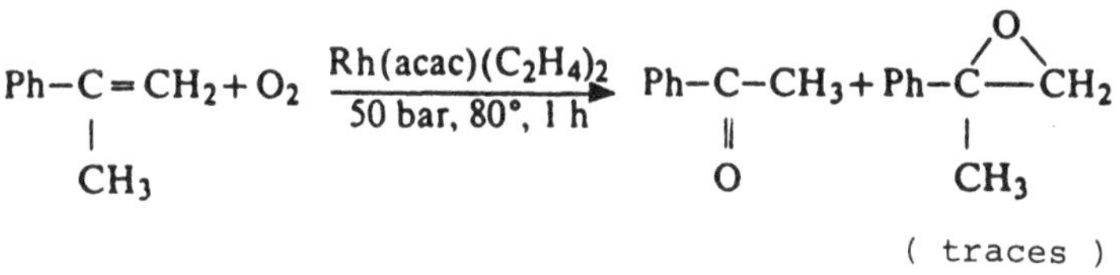

$$\underset{\displaystyle CH_3}{\underset{|}{Ph-C}}=CH_2+O_2 \xrightarrow[50\ bar,\ 80^\circ,\ 1\ h]{Rh(acac)(C_2H_4)_2} \underset{\displaystyle O}{\underset{||}{Ph-C}}-CH_3+\underset{\displaystyle CH_3}{\underset{|}{Ph-\overset{O}{C}}}-CH_2$$

(traces)

Fig. 6.

concentration of oxygen in the solution can direct the rhodium catalyzed oxidation of terminal olefins to give either methyl ketones or C-C bond cleavage we discuss two different possible pathways.

When the oxygen concentration is low (i.e., when air is bubbled through an olefin solution of Rh(I)), excess olefin competes with O_2 for the coordination sites. O_2-Insertion into a rhodium-olefin bond occurs to form a transient peroxometallocycle. The migration of a proton then yields the ketone according to the mechanism proposed by Mimoun.[9] The concentration of oxygen in solution is increased considerably when compressed air is used. In this case it is more likely that oxygen will complex with the rhodium more readily than the olefin leading to a rhodium peroxo-complex, which acts as the propagating catalytic intermediate. In presence of the olefin, dissociation of the O-O bond yields a rhodium dialkoxide which is unstable and decomposes to give the organic carbonyl products with breaking of the C-C bond (Fig. 7). This reaction proceeds surprisingly well even with tetrasubstituted olefins and the Rh-catalyzed oxidation of 2,3-dimethyl-2-butene with compressed air yields acetone at quite reasonable turnover numbers per hour. The oxirane is a side-product of the reaction (Fig. 8).

In contrast, Lyons and Turner[11] have found that when air is bubbled through the solution, the main products of the rhodium-catalyzed oxidation of 2,3-dimethylbutene are epoxybutane and an allylic alcohol. Acetone is only a minor product. These authors proposed a mechanistic pathway involving an allylic hydroperoxide intermediate and this was supported by the fact that the oxirane and the alcohol were produced to approximately the same extent (Fig. 9).

vs.
Rh
R
+ O2
+ O2 - H2C=C
H
R
+ H2C=C
H2C
CH2
H3C-C
O
C=O
O=CH2
Ketonization
C,C-bond cleavage

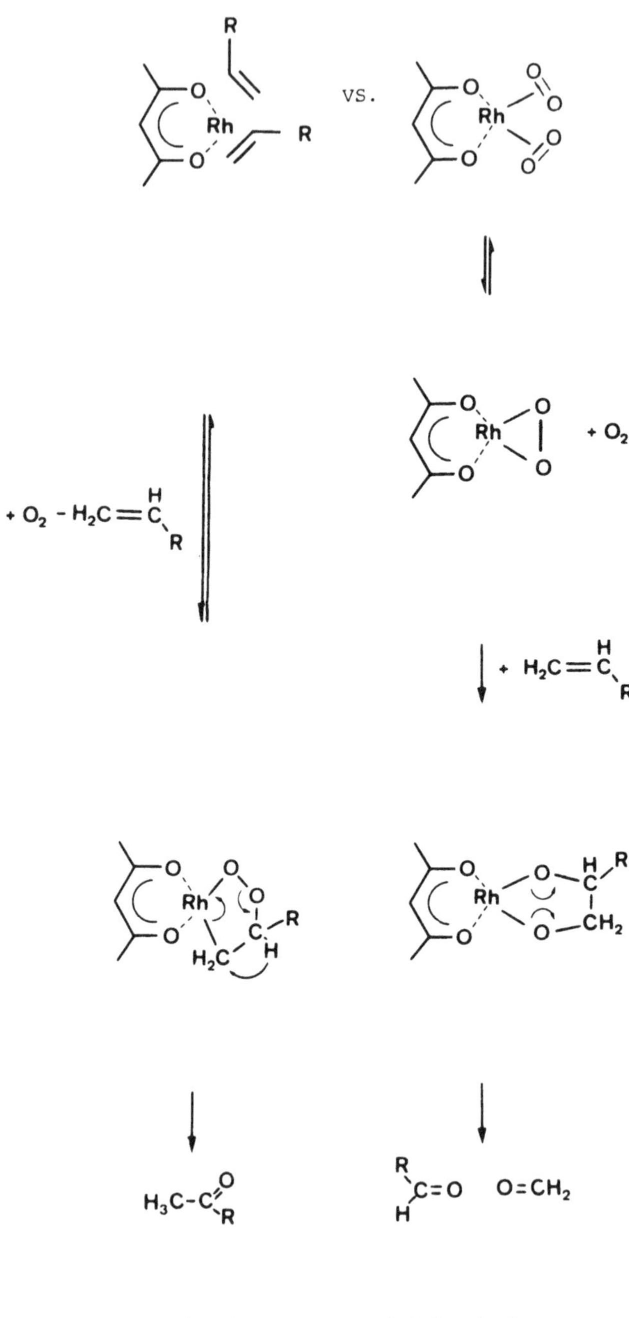

Fig. 7.

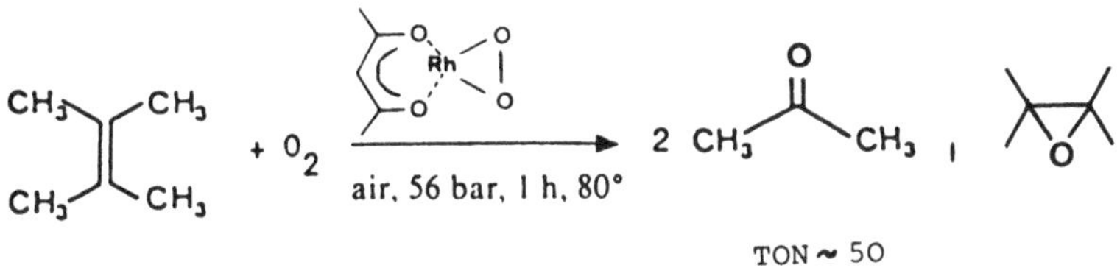

$$(CH_3)_2C=C(CH_3)_2 + O_2 \xrightarrow[\text{air, 56 bar, 1 h, 80°}]{\text{Rh(acac)}(O_2)} 2\ CH_3COCH_3,\ \text{tetramethyloxirane}$$

TON ~ 50

Fig. 8.

The C-C cleavage of olefins by Rh-catalyzed oxidation with compressed air appears to be general. For example, methylidenecyclopentane gave cyclopentanone in presence of the acetylacetonatorhodium catalyst (Fig. 10). 1,3-Butadiene reacted with O_2 in

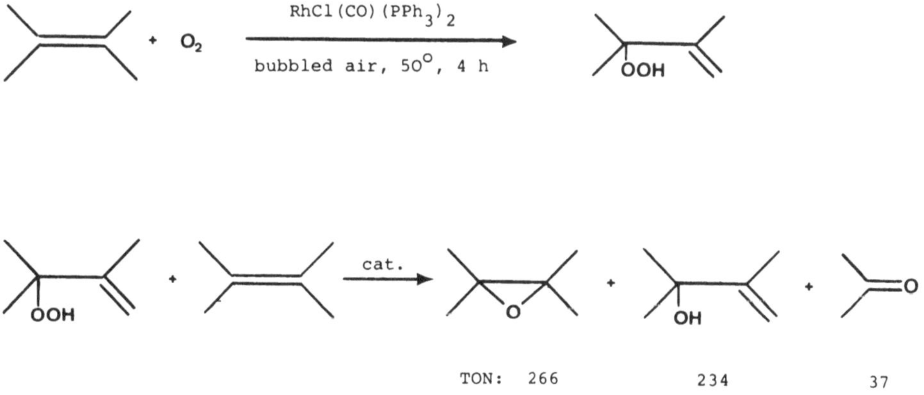

$$\text{tetramethylethylene} + O_2 \xrightarrow[\text{bubbled air, 50}^{\circ}\text{, 4 h}]{RhCl(CO)(PPh_3)_2} \text{allylic hydroperoxide (OOH)}$$

$$\text{hydroperoxide (OOH)} + \text{tetramethylethylene} \xrightarrow{\text{cat.}} \text{epoxide} + \text{allylic alcohol (OH)} + \text{acetone} (=O)$$

TON: 266 234 37

Fig. 9.

toluene in the presence of the Rh-complex to give acrylaldehyde (TON 30) which, under the reaction conditions, was partly transformed to the Diels-Alder adduct. The second C=C double bond in 1,3-dienes appeared to be unreactive against oxidative cleavage and no trace of glyoxal could be detected. In the case of isoprene, almost equimolar amounts of methacrylaldehyde and methylvinylketone were formed (TON 25) indicating that the CH_3 substitutent exerts no directing influence on the C=C double bond cleavage. It should be noted that these unsaturated aldehydes do not undergo further autoxidation reactions in the reaction mixture.

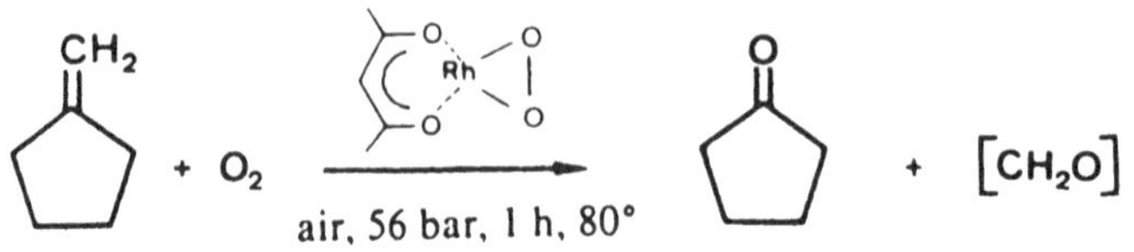

Fig. 10.

The ratio of the products as well as the turnover number depend strongly on the nature of the Rh-complex catalyst. For a catalyst screening we tested a series of Rh-complexes in the oxidative cleavage of 2,3-dimethylbutene under standard conditions. The results are listed in Table I.

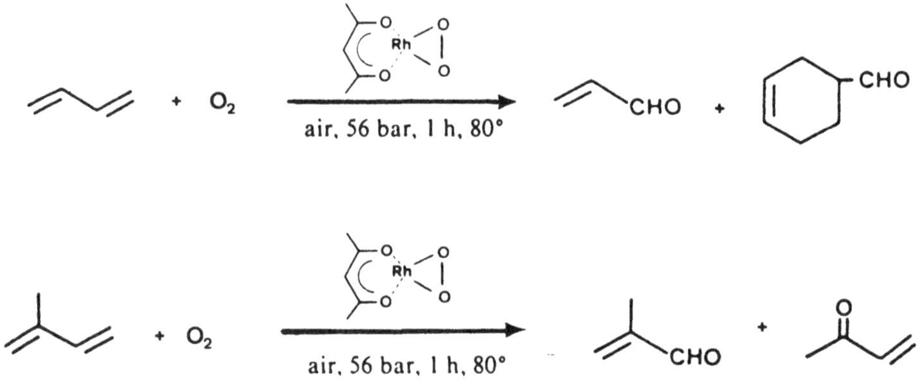

Fig. 11.

Acetylacetonato substituted Rh-alkene compounds, in general, appear to be the most selective catalysts for the oxidative C=C double bond cleavage; cyclopentadienyl complexes as well as Wilkinson's catalyst are also effective. In contrast, the dimeric chlororhodium-alkene systems show almost no catalytic activity. It appears that Rh-complexes containing good leaving groups such as ethylene are more reactive than the corresponding cyclooctadiene or carbonyl complexes which are more stable and consequently show an induction period. From preliminary data it appears that there may be a correlation between the ^{103}Rh NMR data of the different types of complexes and their properties in the oxidation reactions. It is anticipated that ^{103}Rh NMR will find application in the screening of

TABLE I

Catalyst Screening

(40-50 mmole alkene, 0.2 mmole complex, air, 80°, 1 hr)

complex	TON  → 2 O=	δ(Rh)ppm [a]
Acac Rh 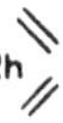	96	+1184,1 ± 0,8
Acac Rh $(C_3H_5)_2$	78	
F_6acac Rh 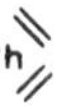	0	
Acac Rh 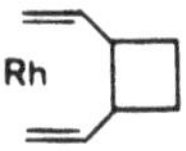	34	
Acac Rh 	14	+1294 ± 0,5
F_6acac Rh	6	
Acac Rh $(CO)_2$	no reaction	
Cp Rh 	50	
Cp Rh 	20	+ 782 ± 0,5
ClRh $(PPh_3)_3$	60	
$[ClRh \ldots]_2$ 	2	
$[ClRh(CO)_2]_2$	no reaction	

a) ^{103}Rh chemical shifts taken from W.v.Philipsborn et al. Helv.Chim.Acta 65, 26-45 (1982)

TABLE II

Solubility of O_2 (25°, normal pressure)

solvent	$[O_2]$ $\frac{\text{Mol } O_2}{\text{Mol solvent}}$	$[O_2]$ $\frac{\text{Mol } O_2}{\text{l solvent}} \cdot 10^{-3}$
Tetradecane	$2.32 \cdot 10^{-3}$	8.9
Dodecane	$2.29 \cdot 10^{-3}$	10.1
Decane	$2.23 \cdot 10^{-3}$	11.4
Heptane	$2.16 \cdot 10^{-3}$	14.7
Hexane	$1.93 \cdot 10^{-3}$	14.8
Toluene	$0.91 \cdot 10^{-3}$	8.6
Ethanol	$0.58 \cdot 10^{-3}$	10.2
Water	$0.023 \cdot 10^{-3}$	1.3

potential catalysts similar to that which we have demonstrated for ^{59}Co complexes.[12]

III. EXPERIMENTAL

It is crucial for the rhodium-catalyzed C=C double bond cleavage that the oxygen concentration in the solution be high enough. Some data on the solubility of O_2 are collected in Table II. Fortunately, organic liquids are much better solvents for O_2 than water or ethanol. Toluene is a good solvent for O_2 and alkanes are even better.

Our earlier experiments on the catalytic olefin oxidation were all carried out batch-wise in an autoclave charged with 50-60 bar of compressed air. Later the reaction was performed continuously in a commercially available loop reactor that had been modified for our purpose

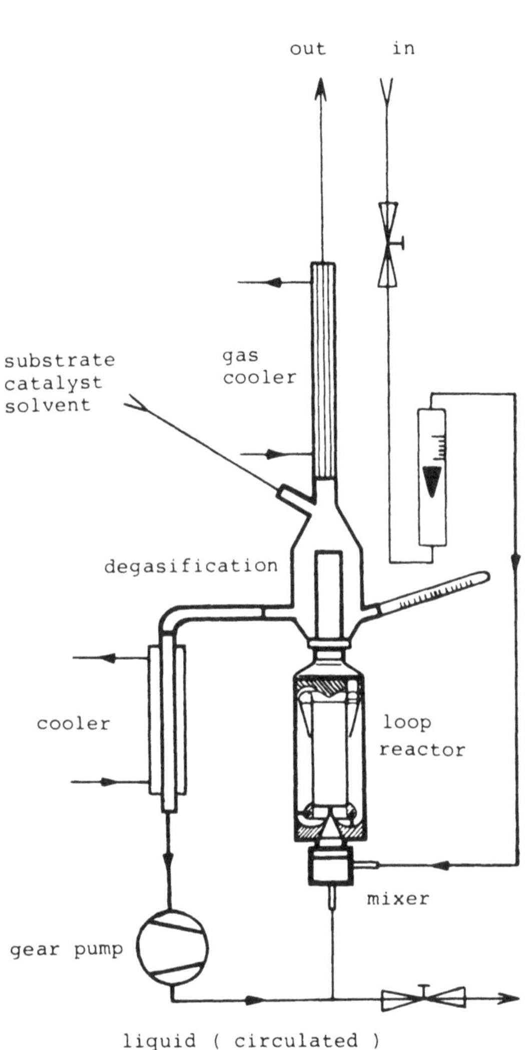

Fig. 12. Loop reactor for catalyzed air oxidation.

(Fig. 12). The olefin and the dissolved catalyst are rapidly circulated by a pump and brought in intense contact with microbubbles of fresh air which is continuously introduced by the mixer. On top of the loop reactor the deoxygenated gas is separated from the liquid in an overflow vessel and the liquid is recirculated.

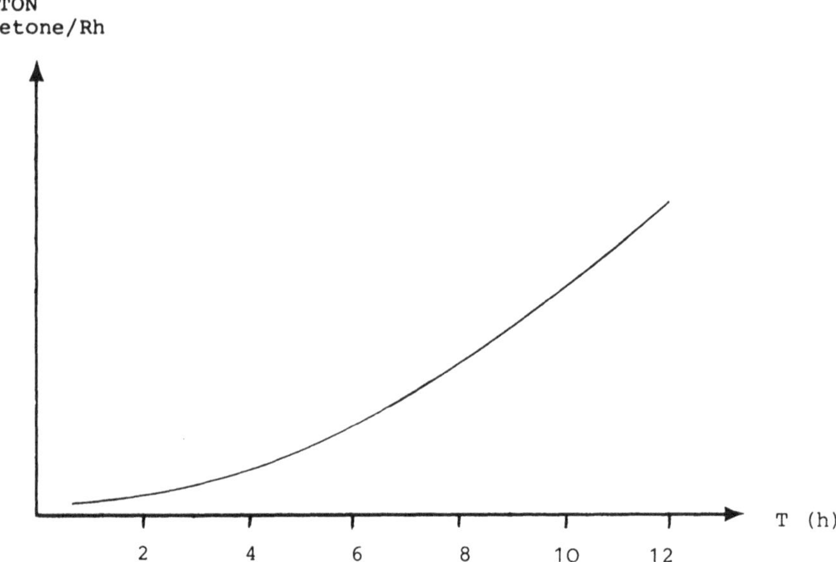

Fig. 13. Acetone TON as a function of time

Since the loop reactor is made of glass, the reaction can be observed directly. As soon as the liquid is oxygenated the color of the pale yellow solution darkens considerably, probably due to the formation of the active rhodium peroxo species. By comparison with the batchwise reactions in an autoclave, the intense mixing of the substrate with fresh air in the loop reactor has several advantages:

(1) The O_2 distribution in the substrate throughout the reaction time is uniform because the system behaves like a back-mix-reactor.

(2) High pressure is not necessary since the recycling of the microbubbles in the loop at 1.1 bar (= 0.11 MPa) produces a sufficiently high O_2 concentration.

(3) The reaction can be followed analytically by taking samples. Fig. 13 shows the oxidation of 2,3-dimethylbutene by air catalyzed by acetylacetonatorhodium(I) in the loop reactor as monitored by gas chromatography over a period of 6 hrs.

Though these data are preliminary it can be seen clearly that acetone is formed catalytically throughout the reaction time showing that the catalyst is still active after six hours.

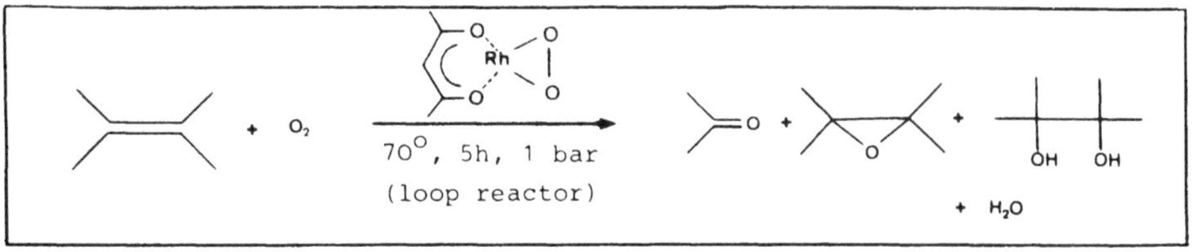

$+ O_2$ → Rh, 70°, 5h, 1 bar (loop reactor) → $>$=O + epoxide + diol (OH OH) $+ H_2O$

epoxide $+ H_2O$ —cat.—//→ diol (OH OH) ← H^+ ← dioxametallacycle RhX

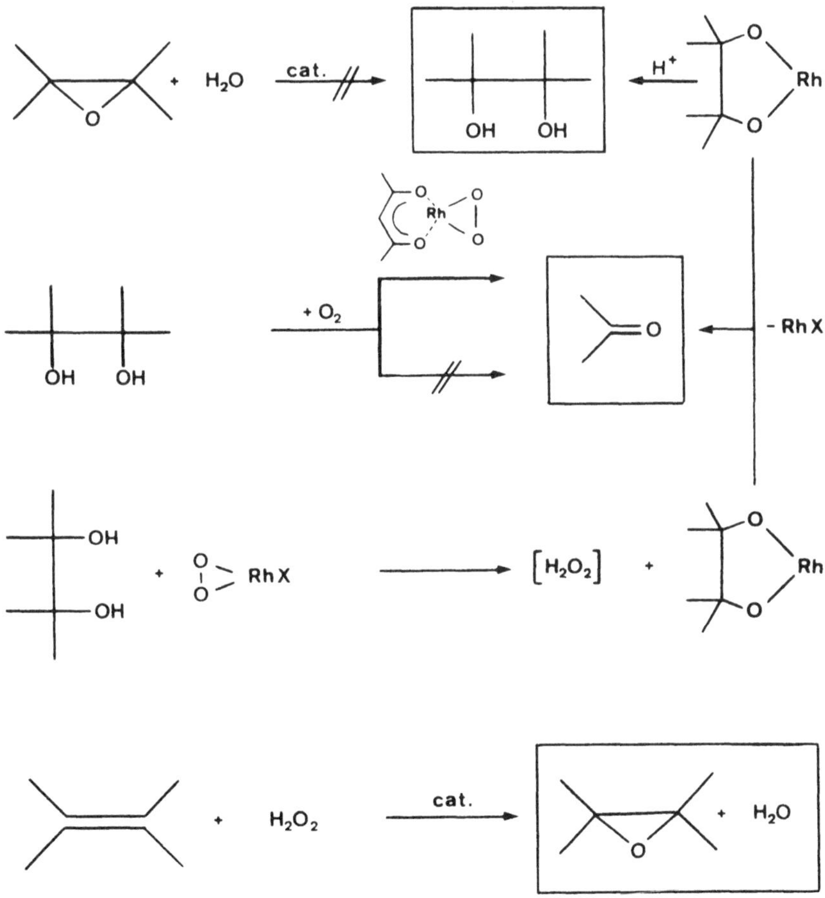

diol $+ O_2$ (Rh) → ketone (=O) ← $-RhX$

diol (OH, OH) $+$ O–O RhX → $[H_2O_2]$ $+$ dioxametallacycle RhX

alkene $+ H_2O_2$ —cat.→ epoxide $+ H_2O$

Fig. 14.

IV. CONCLUSIONS

The complete product spectrum which is obtained in the loop reactor at 70° in 5 hours by air oxidation of 2,3-dimethyl-2-butene using acetylacetonatorhodium as the catalyst is summarized in the upper part of Fig. 14. Besides acetone, which is the main product, smaller amounts of the oxirane and 2,3-butanediol (pinacol) together with some water are detected. It should be noted that no traces of the corresponding allylic alcohol were found. A mechanistic pathway for oxirane formation involving an allylic hydroperoxide is thereby excluded. In order to clarify the origin of the pinacol, we tested whether the subsequent hydration of the oxirane could account for this product. However, no such reaction occurred under these conditions. We assume that the diol byproduct is formed by partial hydrolysis of the intermediate rhodiumalkoxide. Interestingly, pure pinacol is converted to acetone when it reacts with oxygen in the presence of the catalyst under the reaction conditions. No oxidation takes place in absence of the rhodium(I) acetylacetonate. Since it is well established that metal peroxo complexes generally afford hydrogen peroxide on treatment with a proton donor,[13] it is tempting to assume an analogous reaction of the peroxo rhodium species with the diol. The resulting dialkoxide decomposes to give acetone, whereas hydrogen peroxide in turn produces epoxidation of excess starting olefin to form the oxirane and water detected in the product.

It is obvious that the catalyzed cleavage of alkenes by molecular oxygen is still relatively unexplored. Our studies to clarify the scope and merits of the reaction are now in progress.

REFERENCES

1. J.S. Griffith, Proc. Roy. Soc., **A 235**, 23 (1956).
2. L. Vaska, Acc. Chem. Res., **9**, 175 (1976).
3. (a) L. Vaska, Science, **140**, 809 (1963); (b) S.J. LaPlaca and J.A. Ibers, Science, **145**, 920 (1964).
4. (a) J.P. Collman, R.R. Gagne, C.A. Reed, T.R. Halbert, G. Lang, and W.T. Robinson, J. Am. Chem. Soc., **97**, 1427 (1975); (b) J.P. Collman and K.S. Suslick, Pure Appl. Chem., **50**, 951 (1978).

5. F. Fronczek, W.P. Schaefer, and R.E. Marsh, Acta Crystallogr., **B 30**, 117 (1974).
6. G.A. Rodley and W.T. Robinson, Nature, **235**, 438 (1972).
7. J.A. McGinnety, N.C. Payne, and J.A. Ibers, J. Am. Chem. Soc., **91**, 6301 (1969).
8. (a) C.W. Dudley and G. Read, Tetrahedron Lett., 5273, (1973); (b) C.W. Dudley, G. Read, and P.J.C. Walker, J. Chem. Soc., Dalton Trans., 1927 (1974); (c) G. Read and P.J.C. Walker, ibid, 833 (1977).
9. H. Mimoun, J. Mol. Catal., **7**, 1 (1980) and references cited therein.
10. H. Bönnemann, W. Nunez, and D.M.M. Rohe, Helv. Chim. Acta, **66**, 177 (1983).
11. J.E. Lyons and J.O. Turner, J. Org. Chem., **37**, 2881 (1972).
12. H. Bönnemann, W. Brijoux, R. Brinkmann, R. Mynott and W. v. Philipsborn, and T. Egolf, J. Organomet. Chem., (1984) in press.
13. (a) S. Muto, H. Ogata and Y. Kamiya, Chem. Lett., 809 (1975); (b) M. Pizzotti, S. Cenini, and G. LaMonica, Inorg. Chim. Acta, **33**, 161 (1978); (c) B.S. Tovrog, S.E. Diamond, and F. Mares, J. Am. Chem. Soc., **101**, 5067 (1979).

19

Oxidation of Hydrocarbons by Methane-Utilizing Bacteria

Ramesh N. Patel
Exxon Research and Engineering Company
Route 22 East
Annandale, NJ 08807

ABSTRACT

The extensive range of biotransformation catalyzed by aerobic C_1 utilizing (methane, methanol) organisms, methanotrophs, has opened up the possibility of their exploitation for biocatalysis. The broad substrate specificity of enzymes involved in the oxidation of methane is catalytically useful for the production of chemicals. The epoxidation of alkenes (C_2 - C_5) to produce ethylene oxide, propylene oxide, the hyroxylation of alkanes (C_1 - C_8) to produce alcohols, and the production of methylketones such as acetone and 2-butanone from their corresponding n-alkanes (propane, butane) or secondary alcohols (2-propanol, 2-butanol) are industrially important biotransformations catalyzed by methanotrophs. Soluble methane monooxygenase from a methane-utilizing organism, Methylobacterium sp. CRL-26, catalyzed the NAD(P)H-dependent epoxidation/hydroxylation of a variety of hydrocarbons, including terminal alkenes, internal alkenes, substituted alkenes, branch-chain alkenes, alkanes, carbon monoxide, ethers, cyclic, and aromatic compounds. The NAD^+ linked dehydrogenases such as formate dehydrogenase or secondary alcohol dehydrogenase in the presence of formate or secondary alcohol, respectively, regenerated NAD/NADH required for the methane monooxygenase in a coupled enzyme reactions.

Methane monooxygenase was resolved into two components, a hydroxylase and a flavoprotein. In contrast to methane-utilizing bacteria, other hydrocarbon-utilizing bacteria (ethane, propane, butane) catalyzed the epoxidation of alkenes and the hydroxylation of alkanes, except the oxidation of methane, indicating that the oxygenase enzyme system from these microorganisms is different from that of methane monooxygenase. Oxidation of secondary alcohols to the corresponding methylketones in methanotrophs is catalyzed by an NAD^+-dependent, zinc-containing, secondary alcohol dehydrogenase. Primary alcohols were oxidized to the corresponding aldehydes by a phenazine methosulfate-dependent, pyrollo quinoline quinone (methoxatin or PQQ) containing methanol dehydrogenase in methanotrophs.

I. INTRODUCTION

The development of processes for the production of chemicals based on the exploitation of low temperature and atmospheric pressure conditions by specific oxidative catalysis by microorganisms or enzymes derived from them has been an attactive commercial goal. Methane-utilizing organisms, methanotrophs, grown on methane have the ability to oxidize and transform a variety of non-growth substrates to commercially useful chemicals. The methane monooxygenase has attracted considerable industrial interest in this respect because of its ability to insert an oxygen atom into a wide variety of hydrocarbon substrates.[1-3] The products of some of these hydrocarbon oxidations are difficult to make by conventional chemical technology. In this report, the biotechnological application of methanotrophs will be discussed. Particular attention will be paid to their ability to catalyze the oxidative transformation of hydrocarbons. A brief summary of the properties of enzymes involved in catalyzing these biotransformations such as methane monooxygenase, methanol dehydrogenase, and secondary alcohol dehydrogenase will also be presented.

III. RESULTS AND DISCUSSION

A. Oxidation of Methane by Methanotroph

Methanotrophic organisms are characterized by their ability to grow on C_1

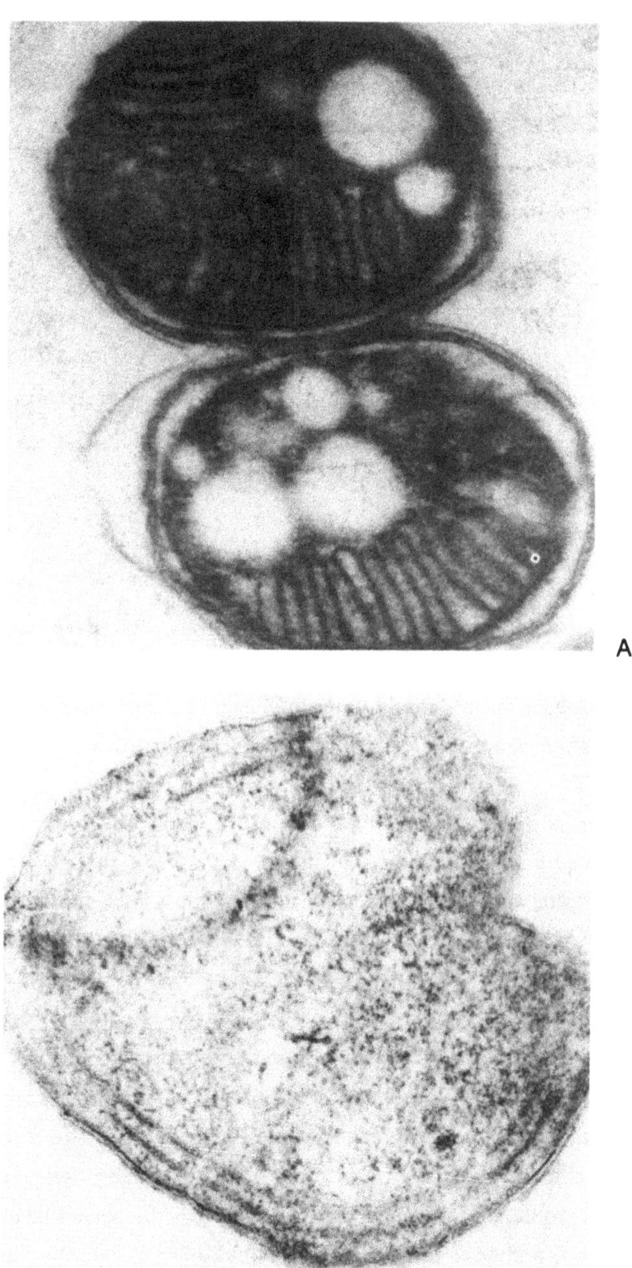

Fig. 1. Electron micrograph of thin-section of methane-utilizing bacteria. A. Methylococcus capsulatus (Type I membranes) Magnification 80,000 X; B. Methylobacterium sp. CRL-26 (Type II membranes) Magnification 81,000 X.

$$CH_4 \xrightarrow[O_2 \quad H_2O]{NADH \quad A \quad NAD} CH_3OH \xrightarrow{PMS \quad B \quad PMSH_2} HCHO \xrightarrow[H_2O]{PMS \quad B \quad PMSH_2} HCOOH \xrightarrow{NAD \quad C \quad NADH} CO_2$$

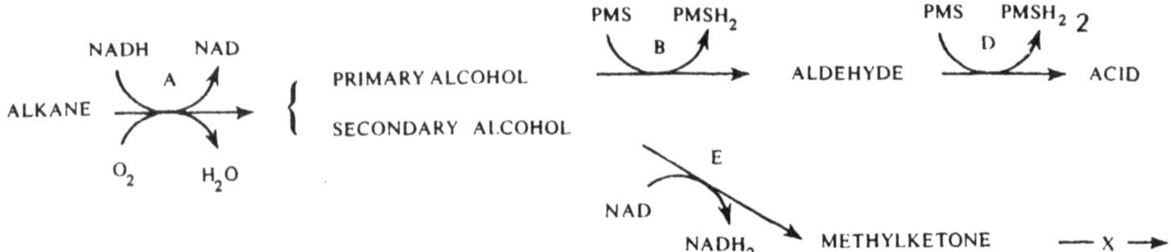

Fig. 2. Pathway of alkane oxidation by methanotrophs. A. methane monooxygenase; B. methanol dehydrogenase; C. formate dehydrogenase; D. aldehyde dehyrogenase; E. secondary alcohol dehydrogenase; PMS. phenazine methosulfate.

compounds (methane and methanol) that are more reduced than carbon dioxide as their sole source of carbon for growth.[4] During the growth on methane, methanotrophs contained extensive intracytoplasmic membranes. On the basis of the organization of intracytoplasmic membranes, carbon assimilation pathway, and the presence or absence of a complete tricarboxylic acid [TCA] cycle, the organisms are divided into two groups.[5,6] The Type I group organisms have bundles of disc-shaped vesicles running across the cells (Fig. 1A), use the ribulose monophosphate pathway for carbon assimilation, and have an incomplete TCA cycle. The Type II group organisms are characterized by a system of paired membranes running parallel to the intracytoplasmic membranes (Fig. 1B), use the serine pathway for carbon assimilation, and have a complete TCA cycle.

The oxidation of methane to carbon dioxide in methanotrophs appears to occur _via_ a series of two-electron oxidation steps involving the enzymes methane monooxygenase, methanol dehydrogenase, and formate dehydrogenase (Fig. 2). Formaldehyde occupies a key position in the oxidative pathway as it is either assimilated into biomass or dissimilated to carbon dioxide to generate energy for growth. Higher alkanes (C_3 to C_8) are oxidized to the corresponding alcohols, both primary and secondary. Primary alcohols are oxidized to the corresponding carboxylic acids by the enzymes methanol dehydrogenase and aldehyde dehydrogenase. Secondary alcohols are oxidized to the corresponding methylketones by secondary alcohol dehydrogenase (Fig. 2).

B. Oxidation of Hydrocarbons by Cell-Suspensions of Methanotrophs

1. Oxidation of Alkenes

We first reported the oxidation of gaseous alkenes such as ethylene, propylene, 1-butene, and butadiene to the corresponding 1,2-epoxides by cell-suspensions of methane-utilizing bacteria grown on methane as the sole carbon source. Internal alkenes, such as 2-butene, were oxidized both terminally at the methyl group and internally across the double bond. Branched chain alkenes (isobutene and isoprene) also were oxidized by the cell-suspension (Table I). Epoxides were identified and estimated by gas chromatography. Epoxides were not metabolized further by the cell suspensions of methanotrophs and accumulated extracellularly in the growth medium.[7,8]

2. Oxidation of Alkanes

Although methanotrophic bacteria are unable to grow on higher alkanes, they have the capability of oxidizing n-alkanes (propane, butane, pentane, hexane) to their corresponding methylketones (acetone, 2-butanone, 2-pentanone, 2-hexanone). Secondary alcohols were detected as intermediates during the oxidation of n-alkanes. Primary alcohols and aldehydes also were detected in very low amounts as products of propane and butane oxidation. These, however, were rapidly oxidized to the corresponding carboxylic acids by the high level of the enzymes, methanol dehydrogenase and aldehyde dehydrogenase. Secondary alcohols were oxidized to the corresponding methylketones by secondary alcohol dehydrogenase. The rate of production of methylketones from their corresponding alkanes is shown in Table II. The product methylketones were not metabolized further and accumulated

TABLE I

Oxidation of Alkenes by Cell Suspensions of Methanotrophs

Substrate	Product	Rate of Product Formation (μmol/h/2.0 mg cell protein)		
		Methylosinus sp. (CRL-15)	*Methylobacterium* sp. (CRL-26)	*Methylococcus capsulatus* (CRL-M1)
Ethylene	Ethylene oxide	1.9	0.9	5.5
Propylene	Propylene oxide	3.6	2.5	5.5
1-Butene	1,2 Epoxybutane	0.45	0.9	1.3
2-Butene	2,3 Epoxybutane	0.66	0.25	0.75
	2-Butene-1-ol	0.38	0.18	0.41
Isobutene	1,2-Epoxy-2-Methyl Propane	0.64	0.84	0.92
Isoprene	1,2-Epoxyisoprene	0.24	0.20	0.26
Butadiene	1,2-Epoxybutene	2.5	2.8	4.8

TABLE II

Production of Secondary Alcohols and Methylketones from n-Alkanes by Cell Suspensions of Methanotrophic Bacteria

	Rate of Product Formation (μmol/h/2.0 mg of Cell Protein)							
	Propane		Butane		Pentane		Hexane	
Organisms	2-Propanol	Acetone	2-Butanol	2-Butanone	2-Pentanol	2-Pentanone	2-Hexanol	2-Hexanone
Methylosinus sp. (CRL-15)	2.1	1.5	1.2	1.2	0.12	0.20	0.10	0.03
Methylococcus capsulatus (Texas)	11	1.2	1.0	0.52	0.15	0.11	0.08	0.02
Methylobacterium sp. (CRL-26)	1.4	1.5	0.8	1.0	0.08	0.15	0.05	0.024

TABLE III

Production of Methylketones from Secondary Alcohols by Cell Suspensions of Methane-or Methanol-grown Organisms

	Rate of Product Formation (μmol/h/mg protein)			
Organisms	2-Propanol → Acetone	2-Butanol → 2-Butanone	2-Pentanol → 2-Pentanone	2-Hexanol → 2-Hexanone
Methane-Grown Bacteria:				
Methylosinus trichosporium (OB3b)	0.30	4.8	2.7	0.09
Methylocystis parvus (OBP)	0.52	1.1	0.70	0.03
Methylosinus sporium(5)	1.0	3.0	2.0	0.07
Methylococcus capsulatus (Texas)	2.0	2.5	0.24	0.08
Methylomonas sp. (CRL-17)	1.2	2.0	0.48	0.04
Methylobacter sp. (CRL-19)	0.42	1.8	0.60	0.03
Methylobacterium organophilum (XX)	0.70	2.5	1.2	0.08
Methylobacterium sp. (CRL-26)	0.78	2.0	1.0	0.09
Methanol-Grown Yeasts:				
Candida utilis ATCC 26387	6.2	6.8	1.5	0.80
Hansenula Polymorpha ATCC 26012	5.9	5.8	1.4	0.72
Pichia sp. NRRL-Y 11328	5.2	6.8	1.2	0.50

extracellularly. Branched chain alkanes such as isobutane and isopentane were oxidized to the corresponding alcohols by cell-suspensions of methanotrophs.[9,10]

3. Oxidation of Secondary Alcohols

Cell-suspensions of methane and methanol grown methanotrophs or methanol-utilizing yeasts catalyzed the oxidation of secondary alcohols to the corresponding methylketones.[11,12] The rate of production of methylketones, which accumulated extracellularly, are shown in Table III.

4. Oxidation of Carbon Monoxide, Ammonia, and Aromatic Compounds

Cell suspensions of methanotrophs grown on methane catalyzed the oxidations of carbon monoxide to carbon dioxide,[13] ammonia to nitrite,[14] ethylbenzene to 4-hydroxyethylbenzene, m- and p-cresol to the corresponding hydroxybenzoic acid, toluene to benzyl alcohol, and benzene to phenol.[15] m-Xylenes were oxidized exclusively at one of the methyl groups to produce the corresponding alcohols.[15]

C. Oxidation by Organisms Grown on Gaseous Alkanes

Organisms grown on other gaseous alkanes (ethane, propane, butane) catalyzed the hydroxylation of alkanes (C_2 to C_8) to the corresponding secondary alcohols and methylketones, the epoxidation of alkenes (C_2 to C_8) to the corresponding epoxides, and the oxidation of secondary alcohols to the corresponding methylketones.[16,17] These organisms are different from methane-utilizing organisms in their inability to grow on methane as a carbon source and the fact that they are unable to catalyze the oxidation of methane to methanol. This indicates that the oxygenase enzyme system in these organisms is different from that of the methane-utilizing organisms.[18,19]

D. Oxidation of Hydrocarbons by Cell-free systems from Methanotrophs

1. Methane Monooxygenase

The initial enzymatic attack upon the methane molecule has been known for some years to involve an oxygenase reaction,[20] however, the enzyme itself has proved difficult to study because of its extremely labile nature. Cell free particulate methane-oxidizing activity, dependent on the presence of the cofactor, nicotinamide adenine dinucleotide [NADH], has been demonstrated in several methanotrophs.[21-23] The particulate preparations from Methylobacterium sp. CRL-26[23] catalyzed the oxygen and NADH-dependent epoxidation of alkenes (C_2 to C_4) and the hydroxylation of alkanes (C_1 to C_4). Activity was directly measured by estimating the products of the reactions by gas chromatography. Attempts to solubilize particulate activity proved unsuccessful and the preparations were unstable.

The particulate enzyme purified from Methylosinus trichosporium (OB3b) could be resolved into three components, a soluble co-binding cytochrome C (MW 13,000) containing 1 atom of iron/mole and a variable amount of copper (0.3 - 0.8 atom/mole), a copper-containing protein (1 atom of copper/mole) with a molecular weight of 47,000, and a small protein with 9,400 molecular weight. The purified enzyme system utilized ascorbate or methanol (in the presence of methanol dehydrogenase) as electron donors. The NADH or NADPH could not act as an electron donor. The purified enzyme system catalyzed the oxidation of ethane, propane, butane, and carbon monoxide.[24]

In contast to Methylosinus trichosporium (OB3b), methane monooxygenase from Methylococcus capsulatus (Bath) was contained mainly in the soluble fraction. Soluble methane monooxygenase catalyzed the oxygenation of alkanes, alkenes, cyclic, and aromatic compounds. Only NAD(P)H was an effective electron donor; ascorbate or methanol could not support methane monooxygenase activity. The soluble methane monooxygenase was resolved into three components.[25] Component A has a molecular weight of about 220,000 and subunits of 68,000 and 47,000 molecular weight. It contains 2 gram atoms of iron and acid-labile sulfide per mole. Component C is an iron flavoprotein of 44,600 molecular weight and contains 1 mole of flavin adenine dinucleotide (FAD), 2 gram atoms of iron and 2 moles of acid-labile sulfur per mole of protein. Component B is a colorless protein of 15,000 molecular weight. All three proteins were required for methane monooxygenase activity.[26] The properties of methane monooxygenase component proteins from Methylococcus capsulatus (Bath)

TABLE IV

Hydroxylation of Alkanes by Soluble Methane Monooxygenase from Methylobacterium sp. CRL-26

Substrate	Product	Specific Activity (nmoles/min/mg of protein)
Methane	Methanol	93
Ethane	Ethanol	64
Propane	1-Propanol	15
	2-Propanol	22
Butane	1-Butanol	43
	2-Butanol	25
Pentane	1-Pentanol	21
	2-Pentanol	45
Hexane	1-Hexanol	39
	2-Hexanol	21
Heptane	1-Heptanol	12
	2-Heptanol	50
Octane	1-Octanol	2.5
	2-Octanol	17
Chloromethane	Formaldehyde	44
Dichloromethane	ND	40
Trichloromethane	ND	21
Tetrachloromethane	ND	0
Bromomethane	Formaldehyde	48
Fluoromethane	Formaldehyde	19
Nitromethane	ND	12
Nitroethane	ND	18
1-Nitropropane	ND	50
2-Nitropropane	ND	19
1-Bromobutane	ND	49
2-Bromobutane	ND	10
Isobutane	Isobutanol	34
	Tert. butanol	40

[a] Products of reactions were identified and estimated by gas chromatography retention time comparisons and co-chromatography with authentic standards. Specific activities were expressed as the rate of products formation/min/mg of protein in S(80) fractions.

ND = Products were not identified. Rate of oxidation was expressed as nmol of substrate utilized per min per mg protein.

thus appear to be quite different from those reported for Methylosinus trichosporium (OB3b). However, recently it was discovered that Methylosinus trichosporium (OB3B) also contained soluble methane monooxygenase activity very similar to that of the Methylococcus capsulatus (Bath) strain.[27] The only effective electron donor for monooxygenase was NAD(P)H and neither ascorbate nor methanol could serve as an electron donor.

Previously, we reported that Methylobacterium sp. CRL-26 grown in shaker flasks with methane (methane and air, 1:1 parts by volume) as the carbon source contained mainly particulate methane monooxygenase activity which was very sensitive to inhibition by metal-chelating and binding compounds such as 1,10-phenanthroline, 2,2-bipyridyl, thiosemicarbazide, and potassium cyanide. Now we have discovered that Methylobacterium sp. CRL-26 obtained by growth in a fermentor with high dissolved oxygen contained mainly soluble methane monooxygenase insensitive to inhibition by various metal-chelating and binding compounds. It is possible, therefore, that particulate methane monooxygenase retains a close association with electron transfer protein, and inhibition by metal-chelating compounds is due to inhibition of some electron carrier protein.[23]

Soluble methane monooxygenase from Methylobacterium sp. CRL-26 catalyzed the NADH-and oxygen-dependent oxygenation of alkanes, alkenes, ethers, cyclic, and aromatic compounds. Alkanes (C_1 to C_8 tested) were hydroxylated to the corresponding alcohols. Both primary and secondary alcohols were produced from the oxidation of C_3 to C_8 alkanes. Among the monohalogenated methane derivatives, bromomethane and chloromethane were oxidized more rapidly than fluoromethane. Formaldehyde was detected as the product of chloromethane and bromomethane oxidation. Dichloromethane and trichloromethane were oxidized more slowly than chloromethane. Tetrachloromethane was not oxidized. 1-Nitropropane and 1-bromobutane were oxidized more rapidly than were 2-nitropropane and 2-bromobutane. The branched chain alkane, isobutane, was oxidized to a mixture of isobutanol and tertiary butanol (Table IV).

Terminal alkenes (ethylene, propylene, 1-butene, and butadiene) were oxidized to the corresponding 1,2-epoxides by the soluble methane

TABLE V

Oxidation of Alkenes, Carbon Monoxide, Ethers, Cyclic and Aromatic Compounds by Soluble Methane Monooxygenase from Methylobacterium sp. CRL-26

Substrate	Product	Specific Activity (nmol/min/mg of protein)
Ethylene	Ethylene Oxide	55
Propylene	Propylene Oxide	100
But-1-ene	1,2-Epoxybutane	87
Butadiene	1,2-Epoxybutene	75
Isobutylene	1,2-Epoxyisobutene	95
Cis-but-2-ene	Cis-2,3-Epoxybutane	22
	Cis-2-Buten-1-ol	15
Trans-But-2-ene	Trans-2,3-Epoxybutane	25
	Trans-2-Buten-1-ol	18
2-Methyl-1-butene	ND	42
2-Methyl-2-butene	ND	16
1-Bromo-1-butene	ND	83
2-Bromo-2-butene	ND	30
Isoprene	1,2-Epoxyisoprene	38
Carbon monoxide	Carbon dioxide	30
Dimethyl ether	Methanol	25
	Formaldehyde	10
Butyl ether	ND	34
Cyclohexane	Cyclohexanol	36
Toluene	Benzyl alcohol	22
	Cresol	15
Benzene	Phenol	20

Products were identified and estimated by gas chromatography retention time comparisons and co-chromatography with authentic standards. Specific activities were expressed as nmol of product formed or substrate utilized per min per mg of protein.

ND = Products were not identified. Rate of oxidation was expressed as nmol of substrate e utilized per min per mg protein.

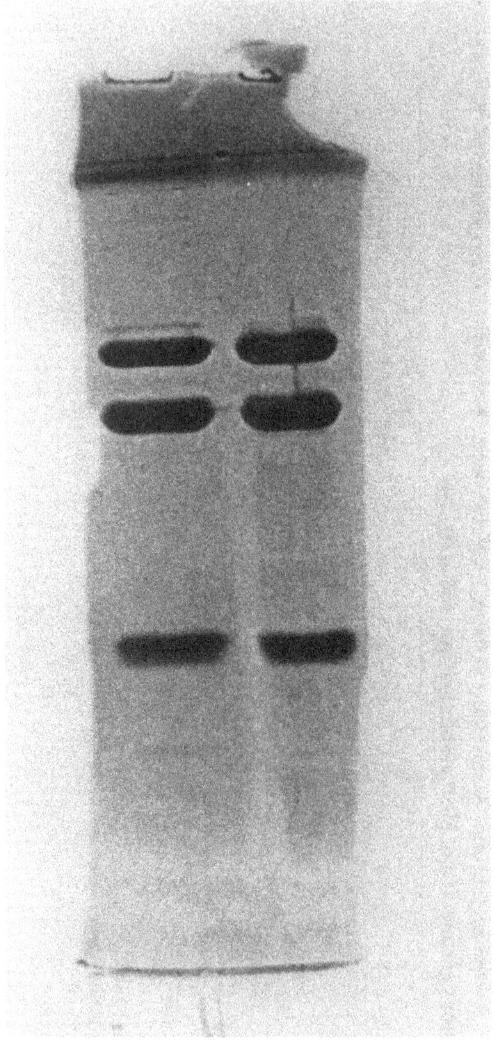

A

Fig. 3. Subunits of the components of the methane monooxygenase system. A. sodium dodecyl sulfate polyacrylamide gel-electrophoresis of the purified hydroxylase; B. Sodium dodecyl sulfate polyacrylamide gel-electrophoresis of the purified flavoprotein:NADH oxidoreductase.

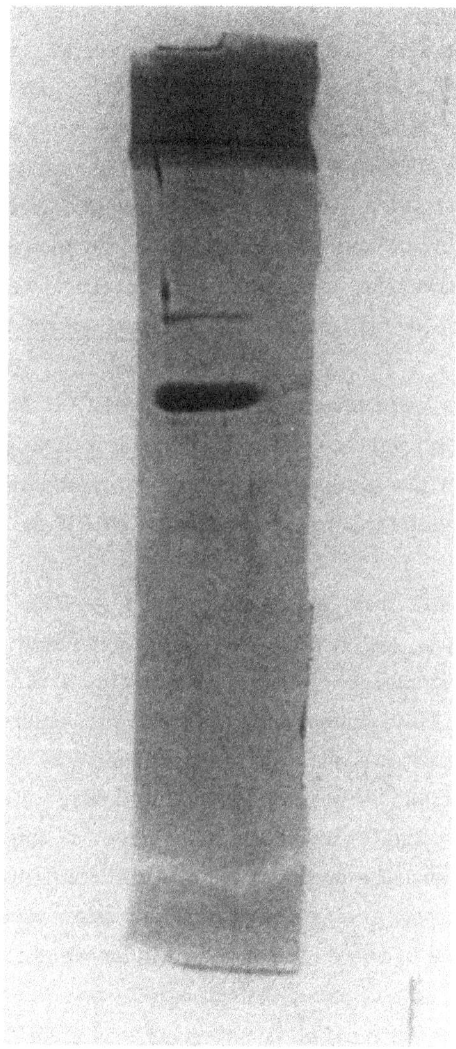

B

monooxygenase from Methylobacterium sp. CRL-26. Internal alkenes such as trans but-2-ene and cis but-2-ene also were oxidized. A mixture of trans 2,3-epoxybutane and trans but-2-ene-1-ol were produced from trans but-2-ene indicating the oxidation of the internal double bond and the terminal methyl group. Substituted alkenes such as 2-methyl-1-butene, 2-methyl-2-butene, 1-bromo-1-butene, 2-bromo-2-butene and the branched chain alkenes, isobutylene and isoprene, were also oxidized (Table IV). Carbon monoxide, dimethylether, dibutylether, cyclohexane, toluene, and benzene were oxidized by a soluble methane monooxygenase from Methylobacterium sp. CR-26 (Table V).

A number of electron donors were unable to replace NAD(P)H as a cofactor for soluble methane monooxygenase from Methylobacterium sp. CRL-26. Ascorbate or methanol in the presence of co-binding cytochrome C or methanol dehydrogenase, respectively, could not replace NADH as an electron donor.[28]

Soluble methane monooxygenase from Methylobacterium sp. CRL-26 was resolved into two components, an hydroxylase component and a flavoprotein:NADH oxidoreductase component by diethylaminoethyl (DEAE) cellulose column chromatography. Both components, which were required for methane monooxygenase activity were purified to homogeneity by ion-exchange, affinity, and molecular sieve column chromatography. The purified hydroxylase and flavoprotein:NADH reductase gave a single precipitin band when treated with antisera prepared against the respective purified components and each gives a single symmetrical peak when analyzed by ultracentrifugation. The hydroxylase has a molecular weight of about 225,000. Since it has subunit size of about 55,000, 40,000, and 20,000 it is an $\alpha_2\beta_2\gamma_2$ protein (Fig. 3A). The purified hydroxylase is a colorless protein. Its UV/visible absorption spectrum shows no peak in the visible but there is a peak around 280nm in UV region. The purified hydroxylase contained about 3.0 gram atoms of iron per mole of protein and probably the iron is a part of its catalytic center. Acid-labile sulfide could not be detected in the hydroxylase by colorimetric analysis. This indicates the presence of a novel iron-containing center(s) in the hydroxylase protein. The flavin:NADH oxidoreductase has a molecular weight of about 40,000 (Fig. 3B) and it is a single polypeptide with an UV/visible absorption

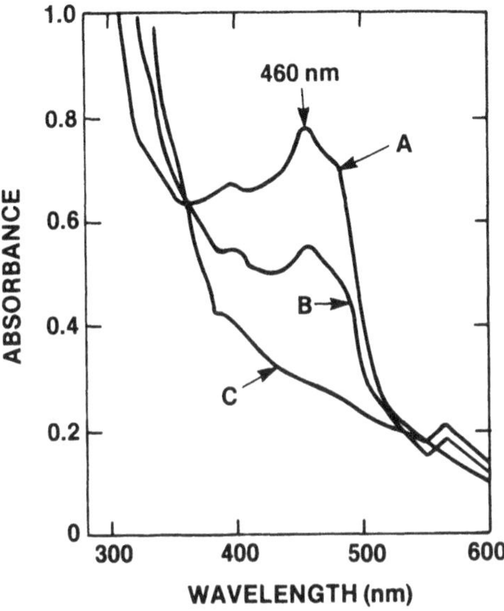

Fig. 4. Anaerobic reduction of the flavoprotein component of methane monooxygenase by NADH. A. Flavoprotein (3mg/ml); B. Flavoprotein + NADH, 2.5×10^{-5}M; C. Flavoprotein + NADH, 5.0×10^{-5}M.

spectrum showing peaks at 460 nm, 390 nm and 280 nm. The purified flavin:NADH oxidoreductase contained 1 mole of flavin adenine dinucleotide (FAD) and 2 moles each of iron and acid-labile sulfide per mole of protein. The purified oxidoreductase is directly reduced by its cofactor, NADH, under anaerobic conditions (Fig. 4). During anaerobic titration of the oxidoreductase with NADH, a spectral species appeared in the 570 nm region indicating the formation of the semiquinone form of FAD. This semiquinone also was detected as a free radical signal during epr spectroscopy of the oxidoreductase when it was partially reduced with NADH (Fig. 5). On the basis of available information the possible mechanism of electron transfer within the methane monooxygenase complex of Methylobacterium sp. CRL-26 is shown in Fig. 6.

2. Secondary Alcohol Dehydrogenase

Methane-utilizing bacteria and methanol-utilizing yeasts contained an NAD^+

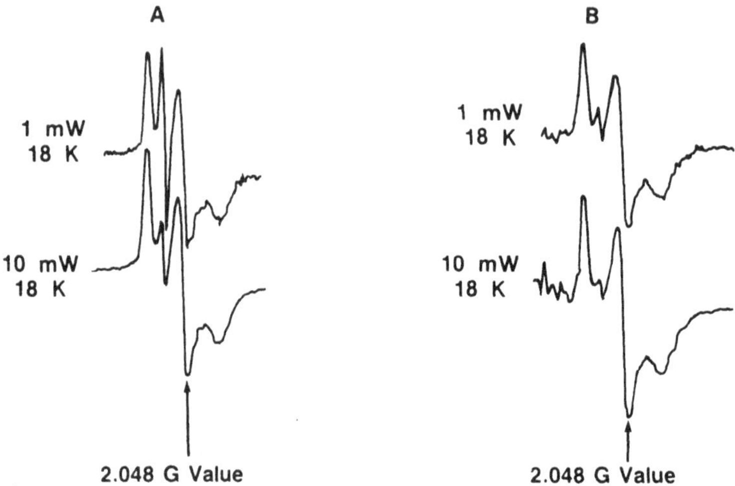

Fig. 5. EPR spectrum of flavoprotein:NADH reductase component of methane monooxygenase in the presence of (A) NADH and (B) sodium dithionite under anaerobic condition.

-dependent dehydrogenase activity which catalyzed the oxidation of secondary alcohols to the corresponding methylketones. The secondary alcohol dehydrogenase was purified to homogeneity from methanol-utilizing yeast, Pichia sp.,[29] and bacterium, Pseudomonas sp..[30] Yeast SADH catalyzed the oxidation of secondary alcohols with the following rates relative to 2-butanol: 2-propanol (35%), 2-butanol (100%), 2-pentanol (67%), 2-hexanol (65%), 2-heptanol (62%), 2-octanol (63%), 2-nonanol (63%), 3-pentanol (50%), 3-hexanol (39%), 3-octanol (27%), 2,3-butanediol (9%), 2,4-pentandiol (7%), 2,5-hexandiol (13%).

Primary alcohols (C_1 to C_6 tested), 1,2-butanediol, isobutanol, tert. butanol, and aldehydes (C_1 to C_6) tested) were not oxidized by the purified

NADH + H → NAD ⇄ FLAVOPROTEIN [OXIDIZED] / FLAVOPROTEIN [REDUCED] ⇄ HYDROXYLASE [REDUCED] / HYDROXYLASE [OXIDIZED] ⇄ $CH_4 + O_2$ → $CH_3OH + H_2O$

Fig. 6. Possible mechanism of electron transfer within the methane monooxygenase complex of Methylobacterium sp. CRL-26.

yeast SADH. In contrast to yeast SADH, the purified bacterial SADH did not oxidize 2-heptanol, 2-octanol, 2-nonanol, 3-pentanol, 3-hexanol, 3-heptanol, and 3-octanol. The purified yeast SADH preferentially oxidized the (-) enantiomers of 2-butanol and 2-octanol. The activity with (+) 2-butanol and (+) 2-octanol was only 36% and 13% of that with (-) 2-butanol and (-) 2-octanol, respectively. The purified yeast SADH also catalyzed the reduction of methylketones to the corresponding secondary alcohols. The apparent K_m values for NAD, NADH, 2-butanol, and 2-butanone are 0.05 mM, 0.1mM, 0.4mM, and 1mM, respectively. Various sulfhydryl inhibitors e.g. p-hydroxymercuribenzoate, 5,5-dithiobis-(2-nitrobenzoate), and acetamide and metal-chelating compounds (1,10-phenanthroline, 2,2-bipyridyl, and imidazole) strongly inhibited both the bacterial and yeast SADH, indicating the involvement of metal(s) and sulfhydryl group(s) in the oxidation of secondary alcohols. The purified yeast and bacterial SADH exhibited optimum activity around pH 8.0. Both have a molecular weight of around 95,000-98,000, contain two subunits of 48,000 molecular weight and have about two atoms of zinc per molecule of enzyme protein (Table VI).

3. Regeneration of NADH for Methane Monooxygenase

Recently, we have demonstrated the regeneration of NADH, the cofactor required for the epoxidation of alkenes, such as propylene, and the hydroxylation of alkanes, such as methane, catalyzed by methane monooxygenase.[31] The NAD^+ linked dehydrogenases were used for the regeneration of the co-factor in the monooxygenase reaction system. Formate in the presence of NAD^+ acts as an electron donor for methane monooxygenase. The soluble fraction from Methylobacterium sp. CRL-26 contained an NAD^+ linked formate dehydrogenase. Hence, methane monooxygenase and formate dehydrogenase in a coupled enzyme system regenerated co-factor NAD^+/NADH (Table VI). We also have used NAD^+ linked secondary alcohol dehydrogenase to regenerate the NAD^+/NADH required for epoxidation/hydroxylation of alkenes/alkanes and dehydrogenation of secondary alcohols, such as 2-propanol, 2-butanol, to the corresponding methylketones (Table VII).

TABLE VI

Properties of Secondary Alcohol Dehydrogenase and Methanol Dehydrogenase

Properties	Secondary Alcohols Dehydrogenase Secondary Alcohols --> Methylketones	Methanol Dehydrogenase Primary Alcohols --> Aldehydes
Substrate specificity	2-Propanol (35%) 2-Butanol (100%) 2-Pentanol to (65%) 2-Decanol 3-Pentanol (50%) 3-Hexanol (39%) 3-Octanol (37%) (-)-2-Butanol (140%) (+)-2-Butanol(40%) 1-Propanol (0%) 1-Butanol (0%)	Methanol (100%) Ethanol (95%) 1-Propanol to (80%) 1-Decanol 2-Phenoxyethanol (72%) 2-Aminoethanol (75%) Formaldehyde (95%) 3-chloro-1-Propanol (80%) 2-Bromomethanol (80%) 2-Propanol (0%) 2-Butanol (0%)
Activator required	none	Ammonia, Methylamine
Electron acceptor	NAD^+	Phenazine methosulfate
Molecular weight	98,000	120,000
Subunit size	48,000	60,000
Metal content	2 atoms zinc/mol	None
Optimum pH for activity	8.0	9.5
Absorption maxima	None	350 nm
Fluorescence maxima	None	440 nm and 470 nm
Prosthetic group	--	Methoxatin [Pyrrolo quinoline quinone]

100% activity for secondary alcohol dehydrogenase was expressed as 6198 nmol of 2-butanol oxidized to 2-butanone per min per mg. of purified protein.

100% activity for methanol dehydrogenase was expressed as 3160 nmol of methanol oxidized to formaldehyde per min per mg. of purified protein.

TABLE VII

Oxidation of Alkanes/Alkenes in Coupled Enzyme Reactions (Methane Monooxygenase and Dehydrogenase) for Regneration of NAD+/NADH

Substrate	Product	Cofactor Generating System	Rate of Product Formation (nmol/min/mg protein)
Methane	Methanol		125
Ethane	Ethanol	NAD^+-linked	147
Ethylene	Ethylene oxide	Formate dehydrogenase[a]	145
Propylene	Propylene oxide		267
Methane	Methanol		151
2-Butanol	2-Butanone		290
Ethane	Ethanol	NAD^+-linked	176
2-Butanol	2-Butanone	Secondary alcohol Dehydrogenase[b]	290
Ethylene	Ethylene oxide		210
2-Butanol	2-Butanone		300
Propylene	Propylene oxide		290
2-Butanol	2-Butanone		300

a Reaction mixture in a total of 1.0 ml contained 50 μmol potassium phosphate buffer pH 7.0; 10 μmol NAD^+; soluble methane monooxygenase from *Methylobacterium* sp. CRL-26, 250 μl; and sodium formate, 20 μmol. Reaction started by addition of substrate and product formation was estimated by gas chromatography.

b Reaction mixture in a total of 1.0 ml contained 50 μmol potassium phosphate buffer, pH 7.0; 10 μmol NAD^+; soluble methane monooxygenase from *Methylobacterium* sp. CRL-26, 250 μl; 2-butanol, 20 μmol; purified yeast secondary alcohol dehydrogenase 100μg. Reaction was started by addition of substrate and the product formation was estimated by gas chromatography.

4. Methanol Dehydrogenase

The NAD(P)+-independent oxidation of primary alcohols by methanol dehydrogenase was studied in a methanol-utilizing organism, Pseudomonas M_{27}[32] and in a methane-utilizing organism, Methylococcus capsulatus.[33] In vitro, methanol dehydrogenase can be coupled to phenazine methosulfate, or oxygen, or to the artificial electron acceptor, 2,6-dichlorophenol indophenol. The purified methanol dehydrogenase from both organisms catalyzed the oxidation of primary alcohols in the presence of ammonia or a primary amine as an activator. Secondary and tertiary alcohols were not oxidized. The methanol dehydrogenases do not contain any metal ions and have two identical subunits of 60,000, with a molecular weight of the native enzyme being 120,000. The purified enzyme has a pH optimum of 9-9.5 and has an absorption peak at 350 nm. Formate was isolated as a product of methanol oxidation and aldehydes were detected as products of primary alcohol (C_2 to C_5) oxidation by the purified methanol dehydrogenase from Methylococcus capsulatus (Table VI). The oxidation of formaldehyde by methanol dehydrogenase is due to the hydration of formaldehyde in aqueous solution.

The structure of the prosthetic group and the mechanism of action of methanol dehydrogenase has recently been elucidated by crystallizing the cofactor of methanol dehydrogease as fluorescent orange needles and determining its structure by X-ray diffraction analysis. The structure obtained from this analysis, 4,5-dihydro-5-hydroxy-4-oxo-5 (2 oxopropyl)-lH-pyrrolo [2,3-f] quinoline quinone (PQQ), was proposed for the methanol dehydrogenase prosthetic group, since the O-quinone structure is essential for enzyme activity.[34,35]

III. CONCLUSIONS

Methane-utilizing bacteria, methanotrophs, have the ability to grow on cheap and available C_1 compounds such as methane and methanol and have the capacity to oxygenate a variety of hydrocarbons including catalyzing the epoxidation of alkenes (propylene to propylene oxide), hydroxylation of alkanes (methane to methanol), cyclic (cyclohexane to cyclohexano), and aromatic compounds (benzene to phenol). Success of such processes for commercialization will depend upon the regeneration of the cofactor, NADH, required to catalyze these reactions by methane monooxygenase enzyme.

Regneration of the co-factor, NADH, in the coupled enzyme reaction with NAD^+-linked dehydrogenase (such as secondary alcohol dehydrogenase) along with methane monooxygenase will produce industrially important products such as methanol from methane or propylene oxide/ethylene oxide from propylene/ethylene, respectively, along with acetone or 2-butanone (MEK) from the dehydrogenation of 2-propanol or 2-butanol. Methanotrophs may find future application in the removal or detoxification of specific chlorinated hydrocarbons. Future research on the genetics of methanotrophs particularly the cloning of the genes responsible for methane monooxygenase will be attractive to increase the rate of the oxygenation reaction by increasing the synthesis of methane monooxygenase component proteins. The mechanism of methane monooxygenase with respect to the function of each component protein, the determination of the structure of the catalytic sites and the active oxygen species produced during the catalysis will lead to the understanding of the biocatalytic oxygenation of hydrocarbons and may reveal ways to avoid the need for the cofactor, NADH, to catalyze the epoxidation and hydroxylation reactions.

REFERENCES

1. H. Dalton in **Hydrocarbons in Biotechnology** (D.E.F. Harrison, I.J. Higgins and R. Watkinson, eds.), Hayden & Son Ltd., London, 1980, p. 85.
2. R.N. Patel and C.T. Hou in **Development in Industrial Microbiology** (L.A. Underkofler, ed.), Keuffel & Essex Co., Virginia, 1983, p. 141.
3. D.J. Best and I.J. Higgins in **Topics in Enzyme and Fermentation Biotechnology** (A. Wiseman ed.), John Wiley & Sons, New York, 1983, p. 38.
4. J. Colby, H. Dalton, and R. Whittenbury, Ann. Rev. Microbiol. **33**, 481 (1979).
5. R. Whittenbury, K.C. Phillips, and J.F. Wilkinson, J. Gen. Microbiol., **61**, 205 (1970).
6. R.N. Patel, D.S. Hoare, and B.F. Taylor, Bacteriol. Proc., 166 (1969).
7. C.T. Hou, R.N. Patel, and A.I. Laskin, Adv. Appl. Microbiol. **26**, 41 (1980).

8. R.N. Patel, C.T. Hou, and A.I. Laskin, **Development in Industrial Microbiology** (L.A. Underkofler, ed.) Keuffel & Essex Co., Virginia, 1981, p. 187.
9. R.N. Patel, C.T. Hou, A.I. Laskin, A. Felix, and P. Derelanko, Appl. Environ. Microbiol. **39**, 720 (1980).
10. R.N. Patel, C.T. Hou, A.I. Laskin, A. Felix, and P. Derelanko, Appl. Environ. Microbiol. **39**, 727, (1980).
11. C.T. Hou, R.N. Patel, A.I. Laskin, N. Barnabe, and I. Barist, Appl. Environ. Microbiol. **38**, 142 (1979).
12. R.N. Patel, C.T. Hou, A.I. Laskin, P. Derelanko, and A. Felix, Appl. Environ. Microbiol., **38**, 219 (1979).
13. T. Ferenci, T. Strom, and J.R. Quayle, J. Gen. Microbiol., **91**, 79 (1975).
14. H. Dalton, Arch. Microbiol., **114**, 272 (1977).
15. I.J. Higins, R.C. Hammond, F.S. Sariaslani, D. Best, M.M. Davis, S.E. Tryhorn, and F. Taylor, Biochem. Biophys. Res. Commun., **89**, 671 (1979).
16. C.T. Hou, R.N. Patel, A.I. Laskin, N. Barnabe, and I. Barist. Appl. Environ. Microbiol. **46**, 178 (1983).
17. R.N. Patel, C.T. Hou, A.I. Laskin, A. Felix, and P. Derelanko, J. Appl. Biochem., **5**, 121 (1983).
18. C.T. Hou, R.N. Patel, A.I. Laskin, N. Barnabe, and I. Barist, Appl. Environ. Microbiol., 46(, 171 (1983).
19. R.N. Patel, C.T. Hou, A.I. Laskin, A. Felix, and P. Derelanko, J. Appl. Biochem., **5**, 107 (1983).
20. I.J. Higgins and J.R. Quayle, Biochem. J., **118**, 210 (1970).
21. D.W. Ribbons, J. Bacteriol., **122**, 1351 (1975).
22. G.M. Tonge, D.E.F. Harrison, C.J. Knowles, and I.J. Higins, FEBS Lett., **58**, 293 (1975).
23. R.N. Patel, C.T. Hou, A.I. Laskin, A. Felix, and P. Derelanko, J. Bacteriol., **139**, 675 (1979).
24. G.M. Tonge, D.E.F. Harrison, and I.J. Higgins, Biochem. J., **161**, 333 (1977).
25. J. Colby and H. Dalton, Biochem. J., **165**, 395 (1978).
26. H. Dalton, Adv. Appl. Microbiol., **26**, 71 (1980).

27. D.I. Stirling and H. Dalton, Eur. J. Biochem., **96**, 205 (1979).

28. R.N. Patel, C.T. Hou, A.I. Laskin, and A. Felix, Appl. Environ. Microbiol., **44**, 1130 (1982).

29. R.N. Patel, C.T. Hou, A.I. Laskin, A. Felix, and P. Derelanko, Eur. J. Biochem., **101**, 401 (1979).

30. C.T. Hou, R.N. Patel, N. Barnabe, and I. Marczak, Eur. J. Biochem., **119**, 395 (1981).

31. R.N. Patel, C.T. Hou, A.I. Laskin and A. Felix, J. Appl. Biochem., **4**, 175 (1982).

32. C. Anthony and L.J. Zatman, Biochem. J., **104**, 953 (1967).

33. R.N. Patel, H.R. Bose, W.J. Mandy, and D.S. Hoare, J. Bacteriol., **110**, 570 (1972).

34. S.A. Salisbury, H.S. Forest, W.B. Cruse, and O. Kennard, Nature, **280**, 848 (1979).

35. J.A. Duine, J.J. Frank, and P.E.J. Verwiel, Eur. J. Biochem., **108**, 187 (1980).

PART V

SELECTED TOPICS

20

Phenol Hydrogenation Process

Jan F. VanPeppen, William B. Fisher, C.H. Chan
Allied Corporation
Petersburg, Virginia

ABSTRACT

The Allied process for the manufacture of caprolactam, the monomer of nylon 6, is based on the hydrogenation of phenol to cyclohexanone. This intrinsically safe hydrogenation is carried out as a continuous process, in the liquid phase below the boiling point, and is catalyzed by palladium on a carbon support. Maximum performance is governed primarily by activity and selectivity characteristics of the catalyst. The activity of the catalyst has to do with the rate of cyclohexanone production per unit time. The selectivity relates to the concurrent formation of the by-product, cyclohexanol, per unit of phenol converted. The more significant of these two is the selectivity. Activity and selectivity are affected inadvertently by extraneous components in the feedstock and intentionally by chemical additives. A computer model was developed to facilitate operational control and to simulate the effect of changes in process variables on system performance.

I. INTRODUCTION

Cyclohexanone is an intermediate in the Allied caprolactam process (Fig. 1).

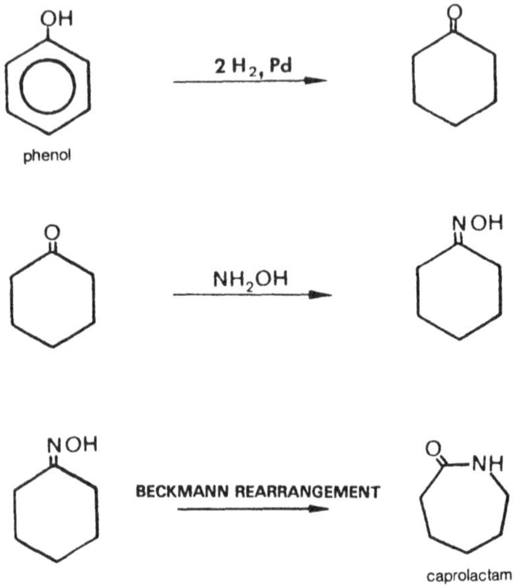

phenol $\xrightarrow{2\,H_2,\,Pd}$

$\xrightarrow{NH_2OH}$

BECKMANN REARRANGEMENT

caprolactam

Fig. 1. Allied caprolactam process.

In this commercial process, the starting material, phenol, is, in the first step, hydrogenated to cyclohexanone. In other caprolactam processes, cyclohexanone is made through the oxidation of cyclohexane. About 80% of the worldwide cyclohexanone production is by the latter process.

In the phenol hydrogenation process, phenol is reacted with hydrogen in the presence of a supported palladium catalyst, either in the vapor phase or in a liquid slurry. The subject of this discussion will be the liquid slurry process practiced by Allied.

In order for this process to be commercially viable, the rate of reactions between phenol and hydrogen must be sufficiently fast at practical temperatures. The rate of reaction is, of course, a function of the catalyst used.

Cyclohexanone, formed from the reaction of one mole of phenol with two moles of hydrogen, can react further with an additional mole of hydrogen to give cyclohexanol. Cyclohexanol is an undesirable byproduct. In order to minimize the formation of cyclohexanol the catalyst must be selective.

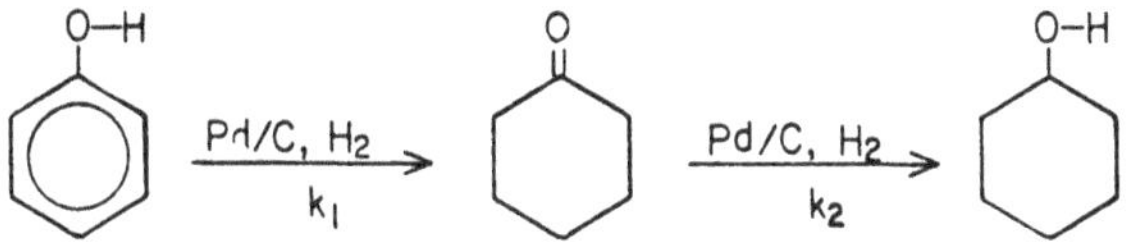

Ea = 16 Kcal/mole
ΔH = −31 Kcal/mole

Selectivity = k_1/k_2

Fig. 2. Hydrogenation of phenol.

For these reasons, the nature and the condition as well as the longevity of activity and selectivity of the catalyst are very critical to the efficiency of the process. Economics dictate that the catalyst can withstand multiple turnovers. Loss of activity or selectivity is usually caused by contaminants in the feedstock.

The Allied phenol hydrogenation process is proprietary technology. This presentation about the manner in which this process is carried out as well as other details serves to further the general knowledge of catalysis.

II. DISCUSSION

The reaction equation for phenol hydrogenation is shown in Fig. 2. The formation of cyclohexanol is believed to occur sequentially through the intermediacy of cyclohexanone. A mathematical model was developed which treats the product formation in this manner.

The operation of the plant process was simulated on a pilot plant scale to investigate process variables, the effect of poisons and additives, and the kinetics.

The following discussion is divided into three parts;

(1) The process
(2) The catalyst
(3) Poisons

A. The Process

The process, illustrated in the simplified flow-sheet shown in Fig. 3, is operated continuously. Feedstock phenol, as purchased, is treated to

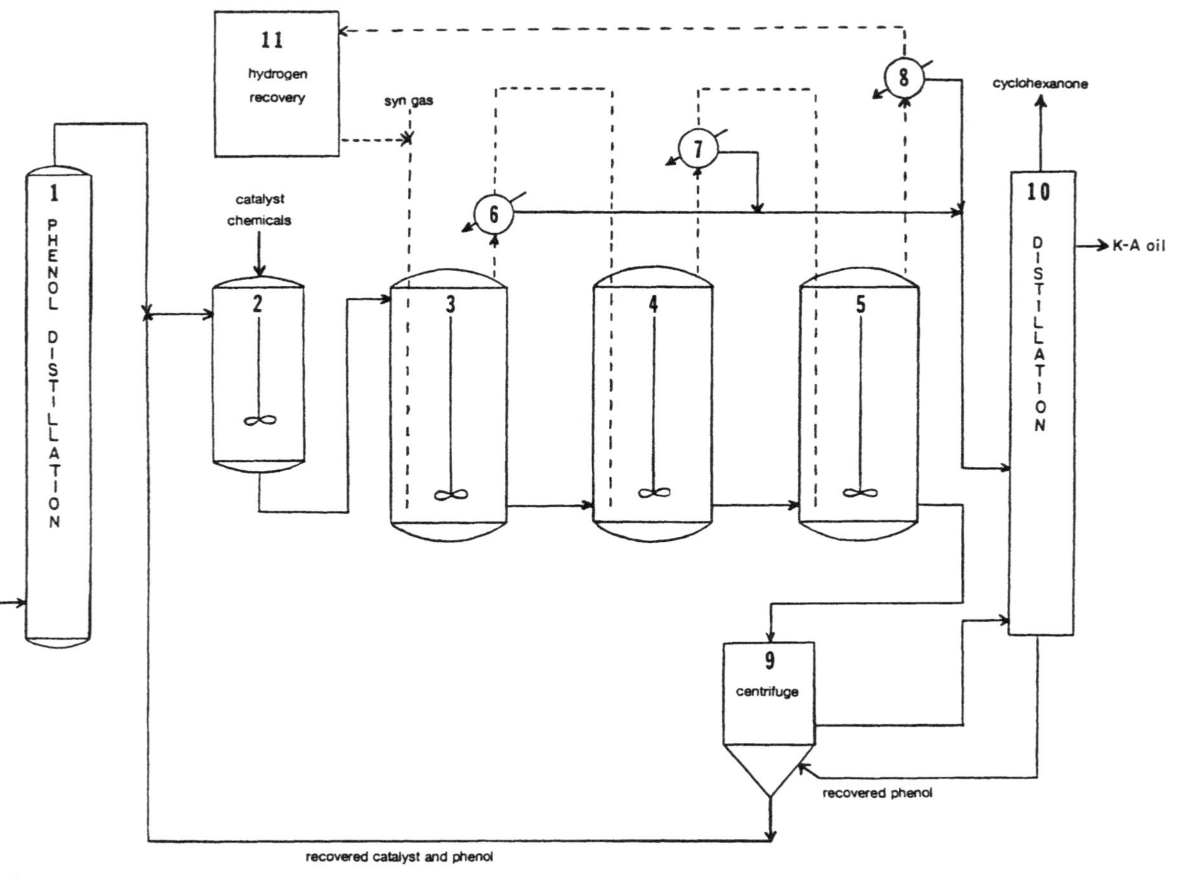

Fig. 3. Phenol hydrogenation process.

remove poisons and distilled in distillation column, **1**. The distilled phenol and a recycle stream containing the catalyst enter the premixing kettle, **2**. Also fed to this kettle are makeup catalyst and, if desired, other chemicals. From the premix kettle the process stream enters the first of a series of reactors, **3**, **4**, and **5**. Although the simplified diagram shows three reactors, this is not necessarily typical of the actual operation. The hydrogenation is effected by bubbling synthesis gas (75% hydrogen, 25% nitrogen) through the stirred vessel, the catalyst is suspended by means of mechanical agitation. Hydrogen is consumed as gas and slurry traverse the reactors. Controlled, stepwise pressure drop between reactors facilitates the forward flow without the need for pumps. As the gas passes through the liquids a large part of the product cyclohexanone, being the lowest boiling component in the reaction mixture, is carried along as vapor. The cyclohexanone vapor is condensed in condensers **6**, **7**, and **8** and is collected away from the process stream before the gas is passed on to the next reactor.

In the last reactor an overall phenol conversion of about 95% is achieved. The reaction slurry emerging from the last reactor is centrifuged in centrifuge **9** to recover the catalyst for recycle. The catalyst free product mixture consisting of cyclohexanone, phenol, and cyclohexanol is rectified in a series of distillation columns shown here as column **10**. The product, cyclohexanone, is taken forward, the byproduct, cyclohexanol, is obtained as a mixture with cyclohexanone, and the unreacted phenol is passed through the centrifuge containing the recovered catalyst. The slurry of recovered catalyst and unreacted phenol are transported to the beginning of the cycle. Unreacted hydrogen, contained in the exit gases, is recovered as a hydrogen rich gas in the hydrogen recovery plant, **11**, and returned to the first reactor.

The temperature of the process is a critical aspect. Complete avoidance of accidental vapor cloud formation requires that the temperature in each reactor does not exceed the atmospheric boiling point of the mixture in the reactor. To maintain a margin of safety the temperatures are held no less than five degrees below the prevailing boiling points. The temperature varies between 155 and 170°C at pressures of 80 to 220 psig.

- HYDROGEN ABSORPTION FROM VAPOR PHASE TO LIQUID PHASE
- ADSORPTION OF DISSOLVED HYDROGEN ONTO CATALYST SURFACE
- ADSORPTION OF PHENOL AND CYCLOHEXANONE ONTO CATALYST SURFACE
- SURFACE REACTION BETWEEN ADSORBED SPECIES
- DESORPTION OF REACTION PRODUCTS
- POISONING OF ACTIVE SITES ON CATALYST SURFACE
- SODIUM EFFECT AND SODIUM EQUILIBRIUM BETWEEN CATALYST SURFACE AND LIQUID PHASE

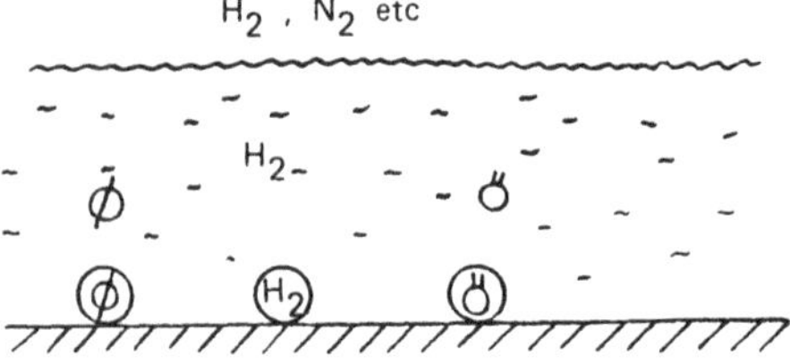

Fig. 4. Phenol hydrogenation process.

The feature of overhead removal of product from each reactor, as mentioned earlier, is an important and unique aspect of the process design. In the first place, removal of the product eliminates the possibility of further reaction to cyclohexanol and thereby the selectivity is increased, the relative catalyst concentration is increased, the residence time per reactor volume is increased, the vaporization of liquid facilitates heat removal from the highly exothermic reaction, and the load on the centrifuge is diminished.

A mathematical model has been developed to describe the continuous hydrogenation process which takes place in a series of continuously stirred reactors. The model is based on successive hydrogenation steps from phenol

TABLE I
Catalyst Characteristics

	Virgin	Recycle
Palladium Content, %	5	1-2
Dispersion, %	20	10
Relative activity	1.0	0.1

to cyclohexanone and then to cyclohexanol as the dominant pathways on the catalyst surface. Included in the model are such factors as catalyst poisoning and the effect of additives. The model also takes into consideration the apparent higher temperature of the catalyst relative to the bulk liquid phase temperature, which is due to heat transfer resistance. Various mechanisms were advanced to account for the different roles of poison and additives. Laboratory as well as plant operating data were used to select the proper model.

Various considerations upon which the mathematical model was based are summarized in Fig. 4.

B. The Catalyst

The catalyst used in the process consists of palladium on carbon. Two distinctly different catalyst species are perceived in the process; (1) a virgin catalyst which is the purchased catalyst added to the first reactor as makeup and (2) the recycle catalyst which is the catalyst in residence in the process loop. In Table I are shown the vastly different characteristics of the two catalyst species. The purchased virgin catalyst contains 5% palladium on carbon support while the palladium content of recycle catalyst is $\sim$2.5%. The reaction conditions appear to redistribute the palladium on the surface since the dispersion on recycle catalyst is only about half that of the virgin catalyst. The activity of recycle catalyst is only about 10% of that of virgin catalyst. This loss of activity is in part due to the lower palladium content and the lesser dispersion. The loss of palladium from the catalyst is a result of palladium desorption, loss of small catalyst particles having high

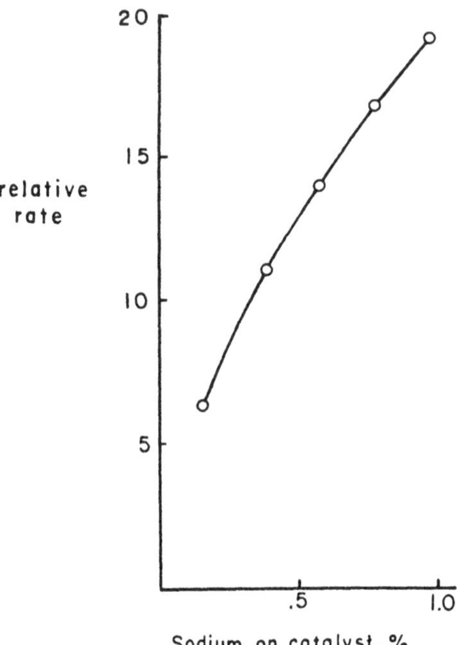

Fig. 5. Effect of base on rate.

palladium content, and chemical dissolution of palladium as soluble complexes of hydride and other complexing agents.

Virgin catalyst used in this process, contains by specification about 0.8% sodium ion as base. Base is an essential ingredient of the catalyst[1] since it acts as a promoter. The response of the rate of cyclohexanone formation to base is shown in Fig. 5. A neutral catalyst performs very poorly in slurry phenol hydrogenation. The role of base was not fully elucidated at this time. The retention of base on the catalyst under the conditions of hydrogenation is relatively poor. In order to maintain adequate catalytic activity, base is added continuously to the process. A level of about 0.5 to 1.0% sodium ion as base on the catalyst is desirable.

One of the reasons for the base requirement might be to neutralize acid groups on the surface of the carbon support. The acidity of activated carbon in the form of pendant carboxylic and phenolic groups is well documented.[2]

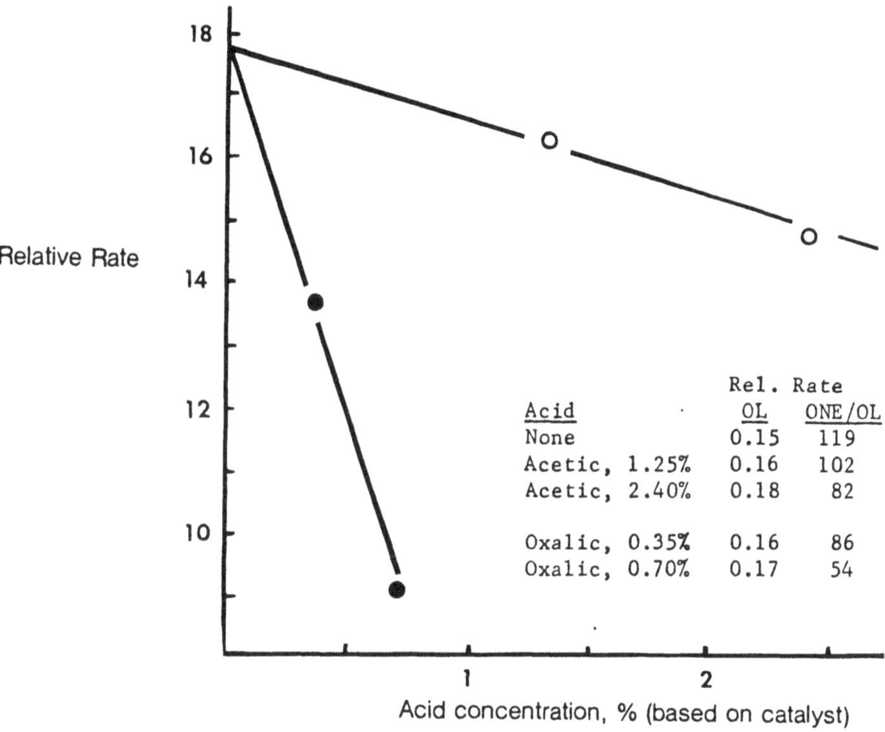

Acid	Rel. Rate OL	Rel. Rate ONE/OL
None	0.15	119
Acetic, 1.25%	0.16	102
Acetic, 2.40%	0.18	82
Oxalic, 0.35%	0.16	86
Oxalic, 0.70%	0.17	54

Fig. 6. Effect of acid on catalyst performance.

The continuous addition of base has a substantial impact on the operation of the process and on plant capacity. Without the addition of base the reaction temperature must be raised by about 30° to 40°C to maintain conversion at same throughput.

C. Poisons

In general, catalyst deactivating agents enter the process with the reactants. To remove poisons, phenol is pretreated with a diamine[3] and undergoes multiple distillation. Poisons which have inadvertently entered the process loop may become enriched in the system if recycle streams are

TABLE II

Effect of Organic Sulfur Compounds on Activity/Selectivity

Sulfur Compound	Concentration % on catalyst	Relative rate Cyclohexanone	Relative rate Cyclohexanol
None	---	14.2	0.17
Thianaphthene	0.02	11.0	0.14
Benzothiazole	0.02	10.4	0.13

not properly controlled. Adsorption of poisons on the catalyst may be either reversible or permanent. At constant levels of poisons in the feedstock, an equilibrium concentration on the catalyst surface is established. Activity of the catalyst and hence, plant capacity are directly related to this concentration of poisons. In the phenol hydrogenation the major poisons are; acids, organic sulfur compounds, iron, and hydroxy-2-propanone.

1. Catalyst Poisoning by Acids

The phenol used in the Allied hydrogenation process is derived from cumene oxidation. In this oxidation process, small amounts of low molecular weight organic acids are coproduced. Complete removal of these organic acids by distillation is difficult. In Fig. 6 is shown the effect of acids on catalyst performance. Acetic and formic acid are found to lower the rate of cyclohexanone formation while the formation of cyclohexanol increased. The dicarboxylic acid, oxalic acid, affected the performance even more severely than did acetic and formic acid. The net effect of the decreased cyclohexanone rate and increased cyclohexanol rate is a loss in activity and, hence, plant capacity, and also a loss of selectivity. Acids can be removed from the catalyst which results in improvement of performance.

2. Poisoning by Organic Sulfur Compounds

Sulfur compounds present in phenol can be traced back to crude oil. Changes in the economies of crude and the consequential supply and worldwide distribution has forced a larger proportion of sour "crude" into the US petrochemical industry. These events and also the emphasis on energy reduction in desulfurization treatments have resulted in a cumene feedstock which contains a higher level of sulfur compounds than it did a decade ago. Oxidation of cumene and subsequent processing to phenol does not remove these sulfur compounds entirely. Two sulfur compounds have been identified as contaminants of phenol; thianaphthene and benzothiazole.

As shown in Table II, phenol spiked with these sulfur compounds (1 ppm in phenol) hydrogenated markedly slower than a control. The rate of cyclohexanol formation decreased proportionally with the rate of cyclohexanone formation. Thus, the selectivity is not changed appreciably by sulfur poisoning.

3. Poisoning by Iron

As the major component of many of the parts used in the construction of the plant, iron is nearly unavoidable as a contaminant of the process streams. In particular, the piping which transports the hydrogen gas to the reactors is a source of iron as iron scales. These scales are usually picked up by the gas whenever surges through the pipes take place, for instance, when the gas flow is turned on or off.

In Table III is shown the effect of iron oxide on catalyst performance. At a level of 0.1% the rate of cyclohexanone formation was not impaired, however, the rate of cyclohexanol formation increased by a factor of 9. The loss of selectivity can be restored by adding a sequestering agent, tetra sodium ethylenediaminetetraacetate (EDTA), which renders the iron ineffective through chelation. The increased rate of cyclohexanone formation in the last entry is a consequence of the basicity of the sodium salt of EDTA.

The effect of iron on palladium was investigated by others. Garten[4] demonstrated a profound change in properties when palladium on alumina was contaminated with iron. He showed evidence of alloy formation in iron contaminated palladium catalyst under conditions similar to those of the phenol hydrogenation process. Alloy formation did not occur in iron contaminated catalysts used in the Allied process. This evidence was based on the energy states of iron as determined by ESCA.

TABLE III
Effect of Iron on Catalyst

	Relative Hydrogenation Rate Constants	
	Phenol	Cyclohexanone
Control	17	0.12
Control + Fe^{3+} (1000 ppm on cat.)	17	1.02
Control + Fe^{3+} (1000 ppm on cat.) + Sequestrene	27	0.20

4. Poisoning by Acetol

Acetol is a byproduct in the cumene phenol process. The probable formation of acetol[5] is shown in Fig. 7. Removal of acetol from phenol by conventional distillation is difficult. In Fig. 8 is shown the severe effect on rate by the presence of acetol. This rate retarding effect is most surprising for a compound such as acetol which does not appear to possess significant chelating properties or other features that would raise concern. Because of the severe potential threat to production capacity, the mechanism of acetol poisoning was investigated.

Acetol was found to be unstable under the conditions of phenol hydrogenation. Thus, when acetol, dissolved in phenol, was heated at 150°C in the presence of palladium on carbon the acetol disappeared. Moreover,

$$C_6H_5-C(CH_3)_2-O-O-H + CH_3-C(OH)=CH_2 \longrightarrow C_6H_5-C(CH_3)_2-O-H + CH_3-\overset{OH}{C}\underset{O}{-}CH_3$$

$$CH_3-\overset{OH}{C}\underset{O}{-}CH_3 \longrightarrow CH_3-\overset{O}{\overset{\|}{C}}-\overset{OH}{C}H_2$$

Fig. 7. Formation of acetol.

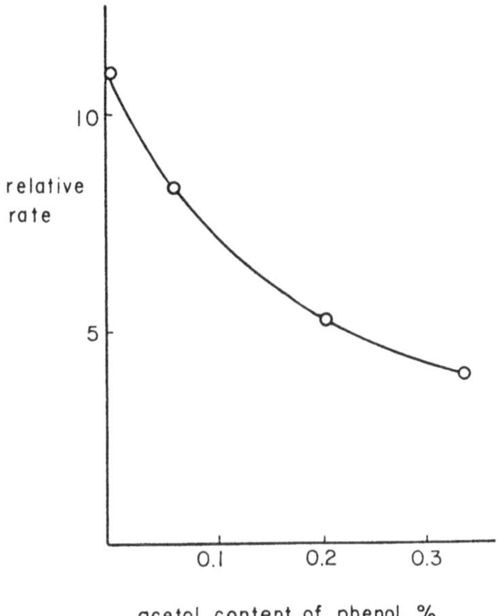

Fig. 8. Effect of acetol on rate.

alpha methylstyrene, present as a contaminant in the phenol, was found to be transformed into cumene. The gas chromatography (GC) traces of this experiment are shown in Fig. 9. When the catalyst was not present, acetol did not disappear noticably nor was cumene formed from alpha-methylstyrene. The reaction equation is shown in Fig. 10. This facile hydrogen transfer process is not unique. For instance, sugars having the same vicinal keto-hydroxy functionality found in acetol, also have this reductive property, which is the basis for the Fehling's test to identify sugars. The expected product resulting from the dehydrogenation, pyruvic aldehyde, could not be detected by GC or mass spectrometry which suggests that it is unstable under the reaction conditions.

Other compounds having the vicinal keto-hydroxy functionality such as, 2-hydroxycyclohexanone and 3-hydroxy-2-butanone (acetoin) also undergo the hydrogen transfer to alpha-methylstyrene. In these instances the reaction products were stable and could be detected in the reaction mixture. The GC trace of the acetoin experiment is shown in Fig. 11.

Contrary to results obtained with acetol and also pyruvic aldehyde,

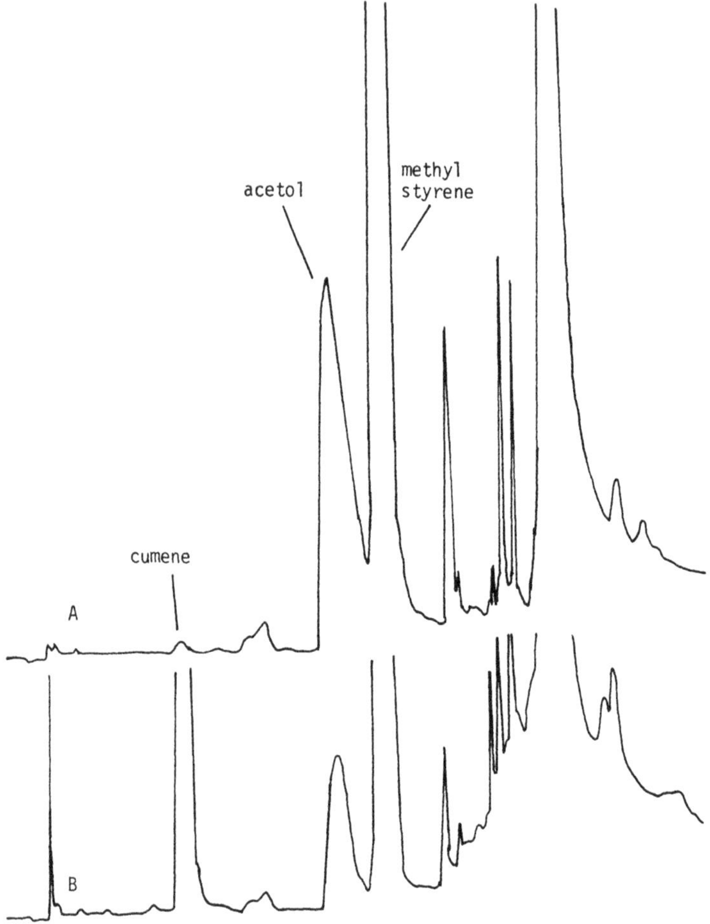

Fig. 9. Hydrogen transfer from acetol to α-methylstyrene. A. G.C. trace of acetol and α-methylstyrene. B. G.C. trace of acetol and α-methylstyrene in phenol heated at 150°C under nitrogen in presence of palladium on carbon.

$$C_6H_5-\overset{\displaystyle CH_3}{\overset{|}{C}}=CH_2 + CH_3-\overset{\displaystyle O}{\overset{\|}{C}}-\overset{\displaystyle O\text{-}H}{\overset{|}{C}}H_2 \xrightarrow[\text{phenol}]{\text{Pd/C}} C_6H_5-\underset{\underset{\displaystyle CH_3}{|}}{\overset{\overset{\displaystyle CH_3}{|}}{C}H} + CH_3-\overset{\displaystyle O}{\overset{\|}{C}}-\overset{\displaystyle O}{\overset{\|}{C}}H$$

methylstyrene acetol cumene pyruvic aldehyde

Fig. 10. Hydrogen transfer.

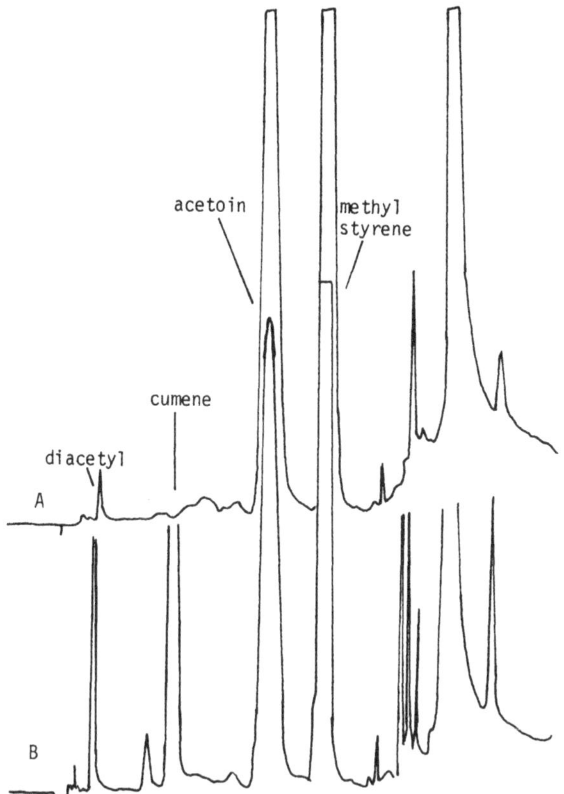

Fig. 11. Hydrogen transfer from acetoin to α-methylstyrene.

hydroxycyclohexanone and acetoin did not affect the rate of phenol hydrogenation in the least.

In view of the apparent instability of the dehydrogenation product of acetol, which is presumably pyruvic aldehyde, the chemistry of this keto aldehyde was studied. It was found that pyruvic aldehyde dissolved in phenol and in presence of palladium on carbon decomposed readily with concommitant formation of carbon monoxide. This reaction proceeds at a reasonably fast rate at about 170°C, the phenol hydrogenation temperature. Under the same reaction conditions, acetol was found to generate carbon monoxide as well and at about the same rate. Hydroxycyclohexanone and acetoin, however, did not form carbon monoxide.

$$CH_3-\overset{O}{\overset{\|}{C}}-\overset{O-H}{\overset{|}{C}H_2} \longrightarrow CH_3-\overset{O}{\overset{\|}{C}}-\overset{O}{\overset{\|}{C}}H + H_2$$

$$CH_3-\overset{O}{\overset{\|}{C}}-\overset{O}{\overset{\|}{C}}H \longrightarrow CH_3-\overset{O}{\overset{\|}{C}}H + CO$$

$$CH_3-\overset{O}{\overset{\|}{C}}H \longrightarrow CH_4 + CO$$

Fig. 12. Decarbonylation of acetol.

The reaction equations involved in the decomposition of acetol are shown in Fig. 12. Except for pyruvic aldehyde, the intermediate, acetaldehyde, and the products methane and carbon monoxide were identified by GC and mass spectrometry. Clearly, the poisoning effect observed when acetol is present is a consequence of the highly poisonous carbon monoxide generated by acetol decomposition.

The chemistry surrounding the poisoning character of acetol can be summarized as follows. The hydroxy-ketone arrangement found in acetol does not have the capacity to affect the palladium catalyst. This is apparent from the lack of such behavior with hydroxycyclohexanone and acetoin. Compounds having the vicinal keto-hydroxy arrangement, can, through a dehydrogenation step, form the corresponding dicarbonyl grouping, which is also incapable of affecting the catalyst. However, whenever one of the carbonyl groups is bonded to hydrogen, such as in pyruvic aldehyde, a palladium catalyzed decarbonylation occurs with formation of carbon monoxide. Thus, the decarbonylation reaction is structure dependent as illustrated in Fig. 13.

III. SUMMARY

Cyclohexanone can be manufactured from phenol in high yields through the process described. In the slurry type process, yields of cyclohexanone

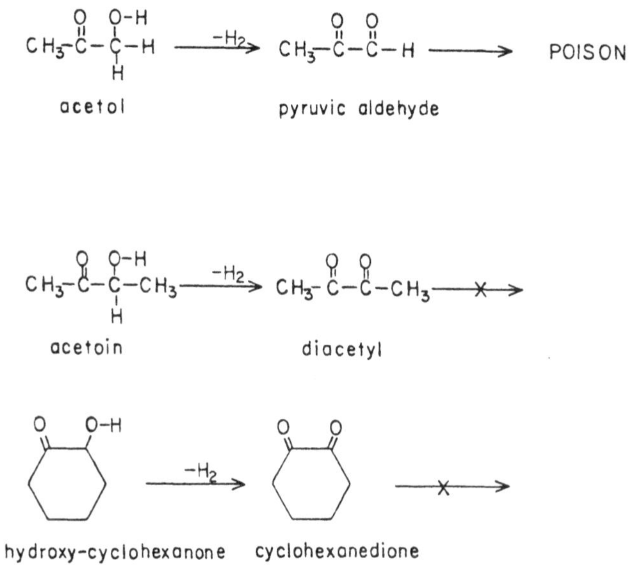

$CH_3-C(=O)-CH(OH)-H \xrightarrow{-H_2} CH_3-C(=O)-C(=O)-H \longrightarrow$ POISON

acetol pyruvic aldehyde

$CH_3-C(=O)-CH(OH)-CH_3 \xrightarrow{-H_2} CH_3-C(=O)-C(=O)-CH_3 \not\longrightarrow$

acetoin diacetyl

$\xrightarrow{-H_2}$ $\not\longrightarrow$

hydroxy-cyclohexanone cyclohexanedione

Fig. 13. Structure limitation on decarbonylation.

exceed 98%. High yields and maximum plant capacity are achieved when the activity and the selectivity of the catalyst are maintained at optimum levels. To that end, the catalyst must be continuously promoted by base. Contaminants present in the feedstock which deactivate the catalyst must be eliminated from the feedstock before hydrogenation. Organic acids, iron, sulfur compounds, and acetol are deactivating contaminants of concern in the liquid phase slurry process using cumene derived phenol. Contamination of the process streams with catalyst poisons is a threat which is recognized, the consequences are understood, and the technology to avoid contamination is in place. The Allied phenol hydrogenation proces produces cyclohexanone in a safe and very efficient manner.

It appears that the process conditions of the slurry type phenol hydrogenation affect the catalyst in a unique way. The findings and knowledge acquired in this process do not apply to other, similar processes. For instance, the seemingly similar vapor phase phenol hydrogenation process appears to have entirely different requirements.

The phenol hydrogenation process contributes to the general knowledge about catalysis that palladium can be an extremely selective catalyst. The ability of this noble metal to nearly ignore the abundant presence of the double bond of the carbonyl group is very remarkable.

BIBLIOGRAPHY

1. J.F. VanPeppen and W.B. Fisher, U.S. Patent 4,200,553.
2. E.D. Little and Z.K. Cheema, U.S. Patent 3,965,187.
3. H.P. Boehm, Adv. Catal., **16**, 179 (1966).
4. R.L. Garten, J. Catal., **43**, 18 (1976).
5. N. Indictor, J. Org. Chem., **30**, 2074 (1965).

21

Preparation of Nitriles by Catalyzed Dehydration of Carboxamides and Aldoximes

John F. Stenberg
Research Laboratories
Eastman Kodak Company
Rochester, NY 14650

ABSTRACT

Aromatic nitriles have been prepared in high yield and low cost from appropriate aromatic primary carboxamides or aromatic aldoximes by dehydration with acetic anhydride in the presence of certain transition metals or metal salts. In the absence of these metals or metal salts, acetic anhydride was ineffective for the dehydration. A number of substituents on the aromatic nucleus could be tolerated.

I. INTRODUCTION

For some of our investigations, a number of substituted aromatic nitriles were needed. Since high-volume demands were anticipated, ease of preparation and ultimate unit costs were of concern. Nitrile syntheses encompass a vast number of reagents, intermediates, and conditions. The approach most useful to our needs seemed to be the use of aromatic primary amides as precursors.

A computer-assisted literature search surveyed the various methods reported for the dehydration of amides and oximes. A number of interesting agents came to light (Table I). Most of these had to be specifically prepared, were very expensive, were consumed during the reaction, or

TABLE I

Some Reagents Used for Dehydration

Reagent	Reference
$(PNCl_2)_3$ (hexachlorocyclotriphosphazene)	[1]
2,4,6-trichloro-1,3,5-triazine	[2]
p-$CH_3C_6H_4SO_2Cl$ + pyridine	[3]
PI_3	[4]
P_2I_4	[5]
$ClSO_2F$, $ClSO_2Cl$	[6]
Ⓟ$-C_6H_4-P(C_6H_5)_2$ + CCl_4	[7]
$(CF_3SO_2)_2O$	[8]
$(CF_3SO_2)_2P(C_6H_5)_3$	[9]

required exotic conditions not amenable to scale-up. Most also applied principally to aliphatic systems only.

Pesson, et al.[10] used a combination of p-toluenesulfonyl chloride and pyridine at reflux for the dehydration shown in Eq. 1. In this method, molar

$$C_6H_5-CH(CONH_2)-CH_2-(\text{2-tetrahydrofuryl}) + p\text{-}CH_3C_6H_4SO_2Cl + C_5H_5N \longrightarrow C_6H_5-CH(C\equiv N)-CH_2-(\text{2-tetrahydrofuryl}) \quad (1)$$

equivalents of p-toluenesulfonic acid and pyridine had to be removed and discarded. Weiss, et al.[11] obtained 1 from the corresponding acetamide in refluxing acetic anhydride (Eq. 2). This was an aliphatic system but did show the utility of acetic anhydride.

$$(p\text{-}ClC_6H_4)_2CH-\overset{O}{\overset{\|}{C}}NH_2 + (CH_3CO)_2O \longrightarrow (p\text{-}ClC_6H_4)_2CHC{\equiv}N \quad (2)$$

1

Campagna, et al.[12] demonstrated that trifluoracetic anhydride/pyridine would dehydrate a number of amides, including some aromatics (Eq. 3). Although this was a good system, it would require

$$R\overset{O}{\overset{\|}{C}}NH_2 + (CF_3CO)_2O + C_5H_5N \longrightarrow RC{\equiv}N \quad (3)$$

extensive cooling capability, and the combined chemical costs of the trifluoracetic anhydride/pyridine system would be high.

II. ACETIC ANHYDRIDE DEHYDRATION

As a starting point for a potentially inexpensive system, we looked for a way to boost the power of acetic anhydride with a low-cost and readily available catalyst. p-Nitrobenzamide was chosen as a promising candidate because all anticipated products were high-melting solids.

Refluxing a mixture of p-nitrobenzamide and acetic anhydride produced no reaction (Table II). Addition of small amounts of pyridine (∿10%) produced some reaction, with most of the product being the result of acetylation at the amide nitrogen. The use of large amounts of pyridine and less acetic anhydride at reflux produced up to 50% of the desired nitrile, plus acetylated amide, p-nitrobenzoic acid, and starting amide.

This was not a grand breakthrough, since the work-up required separation of components as well as either recovery of the pyridine or

TABLE II

Acetic Anhydride + p-Nitrobenzamide

Agent	Product
None	A
Pyridine (catalytic)	A + C
Pyridine (> 2×)	A(10) + B(50) + C(trace) + D(30)
DMF	A(90) + B(10)
Ⓟ–4-pyridyl	A + B(trace) + C(trace)
$CH_3C_6H_4SO_3H$	A + D
Alumina	A + C(major)

A = $O_2N-C_6H_4-CONH_2$

B = $O_2N-C_6H_4-C{\equiv}N$

C = $O_2N-C_6H_4-CONHCOCH_3$

D = $O_2N-C_6H_4-COOH$

absorbing the cost of discarding it. It did, however, show that the dehydration was feasible.

The use of DMF in catalytic amounts with acetic anhydride gave about 10% nitrile. Poly(4-vinylpyridines) promoted the formation of a small amount of nitrile and some acetylated amide. The use of p-toluenesulfonic acid produced mainly p-nitrobenzoic acid on work-up. Alumina was an excellent catalyst for promoting the undesired N-acetylation of the amide.

We recalled that transition metals have found many uses as catalysts. In an experiment in which acetic anhydride at or near the boiling point was used also as solvent and with nickelous acetate hexahydrate used in catalytic amounts, p-nitrobenzamide was completely converted to the nitrile with only a trace of contaminant. The work-up was simple: the cooled reaction mixture was poured into cold water, and the product

TABLE III

Amides Dehydrated to Nitriles by $(CH_3CO)_2O$/Ni

$C_6H_5CONH_2$	$H_2NC(O)-C_6H_4-C(O)NH_2$
$Cl-C_6H_4-CONH_2$	1-$C_{10}H_7CONH_2$
$O_2N-C_6H_4-CONH_2$	2-$C_{10}H_7CONH_2$
$CH_3O-C_6H_4-CONH_2$	

precipitated. If desired, a simple crystallization sufficed. The use of nickelous chloride produced similar results.

The use of cobaltous acetate gave complete conversion of the starting amide; however, about 10% of the product was the N-acetylated amide, the balance being the desired nitrile. Copper acetate was nearly as effective as nickel acetate (1-2% side reaction) and somewhat better than cobalt acetate.

A number of amides were treated in the same fashion with good results (Table III).

We showed that N-acetylation of the amide was a competing reaction and not part of the dehydration sequence. A sample of purified N-acetyl-p-nitrobenzamide, subjected to the same dehydration conditions (acetic anhydride, nickel acetate, reflux), was recovered unchanged.

We had thus demonstrated that nickel or copper salts could catalyze the dehydration. However, the salts were soluble in the aqueous work-up, so the overall cost would include the cost of the small amounts of the salts used.

To address this cost problem, we used finely powdered nickel. This was nearly as effective as nickel acetate. The starting amide was completely consumed, and the isolated product was 97-97% desired nitrile,

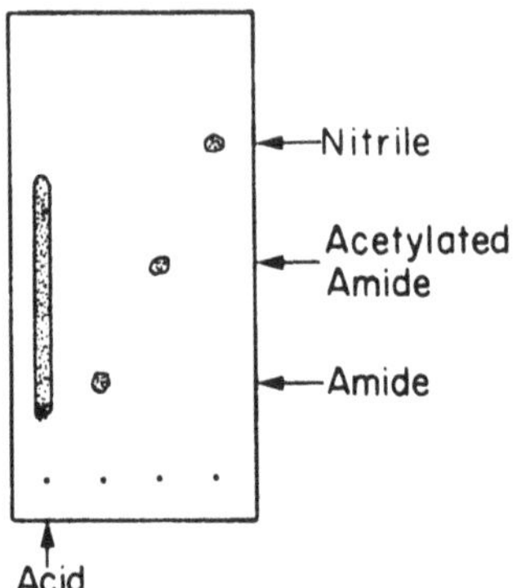

TLC, SiO_2 Plate

Fig. 1. TLC analysis of dehydration mixtures.

the balance being the N-acetylated amide. In the work-up, the reaction mixture was readily decanted from the nickel powder, and the nickel powder could be reused several times with little, if any, loss of activity. Over several runs there was little change in the mass of the nickel powder. Copper powder was useful but less efficient, allowing up to 10% of N-acetylation.

p-Nitrobenzamide was recovered unchanged from stirring in acetic acid at reflux in the presence of nickel(ous) acetate.

To further test the utility of this reaction, we subjected benzaldoxime and p-nitrobenzaldoxime to the same dehydration conditions (acetic anhydride, nickel acetate, reflux). They were converted to the corresponding nitriles with equal facility and with equally high yields. Optimum conditions were not determined.

The dehydration of p-nitrobenzamide was more than 90% complete within 30 minutes at reflux, with complete disappearance of the starting material taking place before 45 minutes.

A convenient way to monitor the reaction was by TLC on silica gel plates with diethyl ether or ether/hexane as the eluting solvent (Fig. 1). The spots were verified by comparison with authentic reference samples and with known mixtures.

The system described has the advantage of using low cost, readily available acetic anhydride and catalytic quantities (from less than 10 wt% to trace amounts) of the chosen metal or metal salt. No attempt was made to optimize amounts or conditions. Nickel(ous) acetate and nickel powder were the best for the reactions described. The reactions were easily run and the isolations were simple.

REFERENCES

1. G. Rosini and A. Mediri, Synthesis, **10**, 665 (1975).
2. G. Olah, S. Narang, A. Fung, and B. Gupta, Synthesis, **8**, 657 (1980).
3. D. Miljkovic and J. Petrovic, J. Org. Chem., **42**, 2101 (1977).
4. J. Denis and A. Krief, J. Chem. Soc., Chem. Commun., 544 (1980).
5. H. Suzuki, T. Fuchita, A. Iwasa, and T. Mishina, Synthesis, **12**, 905 (1978).
6. G. Olah,S. Narang, and A. Garcia-Luna, Synthesis, **8**, 659 (1980).
7. C. Harrison, P. Hodge, and W.Rogers, Synthesis, **1**, 41 (1977).
8. J. Hendrickson, K. Bair, and P. Keehn, Tetrahedron Lett., **8**, 603 (1976).
9. J. Hendrickson and S. Schwartzman, Tetrahedron Lett., **4**, 277 (1975).
10. M. Pesson, et al., Bull. Soc. Chim. Fr., 2263 (1965).
11. P. Weiss, M. Cordasco and L. Reiner, J. Am. Chem. Soc., **71**, 2650 (1949).
12. S. Campagna, A. Carotti, and G. Casini, Tetrahedron Lett., **21**, 1813 (1977).

22

Aminomethylation of Diene Polymers
Novel Route to Polyamines

Felek Jachimowicz and Anders Hansson
W.R. Grace & Co.
Washington Research Center
7379 Route 32
Columbia, MD 21044

ABSTRACT

A general technique for amine functionalization of diene polymers is described. The reaction of various polybutadienes, as model diene polymers, with carbon monoxide, primary or secondary amines, and water or dihydrogen as a hydrogen source in the presence of homogeneous Rh or Ru/Rh catalysts leads to the formation of novel polyamines. The scope of this technique is presented along with a discussion of its dependence on the molecular weight and microstructure of the starting polymers, and a description of the product distributions as a function of the catalyst systems.

I. INTRODUCTION

We have previously described a one-step, general catalytic procedure for the synthesis of a variety of secondary and tertiary amines from olefins, carbon monoxide, water and nitrogen source in the presence of homogeneous catalysts.[1] In this paper we would like to discuss the extension of this technique to the preparation of a collection of polytertiary and polysecondary amines.

The introduction of a functional group onto a polymeric backbone is a widely applied technique for the preparation of polymeric materials not readily available by direct polymerization techniques.[2] It occurred to us that modification of polymers containing olefinic double bonds using our previously developed aminomethylation chemistry would result in a new route for the preparation of unique and novel polyamines.

Since polybutadienes have a high degree of olefinic unsaturation, are readily available in a wide range of molecular weights and microstructures, and can be easily functionalized with terminal groups, they have been selected as the principal model, starting material for these investigations.

II. AMINOMETHYLATION

Aminomethylation of 1,2 and 1,4 polybutadienes results in the formation of polymers with a hydrocarbon backbone containing pendant amino groups (Eq. 1). We have defined the degree of functionalization (AI-Amine

```
                                                      CH2)
                                                     /
                      CH=CH                   CH=CH
                     /     \                 /
 (CH2-CH)       (CH2         CH2)       (CH2
      |
      CH
      ||
      CH2          1,4-cis                 1,4-trans
 1,2 (vinyl)    \_____________________________________/

      |                              |
CO/[H] |  HN<            CO/[H]      |  HN<  [H]≡H2O or H2          (1)
      v                              v

 (CH2-CH)             (CH2-CH2-CH-CH2)
      |                        |
      CH2                      CH2
      |                        |
      CH2                      N
      |                       / \
      CH2
      |
      N
     / \
```

Incorporation) (Eq. 2) as the primary, quantitative measure for the efficiency of this chemistry.

$$AI = \frac{\text{Concentration of Repeating Units Containing Amine Nitrogen}}{\text{Total, Initial Concentration of Olefinic Bonds}} \times 100\% \qquad (2)$$

The results depicted in Table I demonstrate a very good agreement between the various methods employed for determination of Amine Incorporation (AI). Because of the ease of use, good reproducibility and versatility, we have utilized titration techniques as the principal analytical tool. In the case of ^{1}H-NMR, we used phenyl-terminated polybutadienes as starting materials with the five aromatic protons serving as a "built-in" internal standard. The application of IR is based on an empirical relationship correlating the absorption at 1120 cm^{-1} with absorption at 1450 cm^{-1}. All of the techniques will be described in detail in the subsequent paper.[3]

Our previous work on aminomethylation and published data on hydroformylation of olefins[4] indicate that the relative reactivity of polybutadiene units should decrease from vinyl, through 1,4-cis to 1,4-trans units. The data depicted in Table II show the high reactivity observed for material having high percentages of the terminal olefinic double bond.

The most difficult problem in the functionalization of polymers is the control of selectivity, especially when an undesired reaction can lead to the formation of reactive functional groups capable of undergoing uncontrolled,

TABLE I

Amine Incorporation Determination Techniques

	AI [%]			
NITROGEN SOURCE	'H-NMR	TITRATION	IR	ELEMENTAL ANALYSIS
Morpholine	60	57	58	
Dimethylamine		55		54
Pyrrolidine	75	65		

TABLE II
Amine Incorporation as a Function of Vinyl Content

VINYL [%]	MW [g/mol]	AI [%]
90	165,000	78
80	1,000	71
70	6,000	63
45	4,500	60
25	900	55

crosslinking reactions. This problem becomes especially serious for the modification of high molecular weight polymers in which the number of crosslinkages required for gelation becomes very small. Our previous work indicated that this chemistry could be formally divided into three consecutive steps. The first, hydroformylation is followed by a condensation of an aldehyde with a primary or secondary amine. Subsequent hydrogenation of the C=N or C=C-N groups completes the reaction sequence. Accumulation of a significant concentration of the aldehyde would lead in this basic amine medium to crosslinking caused by aldol condensation. Indeed, examination of the IR spectra (Fig. 1) of the unstable products reveals the presence of aldehyde (1727 cm^{-1}) and enamine (1653 cm^{-1}) groups. (These absorption bands are practically absent in the stable product or in the starting materials).

The results summarized in Table III indicate the useful range of molecular weights of polybutadiene that can be used as starting materials in this aminomethylation procedure. It is important to note that in order to control the viscosity of the reaction mixture and to prevent formation of

STARTING
POLYBUTADIENE
994
909
Al = 80%
STABLE
1460
1041
Al = 13%
UNSTABLE
1727
1653
4000
3000
2000
1500
1000
500
cm-1

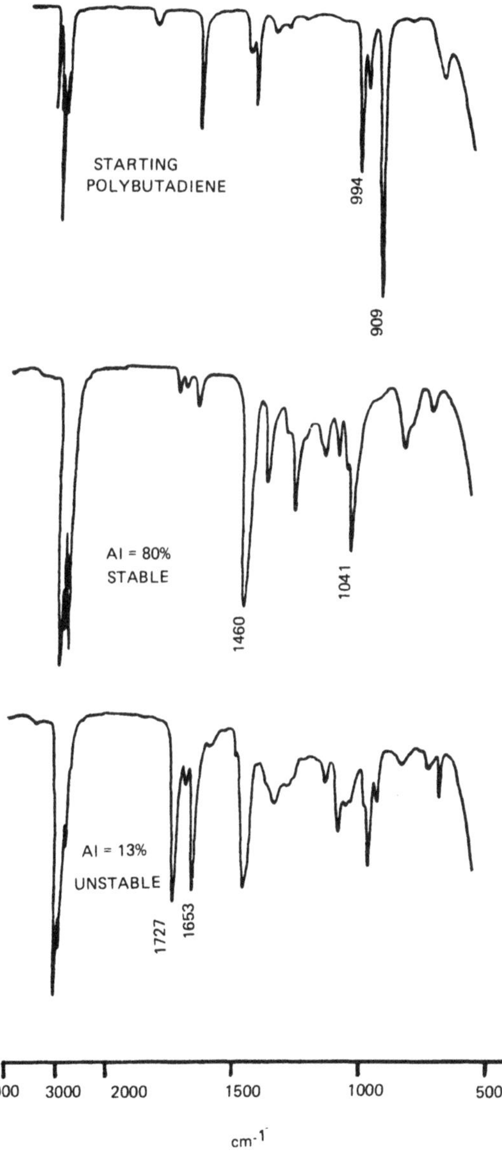

Fig. 1

TABLE III

Amine Incorporation (AI) As a Function of Molecular Weight

MW [g/mol]	Vinyl [%]	Concentration [wt%]	AI [%]
1,000	80	20	75
6,000	70	20	62
30,000	80	15	65
165,000	90	15	68
298,000	93	5	62
625,000	93	1	47

*Hydrogen source - H_2, Nitrogen Source - $HN(CH_3)_2$

infinite networks (gels) when working with higher MW polybutadienes, it is necessary to reduce the concentration of the starting polybutadiene.

A. Catalysts for the Aminomethylation of Polybutadienes

Table IV depicts several examples of catalysts that are capable of promoting the aminomethylation of polybutadienes. Rhodium based catalysts are, by far, the most effective in this transformation. However, since it is not always desirable to obtain a high degree of functionality, Ru or Ir based catalysts can be useful as well. Table V summarizes our investigation on the Rh/Ru mixed metal catalyst system.[5] This mixed catalyst allows the use of relatively small amounts of expensive rhodium (Table VI) while maintaining a high degree of amine functionalization. Polyamines obtained with the Rh/Ru mixed catalyst are very stable and do not exhibit any residual aldehyde or enamine groups.

B. Hydrogen Source

The basic aminomethylation chemistry can utilize water or dihydrogen as the hydrogen source. As the results listed in Tables IV and V point out, the

TABLE IV

A Comparison of Hydrogen Sources in the Aminomethylation Reaction

Catalyst	Hydrogen Source	AI [%]
$RuCl_2(P\emptyset_3)_3$	$H_2(H_2O)$	8 (6)
$RuCl_2(CO)_2(P\emptyset_3)_2$	$H_2(H_2O)$	5 (4)
$IrI(CO)(P\emptyset_3)_2$	$H_2(H_2O)$	37 (14)
$IrH(CO)(P\emptyset_3)_2$	$H_2(H_2O)$	40 (14)
$Rh_6(CO)_{16}$	$H_2(H_2O)$	77 (36)
$RhCl_3(C_5H_5N)_3$	$H_2(H_2O)$	72 (10)

Polybutadiene: MW = 1,000 g/mol, 80% vinyl

$[C{=}C]/[M] = 5 \times 10^2$

TABLE V
Effect of Different Ru Precursors on the Rh/Ru Catalyst System

	$\frac{[C=C]}{[M]} \times 10^{-3}$	Hydrogen Source	AI [%]
$[Rh(NBD)((CH_3)_2P\phi)_3]PF_6$	5	$H_2(H_2O)$	71 (26)
$RuCl_2(P\phi_3)_3$	0.5		
$[Rh(NBD)((CH_3)_2P\phi)_3]PF_6$	5	$H_2(H_2O)$	67 (52)
$Ru_3(CO)_{12}$	0.5		
$[Rh(NBD)((CH_3)_2P\phi)_3]PF_6$	5	H_2	80
$RuCl_2(CO)_2(P\phi_3)_2$	0.5		

TABLE VI

Effect of Concentration on the Rh/Ru - Catalyst System

$\frac{[C=C]}{[Rh_6(CO)_{16}]} \times 10^{-3}$	$\frac{[C=C]}{[RuCl_2(CO)_2(P\emptyset_3)_2]} \times 10^{-3}$	AI [%]
1	-	77
5	0.5	66
8	5	61
16	5	43
30	5	32

TABLE VII

Nitrogen Sources Used in the Aminomethylation Reaction

POLYBUTADIENE + CO/[H] + AMINE ⟶ POLYAMINE

Amine	Product
$HN(CH_3)_2$	POLYAMINE
 HN(morpholine ring)O	POLYMORPHOLINE
$HNCH_2CH_2OH$ / CH_3	POLYETHANOLAMINE
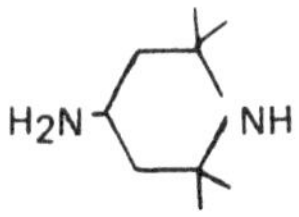 H_2N-(tetramethylpiperidine ring)NH	POLYMERIC HINDERED AMINE
$H_2N\text{-}CH(CH_3)_2$	POLY SECONDARY AMINE

use of dihydrogen results in the formation of polymers with a higher degree of functionalization regardless of the nature of the catalyst.

On examination of the data shown in Tables IV and V it becomes obvious that by a judicious choice of the hydrogen source and catalyst one can obtain polyamines with a wide range of concentration (AI) of pendant amine groups. This feature of the aminomethylation reaction on unsaturated polymers becomes especially important when we consider that modification of polymers usually aims for a specific set of physico-chemical properties of the final product.

Table VII lists several typical secondary and primary amines that can be used in this chemistry. It is especially important to note the feasibility of using primary amines for synthesis of polysecondary amines.

III. CONCLUSION

We have developed a novel, general polymer modification technique for a convenient synthesis of a variety of unique polyamines. By this technique one can prepare a variety of tertiary and secondary polyamines with a wide selection of different pendant amino groups. These novel polyamines can be obtained in various molecular weights and with varying degrees of amine functionality.

REFERENCES

1. F. Jachimowicz and J.W. Raksis, J. Org. Chem., **47**, 445 (1982).
2. **Modification of Polymers**, (C.E. Carraher, Jr., and M. Tsuda, eds.) ACS Symposium Series, Washington, D.C., 1980.
3. F. Jachimowicz, A. Hansson, and F.A. Burnett, in preparation.
4. **Organic Synthesis via Metal Carbonyls**, Vol. 2, (J. Wender and P. Pino, eds.), John Wiley & Sons, New York 1977.
5. A.F.M. Igbal, Helv. Chim. Acta, 1440 (1970) reported on the use of the $Rh_2O_3/Fe(CO)_5$ mixed metal catalyst system for the aminomethylation reaction on monoolefins also R.M. Laine, J. Org. Chem., **45**, 3370 (1980) described the $Ru_3(CO)_{12}/Fe_3(CO)_{12}$ system for the preparation of monoamines. However both of these systems used significantly higher amounts of catalysts, and are less selective.

INDEX

CPSIA information can be obtained
at www.ICGtesting.com
Printed in the USA
LVHW031121071218
599554LV00004B/330/P

9 780824 772635